Computation Engineering
Applied Automata Theory and Logic

Computation Engineering
Applied Automata Theory and Logic

Ganesh Gopalakrishnan
University of Utah

 Springer

Ganesh Gopalakrishnan
University of Utah
Department of Computer Science
50 S. Central Campus Drive
Salt Lake City, UT 84112-9205

Computation Engineering: Applied Automata Theory and Logic

ISBN 978-1-4419-3741-4   e-ISBN 978-0-387-32520-0

Printed on acid-free paper.

9 8 7 6 5 4 3 2 1

springer.com

Cover: Original cartoon art on the front and back covers by Geoff Draper.

Whatever you do will be insignificant,
but it is very important that you do it.
– Mohandas K. Gandhi

Contents

List of Figures

Foreword

It takes more effort to verify that digital system designs are correct than it does to design them, and as systems get more complex the proportion of cost spent on verification is increasing (one estimate is that verification complexity rises as the square of design complexity).

Although this verification crisis was predicted decades ago, it is only recently that powerful methods based on mathematical logic and automata theory have come to the designers' rescue. The first such method was equivalence checking, which automates Boolean algebra calculations. Next came model checking, which can automatically verify that designs have – or don't have – behaviours of interest specified in temporal logic. Both these methods are available today in tools sold by all the major design automation vendors.

It is an amazing fact that ideas like Boolean algebra and modal logic, originating from mathematicians and philosophers before modern computers were invented, have come to underlie computer aided tools for creating hardware designs.

The recent success of 'formal' approaches to hardware verification has lead to the creation of a new methodology: assertion based design, in which formal properties are incorporated into designs and are then validated by a combination of dynamic simulation and static model checking. Two industrial strength property languages based on temporal logic are undergoing IEEE standardisation.

It is not only hardware design and verification that is changing: new mathematical approaches to software verification are starting to be deployed. Microsoft provides windows driver developers with verification tools based on symbolic methods.

Discrete mathematics, logic, automata, and the theory of computability are the basis for these new formal approaches. Although they

have long been standard topics in computer science, the uses made of
them in modern verification are quite different to their traditional roles,
and need different mathematical techniques. The way they are taught
often puts more emphasis on cultivating 'paper and pencil' proof skills,
and less on their practical applications and implementation in tools.
Topics in logic are often omitted, or taught without emphasizing connections with automata, and without explaining the algorithms (e.g.,
fixed-point computation) used in verification.

This classroom-tested undergraduate textbook is unique in presenting logic and automata theory as a single subject. Public domain software is used to animate the theory, and to provide a hands-on taste of
the algorithms underlying commercial tools. It is clearly written and
charmingly illustrated. The author is a distinguished contributor to
both theory and to new tool implementation methods.

I highly recommend this book to you as the best route I know into
the concepts underlying modern industrial formal verification.

Dr. Michael J.C. Gordon FRS
Professor of Computer Assisted Reasoning
The University of Cambridge Computer Laboratory

Preface

Computation Engineering, Applied Automata Theory and Logic:

With the rapidly evolving nature of Computing, and with multiple new topics vying for slots in today's undergraduate and graduate curricula, "classical" topics such as Automata Theory, Computability, Logic, and Formal Specification cannot fill the vast expanses they used to fill in both the undergraduate and the graduate syllabi. This move is also necessary considering the fact that many of today's students prefer learning *theory* as a tool rather than *theory for theory's sake*. This book keeps the above facts in mind and takes the following fresh approach:

- approaches automata theory and logic as the underlying *engineering mathematics* for *Computation Engineering,*
- attempts to restore the Automaton-Logic connection missing in most undergraduate books on automata theory,
- employs many interactive tools to illustrate key concepts,
- employs humor and directly appeals to intuitions to drive points home,
- covers classical topics such as the Rice's Theorem, as well as modern topics such as Model Checking, Büchi Automata, and Temporal Logics.

We now elaborate a bit further on these points, and then provide a chapter-by-chapter description of the book.

Teach Automata Theory and Logic as if they were Engineering Math:

The computer hardware and software industry is committed to using mathematically based (formal) methods, and realizes that the biggest impediment to the large scale adoption of these methods would be the lack of adequately trained manpower. In this context, it is crucial that students who go through automata theory and logic courses *retain* what they have learned, and *know how to use* their knowledge. Today's undergraduate textbooks in the area of this book typically emphasize automata theory, and not logic. This runs the risk of imparting a skewed perspective, as *both* these perspectives are needed in practice. Also most of today's books do not include tool-based experiments. Experience shows that tool based experimentation can greatly enhance one's retention.

Restoring the Missing Automaton/Logic Connection:

Automata theory and logic evolved hand-in-hand - and yet, this connection was severed in the 1970's when the luxury of *separate* automata theory and logic courses became possible. Now, the crowded syllabi that once again forces these topics to co-exist may actually be doing a huge favor: providing the opportunity to bring these topics back together! For example, Binary Decision Diagrams (BDD) - central data structures for representing Boolean functions - can be viewed as minimized DFA. One can introduce this connection and then show how finite state machines can be represented and manipulated either using *explicit* state graphs or *implicitly* using BDD based Boolean transition relations. Another example I've employed with great success is that of the formulation and solution of the logical validity of simple sentences from Presburger arithmetic – such as "$\forall xyz : x + y = z$" – using DFA. Here, the use of automaton operations (such as intersection, complementation, and projection) and corresponding operations in logic (such as conjunction, negation, and existential quantification) in the *same setting* helps illustrate the interplay between two intimately related areas, and additionally helps build strong intuitions.

Teaching Through Interactive Tools:

To the best of my knowledge, none of the present-day undergraduate books in automata theory employ interactive tools in any significant manner. This approach tends to give the false impression to students that these are topics largely of theoretical interest, with the only practical examples coming from the area of compiler parsers. We encourage tool usage in the following ways:

- we illustrate the use of the Grail tools, originally from the University of Western Ontario, to illustrate the application of operations on automata in an interactive manner,

- we illustrate the use of the JFLAP tool kit written by Professor Susan Rodger's group at Duke when discussing nondeterministic Turing machines,
- we employ the BED Binary Decision Diagram package of Andersson and Hulgaard in illustrating several aspects of BDD manipulation and fixed-point computation,
- we illustrate the use of Boolean satisfiability tools such as Zchaff and Minisat in leading up to the concept of NP-completeness,
- we illustrate DNF to CNF conversion and obtaining the Prenex Normal Form using simple programs written in the functional language Ocaml, and finally
- we present simple examples using the model checking tool SPIN.

On page 443, we provides the address of the book website where tool-specific instructions will be maintained.

Letting Intuitions be the Guide:

I have found that introducing diagonalization proofs early on can help students tie together many later ideas more effectively. I employ gentle intuitive introductions (e.g., "the US map has more points than hair on an infinite-sized dog").

Many topics become quite clear if demonstrated in multiple domains. For example, I illustrate fixed-point theory by discussing how context-free grammars are recursive language equations. I also introduce fixed-points by pointing out that if one repeatedly photocopies[1] a document, it will often tend towards one of the fixed points of the image transformation function of the photocopy machine(!).

My example to introduce the fact that there are a countably infinite number of C programs consists of pointing out that the following are legal C programs: main(){}, main(){{}}, main(){{{}}}, ... and then I use the Schröder-Bernstein theorem relating this sequence to the sequence of even numbers ≥ 8 (which can then be bijectively mapped to Natural numbers). In another example, I show that the set of C programs are not regular by applying the Pumping Lemma to main(){{...}} and getting a syntax error when the Pumping Lemma mangles the bracketing structure of such C programs! In my opinion, these examples do not diminish rigor, and may succeed in grabbing the attention of many students.

---

[1] Xerox is still a trademark of Xerox Inc.

Model Checking, Büchi Automata, and Temporal Logics:

From a practical point of view, automata theory and logic play a central role in modeling and verifying concurrent reactive systems. The hardware and software industry employs *model checking* tools to find deep seated bugs in hardware and software. This book closes off with an introduction to the important topic of model checking.

A Tour Through the Book:

Chapter 1 motivates the material in the book, presenting in great detail why the topics it covers are important both in terms of theory and practice.

Chapter 2 begins with the quintessentially important topics of *sets*, *functions*, and *relations*. After going through details such as expressing function signatures, the difference between partial and total functions, we briefly examine the topic of *computable* functions - functions that computers can hope to realize within them. We point out important differences between the terms *procedure* and *algorithm*, briefly touching on the $3x + 1$ problem - a four-line program that confounds scientists despite decades of intense research. In order to permit you to discuss functions concretely, we introduce the *lambda* notation. A side benefit of our introduction to Lambda calculus is that you will be able to study another formal model of computation besides Turing machines (that we shall study later).

Chapter 3 goes through the concept of *cardinality* of sets, which in itself is extremely mentally rewarding, and also reinforces the technique of proof by contradiction. It also sets the stage for defining fine distinctions such as 'all languages,' of which there are "uncountably many" members, and 'all Turing recognizable languages,' of which there are only "countably many" members.

Chapter 4 discusses important classes of binary relations, such as reflexive, transitive, preorder, symmetric, anti-symmetric, partial order, equivalence, identity, universal, equivalence, and congruence (modulo operators).

Chapter 5 provides an intuitive introduction to mathematical logic. We have written this chapter with an eye towards helping you read definitions involving the operators *if, if and only if (iff)*, and quantifiers *for all*, and *there exists*. The full import of definitions laden with *if*s and *if and only if*s, as well as quantifiers, is by no means readily apparent - and so it is essential to cultivate sufficient practice. You will see proof by contradiction discussed at a 'gut level' - we encourage you to play the game of Mastermind, where you can apply this technique quite effectively to win pretty much every time. Nested quantification is reinforced here, as well as while discussing the pumping lemma in Chapter 12. Our message is: *write proofs clearly and lucidly - that way, in case you go wrong, others can spot the mistake and help you.*

Chapter 6 studies the topic of recursion at some depth. Given that a book on automata theory is 'ridden with definitions,' one has to understand carefully when these definitions make sense, and when they end up being 'total nonsense,' owing, say, to being circular or not uniquely defining something. Lambda calculus provides basic notation with which to reason about recursion. We present to you the "friendliest foray into fixed-point theory' that we can muster." We will use fixed-points as a 'one-stop shopping' conceptual tool for understanding a diverse array of topics, including context-free productions and the reachable states of finite-state machines.

In Chapter 7, we begin discussing the notions of *strings* and *languages*. Please, however, pay special attention[2] to "the five most confused objects of automata theory," namely ε, $\emptyset$, $\{\varepsilon\}$, $\{\emptyset\}$, and the equation $\emptyset^* = \{\varepsilon\}$. We discuss the notion of concatenation, exponentiation, 'starring,' complementation, reversal, homomorphism, and prefix-closure applied to languages.

In Chapter 8, we discuss machines, languages, and deterministic finite automata. We construct Deterministic Finite Automata (DFA) for several example languages. One problem asks you to build a DFA that scans a number presented in any number-base b, either most significant digit-first, or least significant digit-first, and shine its green light exactly at those moments when the number scanned so far equals 0 in some modulus k. The latter would be equivalent to division carried out least significant digit-first, with only the modulus to be retained. We will have more occasions to examine the true "power" of DFAs in Chapters 12 and 20. Chapter 8 closes off with a brief study of the limitations of DFA.

Chapter 9 continues these discussions, now examining the crucially important concept of *non-determinism*. It also discusses regular expressions - a syntactic means for describing regular languages. The important topic of Non-deterministic Finite Automaton (NFA) to DFA conversion is also examined.

Automata theory is a readily usable branch of computer science theory. While it is important to obtain a firm grasp of its basic principles

---

[2] Once you understand the fine distinctions between these objects, you will detect a faint glow around your head.

using paper, pencil, and the human brain, it is quite important that one use automated tools, especially while designing and debugging automata. To paraphrase Professor Dana Scott, *computer-assisted tools are most eminently used as the* telescopes *and* microscopes *of learning - to see farther, to see closer, and to see clearer than the human mind alone can discern.* We demonstrate the use of `grail` tools to generate and verify automata.

In Chapter 10, we discuss operations that combine regular languages, most often yielding new regular languages as a result. We discuss the conversion of NFA to DFA - an important algorithm both theoretically and in practice. We also discuss the notion of *ultimate periodicity* which crystallizes the true power of DFAs in terms of language acceptance, and also provide a tool-based demonstration of this idea.

In Chapter 11, we will begin discussing binary decision diagrams (BDD), which are nothing but minimized DFA for the language of satisfying assignments (viewed as sequences, as we will show) for given Boolean expressions. The nice thing about studying BDDs is that it helps reinforce not only automata-theoretic concepts but also concepts from the area of formal logic. It teaches you a technique widely used in industrial practice, and also paves the way to your later study of the theory of NP-completeness.

In Chapter 12, we discuss the Pumping Lemma. We define the lemma in first-order logic, so that the reader can avoid common confusions, and grasp how the lemma is employed. We also discuss *complete* Pumping Lemmas (regular if and only if certain conditions are met).

In Chapter 13, we present the idea of *context-free languages.* Context-free languages are generated by a *context-free grammar* that consists of *production rules.* By reading production rules as recursive equations, we can actually solve for the context-free language being defined. We will also have occasion to *prove*, via induction, that context-free productions are *sound* and *complete* - that they do not generate a string outside of the language, but do generate all strings within the language. The important notions of *ambiguity* and *inherent ambiguity* will also be studied.

Chapter 14 will introduce our first *infinite state* automaton variety - the *push-down automaton (PDA)* - a device that has a finite-control and exactly one unbounded stack. We will study a method to convert PDA to CFG and vice versa. We will also show how to prove PDAs correct using Floyd's Inductive Assertions method.

In Chapter 15, we study Turing machines (TM). Important notions, such as *instantaneous descriptions* (ID) are introduced in this chapter.

Several Turing machine simulators are available on the web - you are encouraged to download and experiment with them. In this chapter, we will also introduce *linear bounded automata* (LBA) which are TMs where one may deposit new values only in that region of the tape where the original input was presented.

Chapter 16 discusses some of the most profound topics in computer science: the *halting* problem, the notion of semi-decision procedures, and the notion of algorithms. With the introduction of unbounded store, many decision problems will become formally undecidable. In a large number of cases, it will no longer be possible to predict what the machine will do. For example, it will become harder or impossible to tell whether a machine will halt when started on a certain input, whether two machines are equivalent, etc. In this chapter, we will formally state and prove many of these undecidability results. We will present three proof methods: (i) through contradiction, (ii) through reductions from languages not known to be decidable, and (iii) through mapping reductions.

Chapter 17 continues with the notion of undecidability, discussing two additional proof techniques: (i) through the computational history method, and (ii) by employing Rice's theorem. These are advanced proof methods of undecidability that may be skipped during the initial pass through the textbook material. One can proceed to Chapter 18 after finishing Chapter 16 without much loss of continuity.

Chapter 18 sets the stage to discuss the theory of NP completeness. It touches on a number of topics in mathematical logic that will help you better appreciate all the nuances. We briefly discuss a "Hilbert style" axiomatization of propositional logic, once again touching on soundness and completeness. We discuss basic definitions including satisfiability, validity, tautology, and contradiction. We discuss the various "orders" of logic including zeroth-order, first-order, and higher-order. We briefly illustrate that the validity problem of first-order logic is only semi-decidable by reduction from the Post Correspondence problem. We illustrate how to experiment with modern SAT tools. We also cover related ideas such as $\neq$-sat, 2-sat, and satisfiability-preserving transformations.

Chapter 19 discusses the notion of polynomial-time algorithms, computational complexity, and the notion of NP-completeness. We reiterate the importance of showing that the problem belongs to NP, in addition to demonstrating NP-hardness.

In Chapter 20, we introduce a small subset of first-order logic called Presburger arithmetic that enjoys the following remarkable property:

given a formula in this logic, we can build a finite automaton such that the automaton has an accepting run if and only if the formula is satisfiable. The neat thing about this technique is that it reinforces the automaton/logic connection introduced in the previous chapter.

Chapters 21 through 23 introduce temporal logic and model checking. These topics are ripe for introduction to undergraduates and graduates in all areas of computer science, but without all the generality found in specialized books. To this end, in Chapter 21, we provide a history of model checking and also a detailed example. Chapter 22 introduces linear-time temporal logic and Computational Tree Logic, contrasting their expressive power, and exactly why *computation trees* matter. Finally, Chapter 23 presents an enumerative as well as a symbolic algorithm for CTL model checking. We also present an enumerative algorithm for LTL model checking through an example. We introduce Büchi automata, discussing how Boolean operations on Büchi automata are performed, and that non-deterministic Büchi automata are not equivalent to deterministic Büchi automata.

Chapter 24 reviews the material presented in the book. Chapter A lists the book website, software tool related information, and possible syllabi based on this book.

Acknowledgements:

This book was written over three years, with numerous interruptions. I have classroom-tested parts of this book in an undergraduate class entitled Models of Computation (CPSC 3100, Spring 2002), a graduate class entitled Foundations of Computing (CPSC 6100, Spring 2005), and a graduate class entitled Model Checking (CPSC 6964, Fall 2005). Many thanks to students who took these classes (with due apologies for having subjected them to half-done material). It was also offered as one of the references by Professor Konrad Slind when he taught CPSC 3100 during Fall 2005, for which I am grateful.

I am very grateful for many friends, colleagues, and students who have proof-read significant parts of this book and provided very valuable comments. As best as I recall, here is a list of people who deserve special mention: Steven Johnson, John Harrison, Rajeev Alur, Michael J.C. Gordon, Mark Aagaard, Paliath Narendran, Ching-tsun Chou, Panagiotis (Pete) Manolios, Konrad Slind, Joe Hurd, Kathi Fisler, Salman Pervez, Xiaofang Chen, Michael Freedman, Geoff Draper, Vijay Durairaj, Shrish Jain, Vamshi Kadaru, Seungkeol Choe, and Sonjong Hwang (and apologies to those I am forgetting). I would also like to thank the authors of the computer-aided design tools I have used in this book (listed on page 443).

Thanks also to the support team at Springer, especially Frank Holzwarth and Deborah Doherty, for their timely help. Many thanks to Michael Hackett, Editor, for helping me launch this book project. Special thanks to Carl W. Harris, Senior Editor, Springer, for patiently leading me through all the major stages of this book, and for very patiently working with the reviewers of this book. Thanks, Carl!

I especially thank Geoff Draper, a student in my CPSC 6100 class (Spring 2005) who provided extensive comments, drew the cartoons you see in this book, as well as kindly gave me permission to employ his cartoons in this book. A reader expecting to read "the usual theory book" is in for a surprise: they will find me joking about at various junctures, as I normally do when I lecture. I keep books such as *The Cartoon Guide to Genetics* by Larry Gonick and Mark Wheelis in mind when I engage in these diversions. Such books are immensely important contributions, in that they make an otherwise dull and dry subject come alive, especially for newcomers. In my humble opinion, many more such books are badly needed to describe computer science theory in a plain and accessible manner to students who might otherwise swallow theory as if it were a bitter pill. I have attempted to slant this book in that direction, without compromising the seriousness of the topic. If I have failed in any way to do so, I offer my advance apologies.[3]

I have benefited immensely by reading many contemporary books - notably one by Michael Sipser and one by Dexter Kozen. I have also drawn from Zohar Manna's book *Mathematical Theory of Computation.* I also like to thank the creators of all the computer based tools employed in this book.

Last but not least, my sincere thanks to my wife Kalpana and daughters Kamala and Kajal for all their love, caring, and cheer. Without their understanding and accommodation, I could not have worked on this book. I also thank my parents, whose names are my middle and last names, for their venerable position of bringing me into this world and seeing to it that I lacked nothing.

Salt Lake City, Utah, USA   *Ganesh Gopalakrishnan*
                                                              April 2006

---

[3] Even Patch Adams might occasionally have overdone his antics!

1

Introduction

Welcome to Computation Engineering - the discipline of applying *engineering principles* to model, analyze, and construct *computing systems*. Human society is, more than ever, reliant on computing systems operating correctly within automobile control systems, medical electronic devices, telephone networks, mobile robots, farms, nuclear power plants, etc. With so much entrusted to computers, how can we ensure that all of these computers are being built and operated in a manner responsible to all flora and fauna? How do we avoid potential disasters - such as mass attacks due to malevolent "viruses," or an avionic computer crashing and rebooting mid-flight? On a deeper, more philosophical note, what exactly is computation? Can we mathematically characterize those tasks that computers are capable of performing, and those they are incapable of performing? The subject-matter of this book is about seeking answers to these deep questions using *automata theory* and *mathematical logic*.

Computation Science and Computation Engineering

We distinguish a computer scientist from a *computation engineer*. We define those people who seek an in-depth understanding of the phenomenon of computation to be a computer scientist. We define those people who seek to *efficiently apply* computer systems to solve real-world problems to be *computation engineers*.[1] The distinction is, in a sense, similar to that between a chemist and a chemical engineer. This

---

[1] We prefer calling the latter activity *computation engineering* as opposed to *computer engineering* because the term "computer engineer" is, unfortunately, applied nowadays to people with a hardware bias in their outlook. In this book, we carry no such bias.

book will expose you to computer *science* through automata theory and logic. It will show you how computation engineers use automata theory and logic as the *engineering mathematics*, much like 'traditional engineers' apply differential and integral calculus to build better-engineered products such as bridges, automobiles, and airplanes.

What is 'Computation?'

By the way of popular analogy,[2] we can offer one possible feeble answer: "if it involves a computer, a program running on a computer, and numbers going in and out, then computation is likely happening.[3]" Such an answer invites a barrage of additional questions such as "what is a computer?"; "what is a program?", etc. There are other tricky situations to deal with as well. Consider another place where computation seems to be happening: within our very body cells. Thanks to modern advances in genetics, we are now able to understand the mind-boggling amount of "string processing" that occurs within our cells - in the process of transcribing the genetic code (which resembles assembly code in a strange programming language), doing all of that wonderful string matching and error correction, and resulting in the synthesis of proteins. Is this also computation?

The short answer is that we *cannot* have either a comprehensive or a permanent definition of what 'computation' means. Unless we employ the *precise* language offered by mathematics, philosophical or emotionally charged discussions are bound to lead nowhere. One must build *formal models* that crystallize the properties observed in real-world computing systems, study these models, and then answer questions about computing and computation in terms of the models. The abstraction must also be at the *right level*. Otherwise, we will end up modeling a computer as a mindless electronic oscillator that hauls bits around.[4]

Given all this, it is indeed remarkable that computer science has been able to capture the essence of computing in terms of a *single formal device*: the so called Turing machine. A Turing machine is a simple device that has finite-state control that interacts with an unbounded storage tape (or, equivalently, a finite-state control that interacts with two unbounded stacks, as we shall show very soon). In fact, several

---

[2] "If it walks like a duck and quacks like a duck, then it is a duck."

[3] I've "laced" this book with several footnotes, hoping to 'break the ice,' and make you believe that you are not reading a theory book.

[4] A similar end-result should we abstract the music of Mozart as a sound pressure waveform.

other formal devices - such as the Lambda calculus, Thue systems, etc. - were proposed around the same time as Turing machines. All these devices were also formally shown to be equivalent to Turing machines. This caused Alonzo Church to put forth his (by now famous) thesis: "All effectively computable functions can be understood in terms of one of these models." In Chapter 15 we shall study Turing machines in great detail; Chapter 6 gives a glimpse of how the Lambda calculus, essentially through the *fixed-point theory*, provides a formal model for computer programs.

A Minimalist Approach

In a minimalist approach, models are created with the smallest possible set of mechanisms. In the case of computational models, it is a bit ironic that the first model proposed - namely Turing machines- was also the most powerful. However, with the increasing usage of computers,[5] two other models born out of practical necessity were proposed, roughly two decades after Turing machines were proposed: finite automata in the late 1950's, and push-down automata shortly thereafter. Rearranging computer history a bit, we will discuss finite automata first, push-down automata next, and finally Turing machines (see Figure 1.1). All these types of machines are meant to carry out *computational procedures* ("procedures" for short,) consisting of instructions. They differ primarily in the manner in which they record data ("state"). A procedure always begins at an initial state which is highlighted by an arrow impinging from nowhere, as in Figure 1.1. The "data input" to a procedure, if any, is provided through the data storage device of the machine. Each instruction, when executed, helps transform the current (data and control) state into the next state. An instruction may also read an input symbol (some view these inputs coming from a *read-only tape*). Also, at every state, one or more instructions may become eligible to execute. A deterministic machine is one that has at most one eligible instruction to execute at any time, while a nondeterministic machine can have more than one eligible instruction.

A procedure halts when it encounters one of the predetermined final states. It is possible for a procedure to never encounter one of its final states; it may loop forever. If a procedure is guaranteed to halt on all inputs, it is called an *algorithm*. Unfortunately, it is impossible to tell

---

[5] Thomas J. Watson, Chairman of IBM in 1943, is said to have remarked, "I think there is a world market for maybe five computers." Well, there are more than five computers in a typical car today. Some cars carry hundreds, in fact!

whether a given procedure is an algorithm - a topic that we shall revisit many times in this book.

One of the central points made in this book is that there are essentially three ways to organize the data (state) recording apparatus of a machine: (i) have *none* at all, (ii) employ one stack to record data, and (iii) employ two stacks to record data (in other words, employ zero, one, or two stacks)! A finite-state control device by itself (*i.e.*, without any additional history recording device) is called a *finite automaton* - either a deterministic finite automaton (DFA) or a nondeterminstic finite automaton (NFA). A finite automaton is surprisingly versatile. However, it is not as powerful as a machine with one stack, which, by the way, is called a *push-down automaton* (PDA). Again there are NPDA and DPDA - a distinction we shall study in Chapters 13 and 14.

A PDA is more powerful than a finite automaton. By employing an unbounded stack, a PDA is able to store an arbitrary amount of information in its state, and hence, is able to refer to data items stacked arbitrarily prior. However, a PDA is not as powerful as a machine with *two* stacks. This is because a PDA is not permitted to "peek" inside its stack to look at some state s held deep inside the stack, unless it is also willing to pop away all the items stacked since s was stacked. Since there could be arbitrarily many such stacked items, a PDA cannot hope to preserve all these items being popped and restore them later.

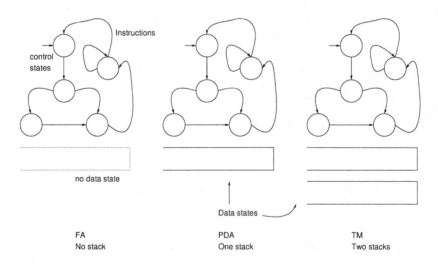

Fig. 1.1. The power of various machines, and how to realize them

The moment a finite-state control device has access to two unbounded stacks, however, it will have the ability to pop items from one stack and push them into the other stack. This gives these two-stack machines the ability to peek inside the stacks; in effect, a finite-state control device with two stacks becomes equivalent to a Turing machine. We can show that adding further stacks does not increase its power! (These two-stack devices were historically first introduced in the form of Turing machines, as we shall see in more detail in Chapter 15, Section 15.2.2.) All computers, starting from the humble low-end computer of a Furbee doll or a digital wrist-watch, through all varieties of desktop and laptop computers, all the way to 'monstrously powerful' computers, (Say, the IBM Blue Gene/L [4] computer) can be modeled in terms of Turing machines.

There is an important point of confusion we wish to avoid early. Real computers only have a finite amount of memory. However, *models* of computers employ an *unbounded* amount of memory. This allows us to study the outcomes possible in *any* finite memory device, regardless of how much memory it has. So long as the finite memory device does not hit its storage limit, it can pretend that it is working with an infinite amount of memory.

How to Measure the Power of Computers?

In comparing computational machines ("computers," loosely), we cannot go by any subjective measure such as their absolute speed, as such "speed records" tend to become obsolete with the passage of time. For instance, the "supercomputers" of the 1950s did not even possess the computational power of many modern hand-held computers and calculators. Instead, we go by the *problem solving ability* of computational devices: we deem two machines M_1 and M_2 to be equivalent if they can solve the same class of problems, *ignoring the actual amount of time taken*. Problem solving, in turn, can be modeled in terms of *algorithmically deciding membership in languages*. Alas, these topics will have to be discussed in far greater detail than is possible now, since we have not set up any of the necessary formal definitions.

Complexity Theory

You may have already guessed this: in studying and comparing computing devices, we cannot ignore time or resources entirely. They do matter! But then one will ask, "how do we measure quantities such as time, space, and energy?" It is very easy to see that using metrics such as minutes and hours is unsatisfactory, as they are *non-robust* measures, being tied to factors with ephemeral significance, such as the clock speed of computers, the pipeline depth, etc. With advances in technology, computations that took hours a few years ago can nowadays be performed in seconds. Again, Turing machines come to our rescue! We define a unit of time to be *one step* of a Turing machine that is running the particular program or computation. We then define time in terms of the asymptotic worst-case complexity notation "$O()$" employed in any book on algorithm analysis (see for instance [29] or [6]). However, such a characterization is often not possible:

- There are many problems (called NP-complete problems) whose best known solutions are exponential time algorithms. It is unknown whether these problems have polynomial time algorithms.
- There are problems for which algorithms are known not to exist; for others, it is not known whether algorithms exist.

However, it is true that researchers engaged in studying even these problems employ Turing machines as one of their important formal models. This is because any result obtained for Turing machines can be translated into corresponding results for real computers.

Automata Theory and Computing

In this book, we approach the above ideas through automata theory. What is automata theory? Should its meaning be cast in concrete, or should it evolve with advances in computer science and computer applications? In this book, we take a much broader meaning for the term 'automata theory.' We will use it to refer to a comprehensive list of closely related topics, including:

- finite and infinite automata (together called "automata"),
- mathematical logic,
- computability and complexity (TR, co-TR, NP-complete, etc.),
- formal proofs,
- the use of automata to decide the truth of formal logic statements,
- automata on infinite words to model the behavior of reactive systems, and last but not least,
- applications of the above topics in formal verification of systems.

Automata theory is a 'living and breathing' branch of theory - not 'fossilized knowledge.' It finds day-to-day applications in numerous walks of life, in the *analysis* of computing system behavior. In this book, we present a number of tools to help understand automata theoretic concepts and also to illustrate how automata theory can be applied in practice. We illustrate the use of automata to describe many real-life finite-state systems, including: games, puzzles, mathematical logic statements, programming language syntax, error-correcting codes, and combinational and sequential digital circuits. We also illustrate tools to describe, compose, simplify, and transform automata.

Why "Mix-up" Automata and Mathematical Logic?

We believe that teaching automata theory hand-in-hand with mathematical logic allows us to not only cover concepts pertaining to formal languages and machines, but also illustrate the deep connections that these topics have to the process of *formalized reasoning* – proving properties about computing systems. Formalized reasoning about computing systems is escalating in importance because of the increasing use of computers in safety critical and life critical systems. The software and hardware used in life and resource critical applications of computers is becoming so incredibly complex that testing these devices for correct operation has become a major challenge. For instance, while performing 100 trillion tests sounds more than adequate for many systems, it is simply inadequate for most hardware/software systems to cover all possible behaviors.

Why Verify? Aren't Computers "Mostly Okay?"

Human society is crucially dependent on software for carrying out an increasing number of day-to-day activities. The presence of *bugs* in software is hardly noticed until, say, one's hand-held computer hangs, when one pokes its reset button and moves on, with some bewilderment. The same is the case with many other computers that we use; in general, the mantra seems to be, "reset and move on!" However, this surely cannot be a general design paradigm (imagine a machine getting stuck in an infinite reboot loop if the problem does not clear following a reset). In the modern context, one truly has to worry about the logical correctness of software and hardware because there have been many "close calls," some real disasters, and countless dollars have been wasted in verifying products before they are released. In 2004, the Mars Spirit Rover's computer malfunctioned when the number of allowed open files in flash memory were exceeded. This caused a shadow of uncertainty to hang over the project for a few days, with scientists wasting their time finding a cure for the problem. Recently, car companies have had recalls due to software bugs in their computerized engine control; the societal costs of such bugs ("software defects") are estimated to be very high [116, 103]. In 1996, Ariane-5, a rocket costing $2B, self-destructed following an arithmetic overflow error that initially caused the rocket nozzles to be pointed incorrectly [84].

A very important human lesson is contained in many software failures. In 1987, a radiation therapy machine called Therac-25 actually caused the death of several patients who came to get radiation therapy for cancer. At the time of the incident, the machine had recently been redesigned and its software was considered so reliable that many of its safety interlock mechanisms had been dispensed with. One of these interlocks was for monitoring the amount of radiation received by patients. To cut a long story short, the software was, after all, buggy, and ended up administering massive radiation overdoses to patients [76]. Every engineer shoulders the societal responsibility to adopt simple and reliable safety measures in the systems they design and deploy to avoid such catastrophes. The approach taken in each project must be: "when in doubt, play it safe, keeping the well being of lives and nature in mind."

The *root cause* of a large majority of bugs is the *ambiguous* and/or *inconsistent* specification of digital system components and subsystems. Software built based on such specifications is very likely to be flawed, hard to debug, and impossible to systematically test. As an example from recent practice, the specification document of a widely used computer interconnect bus called the 'PCI' [95] was shown to be internally inconsistent [28]. Unfortunately, the design community had moved too far along to take advantage of these findings. Many of the disasters of computer science are not *directly* in the form of crashes of rockets or chips (although such disasters have happened[6]). In 2001, the United States Federal Bureau of Investigation (FBI) launched a project to overhaul their software to coordinate terrorist databases. After nearly four years and over $300 million dollars spent, the project has been declared to have an unrealizable set of software requirements, and hence abandoned.

The ability to write precise, as well as, *unambiguous* specifications is central to using them correctly and testing them reliably. Such *formal methods* for hardware and software have already been widely adopted. Organizations such as Intel, AMD, IBM, Microsoft, Sun, HP, JPL, NASA, and NSA employ hundreds of formal verification specialists.

---

[6] The Ariane rocket, worth $2B, was lost because of incorrect version of software running. Intel lost $.5B due to a floating-point bug in their Pentium II microprocessor.

Unfortunately, there are many areas where the power of formal methods has not been demonstrated. This book hopes to expose students to the basics of formal methods, so that they will be able to participate in, as well as contribute to, the formal methods revolution that is happening.

Verifying Computing Systems Using Automaton Models

It turns out that in order to *prove* that even the simplest of computing systems operates correctly, one has to examine so many inputs and their corresponding outputs for correctness. In many cases, it will take *several thousands of centuries* to finish complete testing. For example, if such a computer has one megabit of internal storage, one has to check for correctness over 2^{10^6} states. This is a number of unfathomable magnitude. Chapters 21 through 23 present how automata theory, along with the use of finite-state abstractions and symbolic state representations, helps debug concurrent systems through a powerful approach known as *model checking*.

How does automata theory help ensure that safety critical systems are correct? First of all, it helps create *abstract models* for systems. Many systems are so complex that each vector consists of thousands, if not millions, of variables. If one considers going through all of the 2^{1000} assignments for the bits in the vector, the process will last thousands of millennia. If one runs such tests for several days, even on the fastest available computer, and employs a good randomization strategy in vector selection, one would still have covered only some *arbitrary* sets of behaviors of the system. Vast expanses of its state-space would be left unexamined. A much more effective way to approach this problem in practice is to *judiciously* leave out most of the bits from vectors, and examine the system behavior *exhaustively* over all the remaining behaviors. (In many cases, designers can decide which bits to leave out; in some cases, computer-based tools can perform this activity.) That is, by leaving out the right set of bits, we end up fully covering an abstract model. Experience has repeatedly shown that verifying systems at the level of abstract models can often find serious bugs quicker. The simulation of non-abstracted models is also a necessary part of verification in practice.

It turns out that abstracted models of most systems are *finite automata*. There are a number of techniques being developed that can represent, as well as manipulate, *very large* finite automata. These techniques help minimize the degree to which systems have to be abstracted before they can be exhaustively verified. This, in turn, means that the risk of accidentally missing an error due to overly heavy abstractions is also reduced.

Automaton/Logic Connection

Various branches of mathematical logic are employed to *precisely* and *unambiguously* describe both system *requirements* and system *behaviors*. In modern verification systems, automata theoretic techniques are often employed to process these logical specifications to check whether they are true or false. It is therefore important for students to see these topics treated in a cohesive manner.

The importance of precision and clarity in system descriptions cannot be overstated. Many system description documents rival Los Angeles telephone directories in their size, containing a very large number of subtle assertions that tax the human mind. Each person reading such a document comes up with a different understanding. While engineers are incredibly smart and are able to correct their misunderstandings more often than not, they still waste their time poring over lengthy prose fragments that are ambiguous, and simply too complex to trust one's mind with. It has been widely shown that formal statements in mathematical logic can serve as very valuable augments to text-based documents, supplying the missing precision in the text.

One of the most serious deficiencies of an exclusively natural-language system description is that engineers *cannot mechanically calculate their consequences*. In other words, they cannot ask "what if" (putative) queries about scenarios that are not explicitly discussed in the document. An example from the I/O system world is, "what if I/O transaction x is allowed to overtake transaction y? Does it cause wrong answers to be returned? Does it cause system deadlocks?" The number of potentially interesting "what if" questions pertaining to any real system is extremely large. It is impossible or even counterproductive for a specification document to list all these questions and their answers.[7] On the other hand, an approach where one is able to state the specification of complex systems (such as I/O buses) in a precise language based on mathematical logic, and is able to pose *putative* or *challenge* queries (also expressed in mathematical logic) is highly conducive to gaining a proper understanding of complex systems. Ideally, systems of this kind must either try to show that the posed conjecture or query is true, or provide a clear explanation of *why* it is false. Model checking based verification methods (discussed Chapter 21 onwards) provide

---

[7] Doing so would result in not just a telephone directory, but an entire city library!

this ability in most formal verification based design approaches under development or in actual use today.

The ability to *decide* - provide an answer, without looping, for all possible putative queries - is, of course, a luxury enjoyed when we employ simpler mathematical logics. Unfortunately, simpler mathematical logics are often not as expressive, and we will then have to *deduce*, using partly manual steps and partly automatic (decidable) steps, the answer. In Chapter 18, this book provides a formal proof as to why some of these logics are undecidable.

Avoid Attempting the Impossible

Automata theory and logic often help avoid pursuing the impossible. If one can *prove* that there cannot be a decider for some task, then *there is no point wasting everyone's time in pursuit of an algorithm* for that task. On the other hand, if one does prove that an algorithm exists, finding an *efficient* algorithm becomes a worthwhile pursuit.

As an example, the next time your boss asks you to produce a C-grammar equivalence checker that checks the equivalence between any two arbitrary C-grammar files (say, written in the language Yacc or Bison) and takes no more than "a second per grammar-file line," don't waste your time coding - simply prove that this task is impossible![8]

Solving One Implies Solving All

There is another sense in which automata theory helps avoid work. In many of these cases, researchers have found that while we cannot actually solve a given problem, we can *gang up* or "club together" thousands of problems such that the ability to solve *any one* of these problems gives us the ability to solve *all* of these problems. Often the solvability question will be whether it is tractable to solve the problems - i.e., solve in polynomial time. In many cases, we are simply interested in solvability, without worrying about the amount of time. In these cases also, repeated work is avoided by grouping a collection of problems into an equivalence class and looking for a solution to only *one* of these problems; this solution can be easily modified to solve thousands of practical problems. This is the motivation behind studying NP-completeness and related notions. We will also study the famous Halting problem and problems related to it by taking this approach

---

[8] You may wish to state that the difference between Bison and Yacc is that one cannot wash one's face in a Yacc!

of clubbing problems together through the powerful idea of *mapping reductions.*

Automata Theory Demands a Lot From You!

The study of automata theory is a challenging. It exposes students to a variety of mathematical models and helps build confidence in them, thus encouraging them to be creative, to take chances, and to tread new ground while designing computing systems. Unfortunately, the notion of formally characterizing designs is not emphasized in traditional systems classes (architecture, operating systems, etc.), where the goal has, historically, been high performance and not high reliability. Therefore, it takes an extra bit of effort to put formal specification and verification into practice. Fortunately, the computer industry has embraced formal methods, and sees it as the main hope for managing the complexity and ensuring the reliability of future designs.

It is impossible to learn automata theory "in a hurry." While the subject is quite simple and intuitive in hindsight, to get to that stage takes patience. You must allow enough time for the problems to gestate in your minds. After repeatedly trying and failing, you will be able to carry the problems in your minds. You may end up solving some of them in the shower.[9]

A Brief Foray Into History

Let us take a historical perspective, to look back into how this subject was viewed by two of the originators of this subject - Michael Rabin and Dana Scott - in their 1959 paper (that, incidentally, is cited in their ACM Turing award citation [100]):

> Turing machines are widely considered to be the abstract prototype of digital computers; workers in the field, however, have felt more and more that the notion of a Turing machine is too general to serve as an accurate model of actual computers. It is well known that even for simple calculations, it is impossible to give a prior upper bound on the amount of tape a Turing machine will need for any given computation. It is precisely this feature that renders Turing's concept unrealistic.
>
> In the last few years, the idea of a finite automaton has appeared in the literature. These are machines having only a finite number

---

[9] You may wish to warn your family members that you may one day bolt out of the shower, all soaking wet, shouting "Eureka!"

of internal states that can be used for memory and computation. The restriction of finiteness appears to give a better approximation to the idea of a physical machine. [...]. Many equivalent forms of the idea of finite automata have been published. One of the first of these was the definition of "nerve-nets" given by McCulloch and Pitts. [...]

In short, Rabin and Scott observe that the theory of Turing machines, while all encompassing, is "too heavy-weight" for day-to-day studies of computations. They argue that perhaps finite automata are the right model of most real-world computations. From a historical perspective, the more complex machine form (Turing machine) was proposed much before the much simpler machine form (finite automata). All this is clearly not meant to say that Turing machines are unimportant—far from it, in fact! Rather, the message is that a more balanced view of the topics studied under the heading of automata will help one better appreciate how this area came about, and how the priorities will shift over time.

Disappearing Formal Methods

In the article "disappearing formal methods [104]," John Rushby points out how throughout the history of engineering, various new technologies had been [over] advertised, until they became widely accepted and taken for granted. For instance, many olden-day radios had a digit ('6' or '8') boldly written on their aluminum cases, advertising the fact that they actually employed 6 or 8 (as the case may be) transistors.[10] Centuries ago, differential and integral calculus were widely advertised as the "magic" behind studying planetary motions, as well as building everyday objects, such as bridges. Today, nobody hangs a sign-board on a bridge proclaiming, "this bridge has been designed using differential and integral calculus!" In the same vein, the term *formal methods* is nowadays used as a "slogan" to call attention to the fact that we are really using ideas based on logic and automata theory to design products, as opposed to previously, when we used no such disciplined approach. While the explicit advertisement of the technology behind modern developments is inevitable, in the long run when such applications are well understood, we would no longer be pointing out explicitly

---

[10] Bob Colwell recently told me the story of a radio he opened in his younger days, only to find that it had two transistors connected to nowhere - apparently serving the only purpose of jacking up the number advertised outside!

that we are actually designing, say, floating-point units, using higher order logic.

Said another way, in a future book on automata theory, perhaps one could dispense with all these motivational remarks that have prevented us from heading towards the core of this book. Luckily, that is precisely where we are headed now.

Exercises

1.1. Read the 1972 Turing Award Lecture by Edsger W. Dijkstra entitled "The Humble Programmer" [36]. This, and a significant number of other articles referred to in this book, is available through the ACM Digital Library, and often through the Google search engine.

1.2. Read and summarize the article "The Emperor's Old Clothes" by C. A. R. Hoare [57].

1.3. Read and summarize the article "Computational Thinking" by Jeanette M. Wing [123]. Computational thinking represents a universally applicable attitude and skill set everyone, not just computer scientists, would be eager to learn and use.

2

Mathematical Preliminaries

In this chapter, we introduce many of the fundamental mathematical ideas used throughout the book. We first discuss *sets*, which help organize "things" into meaningful unordered collections. We then discuss functions which help map things to other things. Next, we discuss relations that relate things. We provide a concrete syntax, that of *Lambda expressions*, for writing down function definitions. We then present ways to "count" infinite sets through a measure known as *cardinality*.

2.1 Numbers

We will refer to various classes of numbers. The set *Nat*, or *natural numbers*, refers to whole numbers greater than or equal to zero, i.e., $0, 1, 2, \ldots$. The set *Int*, or *integers*, refers to whole numbers, i.e., $0, 1, -1, 2, -2, 3, -3 \ldots$. The set *Real*, or *real numbers*, refers to both rational as well as irrational numbers, including 0.123, $\sqrt{2}$, π, 1, and -2.

2.2 Boolean Concepts, Propositions, and Quantifiers

We assume that the reader is familiar with basic concepts from Boolean algebra, such as the use of the Boolean connectives *and* ($\wedge$), *or* ($\vee$), and *not* ($\neg$). We will be employing the two *quantifiers* "for all" ($\forall$) and "for some" or equivalently "there exists" ($\exists$) in many definitions. Here we provide preliminary discussions about these operators; more details are provided in Section 5.2.4.

In a nutshell, the quantifiers $\forall$ and $\exists$ are iteration schemes for $\wedge$, $\vee$, and $\neg$, much like Σ (summation) and Π (product) are iteration schemes for *addition* $(+)$ and *multiplication* $(\times)$. The universal quantification operator $\forall$ is used to make an assertion about all the objects in the domain of discourse. (In mathematical logic, these domains are assumed to be non-empty). Hence,

$$\forall x : Nat : \ P(x)$$

is equivalent to an infinite conjunction

$$P(0) \ \wedge \ P(1) \ \wedge \ P(2) \ \ldots$$

or equivalently $\wedge_{x \in Nat} : \ P(x)$.

2.3 Sets

A *set* is a collection of things. For example, $A = \{1, 2, 3\}$ is a set containing three natural numbers. The order in which we list the contents of a set does not matter. For example, $A = \{3, 1, 2\}$ is the same set as above. A set cannot have duplicate elements. For example, $B = \{3, 1, 2, 1\}$ is not a set.[1]

A set containing no elements at all is called the *empty set*, written $\{\}$, or **equivalently**, $\emptyset$. A set may also consist of a collection of other sets, as in

$$P = \{\{\}, \{1\}, \{2\}, \{3\}, \{1, 2\}, \{1, 3\}, \{2, 3\}, \{1, 2, 3\}\}.$$

P has a special status; it contains every *subset* of set A. P is in fact the *powerset* of A. We will have more to say about powersets soon.

2.3.1 Defining sets

Sets are specified using the *set comprehension* notation

$$S = \{x \in D \mid p(x)\}.$$

Here, S includes all x from some universe D such that $p(x)$ is true. $p(x)$ is a Boolean formula called *characteristic formula*. p by itself is called the *characteristic predicate*. We can leave out D if it is clear from the context.

---

[1] An unordered collection with duplicates, such as B, is called a *multi-set* or *bag*.

Examples:

- Set A, described earlier, can be written as

$$A = \{x \mid x = 1 \vee x = 2 \vee x = 3\}.$$

- For any set D,

$$\{x \in D \mid true\} = D.$$

  Notice from this example that the characteristic formula can simply be *true*, or for that matter *false*.

- For any set D,

$$\{x \in D \mid false\} = \emptyset.$$

The next two sections illustrate that care must be exercised in writing set definitions. The brief message is that by writing down a collection of mathematical symbols, one does not necessarily obtain something that is well defined. Sometimes, we end up defining more than one thing without realizing it (the definitions admit multiple solutions), and in other cases we may end up creating contradictions.

2.3.2 Avoid contradictions

Our first example illustrates the famous *Russell's Paradox*. This paradox stems from allowing expressions such as $x \in x$ and $x \notin x$ inside characteristic formulas. Consider some arbitrary domain D. Define a set S as follows:

$$S = \{x \in D \mid x \notin x\}.$$

Now, the expression $x \notin x$ reveals that x itself is a set. Since S is a set, we can now ask, "is S a member of S?"

- If S is a member of S, it cannot be in S, because S cannot contain sets that contain themselves.
- However, if S is not a member of S, then S must contain S!

Contradictions are required to be complete, *i.e.*, apply to all possible cases. For example, if $S \notin S$ does not result in a contradiction, that, then, becomes a consistent solution. In this example, we fortunately obtain a contradiction in all the cases. The proposition $S \in S$ must produce a definite answer - true or false. However, *both* answers lead to a contradiction.

We can better understand this contradiction as follows. For Boolean quantities a and b, let $a \Rightarrow b$ stand for "a implies b" or "if a then b;" in other words, $\Rightarrow$ is the *implication* operator. Suppose $S \in S$. This allows us to conclude that $S \notin S$. In other words, $(S \in S) \Rightarrow (S \notin S)$ is true. In other words, $\neg(S \in S) \lor (S \notin S)$, or $(S \notin S) \lor (S \notin S)$, or $(S \notin S)$ is true. Likewise, $(S \in S) \Rightarrow (S \notin S)$. This allows us to prove $(S \in S)$ true. Since we have proved $S \in S$ as well as $S \notin S$, we have proved their conjunction, which is *false*! With *false* proved, anything else can be proved (since *false* $\Rightarrow$ *anything* is $\neg(false) \lor anything$, or *true*). Therefore, it is essential to avoid contradictions in mathematics.

Russell's Paradox is used to conclude that a "truly universal set" – a set that contains *everything* – cannot exist. Here is how such a conclusion is drawn. Notice that set S, above, was defined in terms of an *arbitrary* set called D. Now, if D were to be a set that contains "everything," a set such as S must clearly be present inside D. However, we just argued that S must not exist, or else a contradiction will result. Consequently, a set containing *everything* cannot exist, for it will lack at least S. This is the reason why the notion of a *universal* set is not an absolute notion. Rather, a universal set specific to the domain of discourse is defined each time. This is illustrated below in the section devoted to universal sets. In practice, we disallow sets such as S by banning expressions of the form $x \in x$. In general, such restrictions are handled using *type theory* [48].

2.3.3 Ensuring uniqueness of definitions

When a set is defined, it must be *uniquely* defined. In other words, we cannot have a definition that does not pin down the exact set being talked about. To illustrate this, consider the "definition" of a set

$$S = \{x \in D \mid x \in S\},$$

where D is some domain of elements. In this example, the set being defined depends on itself. The circularity, in this case, leads to S not being uniquely defined. For example, if we select $D = Nat$, and plug in $S = \{1, 2\}$ on both sides of the equation, the equation is satisfied. However, it is also satisfied for $S = \emptyset$, $S = \{3, 4, 5\}$. Hence, in the above circular definitions, we cannot *pin down* exactly what S is.

The message here is that one must avoid using purely circular definitions. However, sets are allowed to be defined through *recursion* which, at first glance, is "a sensible way to write down circular definitions." Chapter 7 explains how recursion is understood, and how sets can be uniquely defined even though "recursion seems like circular definition."

Operations on Sets

Sets support the usual operations such as membership, *union, intersection, subset, powerset, Cartesian product,* and *complementation*. $x \in A$ means x is a member of A. The *union* of two sets A and B, written $A \cup B$, is a set such that $x \in (A \cup B)$ *if and only if* $x \in A$ *or* $x \in B$. In other words, $x \in (A \cup B)$ implies that $x \in A$ or $x \in B$. Also, $x \in A$ or $x \in B$ implies that $x \in (A \cup B)$. similarly, the *intersection* of two sets A and B, written $A \cap B$, is a set such that $x \in (A \cap B)$ if and only if $x \in A$ *and* $x \in B$.

A *proper* subset A of B, written $A \subset B$, is a subset of B different from B. $A \subseteq B$, read 'A is a subset of B', means that $A \subset B$ or $A = B$. Note that the empty set has no proper subset.

Subtraction, Universe, Complementation, Symmetric Difference

Given two sets A and B, set subtraction, '$\setminus$', is defined as follows:

$$A \setminus B = \{x \mid x \in A \land x \notin B\}.$$

Set subtraction basically removes all the elements in A that are in B. For example, $\{1, 2\} \setminus \{2, 3\}$ is the set $\{1\}$. 1 survives set subtraction because it is not present in the second set. The fact that 3 is present in the second set is immaterial, as it is not present in the first set.

For each type of set, there is a set that contains all the elements of that type. Such a set is called the *universal* set. For example, consider the set of *all strings* over some alphabet, such as $\{a, b\}$. This is universal set, as far as sets of strings are concerned. We can write this set as

$$\text{SigmaStar} = \{\varepsilon, a, b, aa, ab, ba, bb, aaa, aab, \ldots\}.$$

(The reason why we name the above set SigmaStar will be explained in Chapter 7.) Here, ε is the empty string, commonly written as "", a and b are strings of length 1, aa, ab, ba, and bb are strings of length 2, and so on. While discussing natural numbers, we can regard $Nat = \{0, 1, 2, \ldots\}$ as the universe.

> The symbol ε is known to confuse many students. Think of it as the "zero" element of strings, or simply read it as the empty string "". By way of analogy, the analog of the arithmetic expression $0 + 1$, which simplifies to 1, is ε concatenated with a, which simplifies to a. (We express string concatenation through juxtaposition). Similarly, $0 + 2 + 0 = 2$ is to numbers as ε aa $\varepsilon = aa$ is to strings. More discussions are provided in Section 7.2.4.

Universal sets help define the notion of *complement* of a set. Consider the universal set (or "universe") SigmaStar of strings over some alphabet. The complement of a set of strings such as $\{a, ba\}$ is SigmaStar $\setminus \{a, ba\}$. If we now change the alphabet to, say, $\{a\}$, the universal set of strings over this alphabet is

$$\text{SigmaStar1} = \{\varepsilon, a, aa, aaa, aaaa, \ldots\}.$$

Taking the complement of a set such as $\{a, aaa\}$ with respect to SigmaStar1 yields a set that contains strings of a's such that the number of occurrences of a's is neither 1 nor 3.

Given two sets A and B, their *symmetric difference* is defined to be

$$(A \setminus B) \cup (B \setminus A).$$

For example, if $A = \{1, 2, 3\}$ and $B = \{2, 3, 4, 5\}$, their symmetric difference is the set $\{1, 4, 5\}$. The symmetric difference of two sets produces, in effect, the XOR (exclusive-OR) of the sets.

For any alphabet Σ and its corresponding universal set SigmaStar, the complement of the empty set $\emptyset$ is SigmaStar. One can think of $\emptyset$ as *the* empty set with respect to *every* alphabet.

Types versus Sets

The word *type* will be used to denote a *set* together with its associated operations. For example, the type *natural number*, or *Nat*, is associated with the set $\{0, 1, 2, \ldots\}$ and operations such as successor, +, etc. $\emptyset$ is an overloaded symbol, denoting the empty set of every type. When we use the word "type," most commonly we will be referring to the underlying set, although strictly speaking, types are "sets plus their operations."

Numbers as Sets

In mathematics, it is customary to regard natural numbers themselves as sets. Each natural number essentially denotes the set of natural numbers below it. For example, 0 is represented by $\{\}$, or $\emptyset$, as there are no natural numbers below 0. 1 is represented by $\{0\}$, or (more graphically) $\{\{\}\}$, the only natural number below 1. Similarly, 2 is the set $\{0, 1\}$, 3 is the set $\{0, 1, 2\}$, and so on. This convention comes in quite handy in making formulas more readable, by avoiding usages such as

$$\forall i \, : \, 0 \leq i \leq N - 1 \, : \, ..something..$$

and replacing them with

$$\forall i \in N \ : \ ..something..$$

Notice that this convention of viewing sets as natural numbers is exactly similar to how numbers are defined in set theory textbooks, e.g., [50]. We are using this convention simply as a labor-saving device while writing down definitions. We do not have to fear that we are suddenly allowing sets that contain other sets.

As an interesting diversion, let us turn our attention back to the discussion on Russell's Paradox discussed in Section 2.3.2. Let us take D to be the set of natural numbers. Now, the assertion $x \notin x$ evaluates to *true* for every $x \in D$. This is because no natural number (viewed as a set) contains itself - it only contains all natural numbers strictly *below* it in value. Hence, no contradiction results, and S ends up being equal to Nat.

2.4 Cartesian Product and Powerset

The *Cartesian product* operation '×' helps form *sets of tuples* of elements over various types. The terminology here goes as follows: 'pairs' are 'two-tuples,' 'triples' are 'three-tuples,' 'quadruples' are 'four-tuples,' and so on. After 5 or so, you are allowed to say 'n-ple' - for instance, '37-ple' and so on. For example, **the set** $Int \times Int$ denotes the sets of pairs of all integers. Mathematically, the former set is

$$Int \times Int = \{\langle x, y \rangle \ | \ x \in Int \ \wedge y \in Int\}.$$

Given two sets A and B, the **Cartesian product** of A and B, written $A \times B$, is defined to be the set

$$\{\langle a, b \rangle \ | \ a \in A \ \wedge \ b \in B\}.$$

We can take Cartesian product of multiple sets also. In general, the Cartesian product of n sets A_i, $i \in n$ results in a set of "n-tuples"

$$A_0 \times A_1 \times \ldots A_{n-1} = \{\langle a_0, a_1, \ldots, a_{n-1} \rangle \ | \ a_i \in A_i \ for \ every \ i\}.$$

If one of these sets, A_i, is $\emptyset$, the Cartesian product results in $\emptyset$ because it is impossible to "draw" any element out of A_i in forming the n-ples. Here are some examples of Cartesian products:

- $\{2, 4, 8\} \times \{1, 3\} \times \{100\} =$
  $\{\langle 2, 1, 100 \rangle, \langle 2, 3, 100 \rangle, \langle 4, 1, 100 \rangle, \langle 4, 3, 100 \rangle, \langle 8, 1, 100 \rangle, \langle 8, 3, 100 \rangle\}.$

- SigmaStar1×SigmaStar1 = $\{\langle x, y \rangle \mid x \text{ and } y \text{ are strings over } \{a\}\}$.

In taking the Cartesian product of n sets A_i, $i \in n$, it is clear that if $n = 1$, we get the set A_0 back. For example, if $A_0 = \{1, 2, 3\}$, the 1-ary Cartesian product of A_0 is itself. Note that $A_0 = \{1, 2, 3\}$ can also be written as $A_0 = \{\langle 1 \rangle, \langle 2 \rangle, \langle 3 \rangle\}$ because, in classical set theory [50], 1-tuples such as $\langle 0 \rangle$ are the same as the item without the tuple sign (in this case 0).

It is quite common to take the Cartesian product of different *types* of sets. For example, the set *Int* × *Bool* denotes the sets of pairs of integers and Booleans. An element of the above set is $\langle 22, true \rangle$, which is a pair consisting of one integer and one Boolean.

2.4.1 Powersets and characteristic sequences

The **powerset** of a set S is the set of all its subsets. As is traditional, we write 2^S to denote the powerset of S. In symbols,

$$2^S = \{x \mid x \subseteq S\}.$$

This "exponential" notation suggests that the size of the powerset is 2 raised to the size of S. We can argue this to be the case using the notion of *characteristic sequences*. Take $S = \{1, 2, 3\}$ for example. Each subset of S is defined by a bit vector of length three. For instance, 000 represents $\emptyset$ (include none of the elements of S), 001 represents $\{3\}$, 101 represents $\{1, 3\}$, and 111 represents S. These "bit vectors" are called *characteristic sequences*. All characteristic sequences for a set S are of the same length, equal to the size of the set, $|S|$. Hence, the number of characteristic sequences for a *finite* set S is exponential in $|S|$.

2.5 Functions and Signature

A function is a mathematical object that expresses how items called "inputs" can be turned into other items called "outputs." A function maps its *domain* to its *range*; and hence, the inputs of a function belong to its *domain* and the outputs belong to its *range*. **The domain and range of a function are always assumed to be non-empty.** The expression "$f : T_D \to T_R$" is called the signature of f, denoting that f maps the domain of type T_D to the range of type T_R. Writing signatures down for functions makes it very clear as to what the function "inputs" and what it "outputs." Hence, this is a highly recommended practice.

As a simple example, $+ : Int \times Int \rightarrow Int$ denotes the signature of integer addition.

Function signatures must attempt to capture their domains and ranges as tightly as possible. Suppose we have a function g that accepts subsets of $\{1, 2, 3\}$, outputs 4 if given $\{1, 2\}$, and outputs 5 given anything else. How do we write the signature for g? Theoretically speaking, it is correct to write the signature as $2^{Nat} \rightarrow Nat$; however, in order to provide maximum insight to the reader, one must write the signature as

$$2^{\{1,2,3\}} \rightarrow \{4, 5\}.$$

If you are unsure of the exact domain and range, try to get as tight as possible. Remember, you must help the reader.

The *image* of a function is the set of range points that a function actually maps onto. For function $f : T_D \rightarrow T_R$,

$$image(f) = \{y \in T_R \mid \exists x \in T_D : y = f(x)\}.$$

2.6 The λ Notation

The Lambda calculus was invented by Alonzo Church[2] as a formal representation of computations. Church's thesis tells us that the lambda-based evaluation machinery, Turing machines, as well as other formal models of computation (Post systems, Thue systems, ...) are all formally equivalent. Formal equivalences between these systems have all been worked out by the 1950s.

More immediately for the task at hand, the lambda notation provides a *literal* syntax for *naming* functions. First, let us see in the context of numbers how we name them. The sequence of numerals (in programming parlance, the *literal*) '1' '9' '8' '4' names the number 1984. We do not need to give alternate names, say 'Fred', to such numbers! A numeral sequence such as **1984** suffices. In contrast, during programming one ends up giving such alternate names, typically derived from the domain of discourse. For example,

function `Fred`(x) {return 2;}.

Using Lambda expressions, one can write such function definitions without using alternate names. Specifically, the Lambda expression $\lambda x.2$ captures the same information as in the above function definition.

---

[2] When once asked how he chose λ as the delimiter, Church replied, "Eenie meenie mynie mo!"

We think of strings such as $\lambda x.2$ as a *literal* (or *name*) that describes functions. In the same vein, using Lambda calculus, one name for the successor function is $\lambda x.(x+1)$, another name is $\lambda x.(x+2-1)$, and so on.[3] We think of *Lambda expressions* as *irredundant* names for functions (irredundant because redundant strings such as 'Fred' are not stuck inside them). We have, in effect, "de-Freded" the definition of function Fred. In Chapter 6, we show that this idea of employing irredundant names works even in the context of *recursive* functions. While it may appear that performing such 'de-Fredings' on recursive definitions appears nearly impossible, Chapter 6 will introduce a trick to do so using the so-called Y operator. Here are the only two rules pertaining to it that you need to know:

- *The Alpha rule*, or "the name of the variable does not matter." For example, $\lambda x.(x+1)$ is the same as $\lambda y.(y+1)$. This process of renaming variables is called *alpha conversion*. Plainly spoken, the Alpha rule simply says that, in theory,[4] the formal parameters of a function can be named however one likes.
- *The Beta rule*, or "here is how to perform a function call." A function is applied to its argument by writing the function name and the argument name in juxtaposition. For example, $(\lambda x.(x+1))\ 2$ says "feed" 2 in place of x. The result is obtained by substituting 2 for x in the body $(x+1)$. In this example, $2+1$, or 3 results. This process of simplification is called *beta reduction*.

The formal arguments of Lambda expressions associate to the right. For example, as an abbreviation, we allow cascaded formal arguments of the form $(\lambda xy.(x+y))$, as opposed to writing it in a fully parenthesized manner as in $(\lambda x.(\lambda y.(x+y)))$. In addition, the arguments to a Lambda expression associate to the left. Given these conventions, we can now illustrate the simplification of Lambda expressions. In particular,

$$(\lambda zy.(\lambda x.(z+x)))\ 2\ 3\ 4$$

can be simplified as follows (we show the bindings introduced during reduction explicitly):

$$= (\lambda zy.(\lambda x.(z+x)))\ 2\ 3\ 4$$

---

[3] You may be baffled that I suddenly use "23" and "+" as if they were Lambda terms. As advanced books on Lambda calculus show [48], such quantities can also be encoded as Lambda expressions. Hence, anything that is *effectively computable*—computable by a machine—can be formally defined using only the Lambda calculus.

[4] In practice, one chooses mnemonic names.

$$= \text{(using the Beta rule) } (\lambda \underline{z = 2} \ y = 3.(\lambda x.(z + x)))4.$$
$$= (\lambda x.(2 + x))4$$
$$= \text{(using the Beta rule) } (\lambda \underline{x = 4}.(2 + x))$$
$$= 2 + 4$$
$$= 6$$

The following additional examples shed further light on Lambda calculus:

- $(\lambda x.x)$ 2 says apply the identity function to argument 2, yielding 2.
- $(\lambda x.x) (\lambda x.x)$ says "feed the identity function to itself." Before performing beta reductions here, we are well-advised to perform alpha conversions to avoid confusion. Therefore, we turn $(\lambda x.x) (\lambda x.x)$ into $(\lambda x.x) (\lambda y.y)$ and then apply beta reduction to obtain $(\lambda y.y)$, or the identity function back.
- As the Lambda calculus seen so far does not enforce any "type checking," one can even feed $(\lambda x.(x + 1))$ to itself, obtaining (after an alpha conversion) $(\lambda x.(x+1))+1$. Usually such evaluations then get "stuck," as we cannot add a number to a function.

2.7 Total, Partial, 1-1, and Onto Functions

Functions that are defined over their entire domain are *total*. An example of a total function is $\lambda x.2x$, where $x \in Nat$. A *partial* function is one undefined for some domain points. For example, $\lambda x.(2/x)$ is a partial function, as it is undefined for $x = 0$.

The most common use of partial functions in computer science is to model *programs that may go into infinite loops for some of their input values.* For example, the recursive program over Nat,

$$f(x) = if \ (x = 0) \ then \ 1 \ else \ f(x)$$

terminates only for $x = 0$, and loops for all other values of x. Viewed as a function, it maps the domain point 0 to the range point 1, and is undefined everywhere else on its domain. Hence, function f can be naturally modeled using a *partial function*. In the Lambda notation, we can write f as $\lambda x.if \ (x = 0) \ then \ 1$. Notice that we use an "if-then" which leaves the "else" case undefined.

One-to-one (1-1) functions f are those for which every point $y \in image(f)$ is associated with exactly one point x in T_D. A function that is not 1-1 is 'many-to-one.' One-to-one functions are also known as *injections*. An example of an injection is the predecessor function $pred : Nat \rightarrow Nat$, defined as follows:

$$\lambda x.if\ (x > 0)\ then\ (x - 1).$$

We have $pred(1) = 0$, $pred(2) = 1$, and so on. This function is partial because it is undefined for 0.

Onto functions f are those for which $image(f) = T_R$. Onto functions are also known as *surjections*. While talking about the type of the range, we say function f maps *into* its range-type. Hence, *onto* is a special case of *into* when the entire range is covered. **One-to-one, onto, and total functions are known as bijections.** Bijections are also known as **correspondences.**

Examples: We now provide some examples of various types of functions. In all these discussions, assume that $f : Nat \rightarrow Nat$.

- An example of a one-to-one (1-1) function ("injection") is $f = \lambda x.2x$.
- An example of a *many-to-one* function is $f = \lambda x.(x\ mod\ 4)$.
- An example of an onto function is $f = \lambda x.x$.
- An example of a partial function is $f = \lambda x.if\ even(x)\ then\ (x/2)$.
- An example of a bijection is $f = \lambda x.if\ even(x)\ then\ (x+1)\ else\ (x-1)$. All bijections $f : T_D \rightarrow T_R$ where $T_D = T_R = T$ are the same type, are *permutations* over T.
- Into means *not necessarily onto*. A special case of *into* is *onto*.
- Partial means *not necessarily total*. A special case of *partial* is *total*.

For any given one-to-one function f, we can define its *inverse* to be f^{-1}. This function f^{-1} is defined at all its *image* points. Therefore, whenever f is defined at x,

$$f^{-1}(f(x)) = x.$$

For $f : T_D \rightarrow T_R$, we have $f^{-1} : T_R \rightarrow T_D$. Consequently, if f is *onto*, then f^{-1} is *total*—defined everywhere over T_R. To illustrate this, consider the predecessor function, *pred*. The image of this function is Nat. Hence, *pred* is onto. Hence, while $pred : Nat \rightarrow Nat$ is not total, $pred^{-1} : Nat \rightarrow Nat$ is total, and turns out to be the *successor* function *succ*.

Given a bijection f with signature $T_D \rightarrow T_R$, for any $x \in T_D$, $f^{-1}(f(x)) = x$, and for any $y \in T_R$, $f(f^{-1}(y)) = y$. This shows that if f is a bijection from T_D to T_R, f^{-1} is a bijection from T_R to T_D. For this reason, we tend to call f a bijection **between** T_D and T_R - given the forward mapping f, the existence of the backward mapping f^{-1} is immediately guaranteed.

Composition of functions

The composition of two functions $f : A \to B$ and $g : B \to C$, written $g \circ f$, is the function $\lambda x \,.\, g(f(x))$.

2.8 Computable Functions

Computational processes map their inputs to their outputs, and therefore are naturally modeled using functions. For instance, given two matrices, a computational process for matrix multiplication yields the product matrix. All those functions whose mappings may be obtained through a *mechanical process* are called *computable* functions, *effectively computable* functions, or *algorithmically computable* functions. For practical purposes, another equivalent definition of a computable function is one whose definition can be expressed in a general-purpose programming language. By 'mechanical process,' we mean a sequence of elementary steps, such as bit manipulations, that can be carried out on a machine. Such a process must be finitary, in the sense that for any input for which the function is defined, the computational process producing the mapping must be able to read the input in finite time and yield the output in a *finite* amount of time. Chapter 3 discusses the notion of a 'machine' a bit more in detail; for now, think of computers when we refer to a machine.

Non-computable functions are well-defined mathematical concepts. These are genuine mathematical functions, albeit those whose mappings cannot be obtained using a machine. In Section 3.1, based on cardinality arguments, we shall show that non-computable functions do exist. We hope that the intuitions we have provided above will allow you to answer the following problems intuitively. The main point we are making in this section is that just because a function "makes sense" mathematically doesn't necessarily mean that we can code it up as a computer program!

2.9 Algorithm versus Procedure

An *algorithm* is an *effective procedure*, where the word 'effective' means 'can be broken down into elementary steps that can be carried out on a computer.' The term *algorithm* is reserved to those procedures that come with a guarantee of termination on every input. If such a guarantee is not provided, we must not use the word 'algorithm,' but instead use the word *procedure*. While this is a simple criterion one can often apply, sometimes it may not be possible to tell whether to call something an algorithm or a procedure. Consider the celebrated "$3x + 1$ problem," also known as "Collatz's problem," captured by the following program:

```
function three_x_plus_one(x)
  { if (x==1) then return 1;
    if even(x) then three_x_plus_one(x/2);
    else three_x_plus_one(3x+1); }
```

For example, given 3, the `three_x_plus_one` function obtains 10, 5, 16, 8, 4, 2, 1, and halts. Will this function halt for all x? Nobody knows! It is still open whether this function will halt for all x in Nat [24]! Consequently, if someone were to claim that the above program is their actual implementation of an *algorithm* (not merely a *procedure*) to realize the constant function $\lambda x.1$, not even the best mathematicians or computer scientists living today would know how to either confirm or to refute the claim! That is, nobody today is able to prove or disprove that the above program will halt for all x, yielding 1 as the answer.[5]

2.10 Relations

Let S be a set of k-tuples. Then a k-ary relation R over S is defined to be a subset of S. It is also quite common to assume that the word 'relation' means 'binary relation' ($k = 2$); we will not follow this convention, and shall be explicit about the arity of relations. For example, given $S = Nat \times Nat$, we define the binary relation $<$ over S to be

$$< \ = \{\langle x, y \rangle \mid x, y \in Nat \ and \ x < y\}.$$

It is common to overload symbols such as $<$, which can be used to denote binary relations over Int (the set of positive and negative numbers) or $Real$. For the sake of uniformity, we permit the arity of a relation to be 1. Such relations are called *unary* relations, or *properties*. For example, *odd* can be viewed as a unary relation

$$odd = \{x \mid x \in Nat \ and \ x \ is \ odd\},$$

or, equivalently,

$$odd = \{x \mid x \in Nat \ and \ \exists y \in Nat : x = 2y + 1\}.$$

---

[5] Our inability to deal with "even a three-line program" perhaps best illustrates Dijkstra's advice on the need to be *humble programmers* [36].

Much like we defined binary relations, we can define *ternary* relations (or 3-ary relations), 4-ary relations, etc. An example of a 3-ary relation over *Nat* is *between*, defined as follows:

$$between = \{\langle x, y, z \rangle \mid x, y, z \in Nat \wedge (x \leq y) \wedge (y \leq z)\}.$$

Given a *binary* relation R over set S, define the *domain* of R to be

$$domain(R) = \{x \mid \exists y : \langle x, y \rangle \in R\},$$

and the *co-domain* of R to be

$$codomain(R) = \{y \mid \exists x : \langle x, y \rangle \in R\}.$$

Also, the *inverse* of R, written R^{-1} is

$$R^{-1} = \{\langle y, x \rangle \mid \langle x, y \rangle \in R\}.$$

As an example, the inverse of the 'less than' relation, '<,' is the greater than relation, namely '>.' Similarly, the inverse of '>' is '<.' Please note that the notion of *inverse* is defined only for *binary* relations - and not for *ternary* relations, for instance. Also, *inverse* is different from *complement*. The *complement* of < is $\geq$ and the complement of '>' is '$\leq$,' where the complementations are being done with respect to the universe $Nat \times Nat$.

For a binary relation R, let $elements(R) = domain(R) \cup codomain(R)$. The restriction of R on a subset $X \subseteq elements(R)$ is written

$$R \mid_X = \{\langle x, y \rangle \mid \langle x, y \rangle \subseteq R \wedge x, y \in X\}.$$

Restriction can be used to specialize a relation to a "narrower" domain. For instance, consider the binary relation < defined over *Real*. The restriction $< \mid_{Nat}$ restricts the relation to natural numbers.

> *Putting these ideas together,* the symmetric difference of '<' and '>' is the '$\neq$' (not equal-to) relation. You will learn a great deal by proving this fact, so please try it!

2.11 Functions as Relations

Mathematicians seek conceptual economy. In the context of functions
and relations, it is possible to express all functions as relations; hence,
mathematicians often view functions as special cases of relations. Let
us see how they do this.

For $k > 0$, a k-ary relation $R \subseteq T_D^1 \times T_D^2 \times \ldots \times T_D^k$ is said to be *single-
valued* if for any $\langle x_1, \ldots, x_{k-1} \rangle \in T_D^1 \times T_D^2 \times \ldots \times T_D^{k-1}$, there is at most
one x_k such that $\langle x_1, \ldots, x_{k-1}, x_k \rangle \in R$. Any single-valued relation R
can be viewed as a $k-1$-ary function with domain $T_D^1 \times T_D^2 \times \ldots \times T_D^{k-1}$
and range T_D^k. We also call single-valued relations *functional relations*.
As an example, the ternary relation

$$\{ \langle x, y, z \rangle \mid x, y, z \in Nat \wedge (x + y = z) \}$$

is a functional relation. However, the ternary relation *between* defined
earlier is *not* a functional relation.

How do partial and total functions "show up" in the world of relations? Consider a $k-1$-ary function f. If a x_k exists for any input
$\langle x_1, \ldots, x_{k-1} \rangle \in T_D^1 \times T_D^2 \times \ldots \times T_D^{k-1}$, the function is total; otherwise,
the function is partial.

To summarize, given a single-valued k-ary relation R, R can be
viewed as a function f_R such that the "inputs" of this function are
the first $k-1$ components of the relation and the output is the last
component. Also, given a k-ary function f, the $k+1$-ary single-valued
relation corresponding to it is denoted R_f.

2.11.1 More λ syntax

There are two different ways of expressing two-ary functions in the
Lambda calculus. One is to assume that 2-ary functions take a *pair*
of arguments and return a result. The other is to assume that 2-ary
functions are 1-ary functions that take an argument and return a result,
where the result is *another* 1-ary function.[6] To illustrate these ideas,
let us define function RMS which stands for *root mean squared* in both
these styles, calling them rms_1 and rms_2 respectively:

$rms_1 : \lambda \langle x, y \rangle . \sqrt{x^2 + y^2}$
$rms_2 : \lambda x . \lambda y . \sqrt{x^2 + y^2}$

---

[6] The latter style is known as the *Curried* form, in honor of Haskell B. Curry. It
was also a notation proposed by Schönfinkel; perhaps one could have named it
the 'Schönfinkeled' form, as well.

Now, $rms_1\langle 2,4\rangle$ would yield $\sqrt{20}$. On the other hand, we apply rms_2 to its arguments in succession. First, $rms_2(2)$ yields $\lambda y.\sqrt{2^2+y^2}$, i.e., $\lambda y.\sqrt{4+y^2}$, and this function when applied to 4 yields $\sqrt{20}$. Usually, we use parentheses instead of angle brackets, as in programming languages; for example, we write $rms_1(2,4)$ and $\lambda(x,y).\sqrt{x^2+y^2}$.

The above notations can help us write characteristic predicates quite conveniently. The characteristic predicate

$$(\lambda(z,y).(odd(z) \wedge (4 \le z \le 7) \wedge \neg y))$$

denotes (or, 'defines') the relation $= \{\langle 5, false\rangle, \langle 7, false\rangle\}$. This is different from the characteristic predicate

$$(\lambda(x,z,y).(odd(z) \wedge (4 \le z \le 7) \wedge \neg y))$$

which, for x of type $Bool$, represents the relation

$$R' \subseteq Bool \times Nat \times Bool$$

equal to

$$\{\langle false, 5, false\rangle, \langle true, 5, false\rangle, \langle false, 7, false\rangle\}, \langle true, 7, false\rangle\}.$$

Variable x is not used (it is a "don't care") in this formula.

Chapter Summary

This chapter provided a quick tour through sets, numbers, functions, relations, and the lambda notation. The following exercises are designed to give you sufficient practice with these notions.

| Exercises |

2.1. Given the characteristic predicate $p = \lambda x.\ (x > 0 \wedge x < 10)$, describe the unary relation defined by p as a set of natural numbers.

2.2. Given the characteristic formula $f = (x > 0 \wedge x < 10)$, describe the unary relation defined by f as a set of natural numbers.

2.3. Given the characteristic predicate

$$r = \lambda(x,y,z).\ (x \subseteq y \wedge y \subseteq z \wedge x \subseteq \{1,2\} \wedge y \subseteq \{1,2\} \wedge z \subseteq \{1,2\})$$

write out the relation described by r as a set of triples.

2.4. Repeat Exercise 2.3 with the conjunct $x \subseteq y$ removed.

2.5. What is the set defined by $P = \{x \in Nat \mid 55 < 44\}$?

2.6. The powerset, P, introduced earlier can also be written as

$$P = \{x \mid x \subseteq \{1, 2, 3\}\}$$

What set is defined by replacing $\subseteq$ by $\subset$ above?

2.7.
1. What is the set described by the expression

$$\{1, 2, 3\} \cap \{1, 2\} \cup \{2, 4, 5\}.$$

   Here, $\cap$ has higher precedence than $\cup$.
2. What is the *symmetric difference* between $\{1, 2, 3, 9\}$ and $\{2, 4, 5, -1\}$?
3. How many elements are there in the following set:
   $\{\emptyset\} \cup \emptyset \cup \{\{2\}, \emptyset\}$? $\emptyset$ denotes the empty set. It is assumed that sets may contain other sets.

2.8. Formally define the set S of divisors of 64. Either show the set explicitly or define it using comprehension.

2.9. Formally define the set S of divisors of 67,108,864. Either show the set explicitly (!) or define it using comprehension.

2.10. What is the set defined by $\{x \mid x \geq 0 \wedge prime(x) \wedge x \leq 10\}$?

2.11. What is the set defined by

$$\{x \mid x \in 13 \wedge composite(x) \wedge x \geq 1\}?$$

A composite number is one that is not prime.

2.12. What is the set defined by

$$\{x \mid x \in 25 \wedge square(x)\}$$

$square(x)$ means x is the square of a natural number.

2.13. What is the set $S = \{x \mid x \subset Nat \wedge 23 = 24\}$?

2.14. Take $S = Nat$, which contains an infinite number of elements. How many elements are there in the powerset of S? Clearly it also contains an infinite number of elements; but is it the "same kind of infinity?" Think for five minutes and write down your thoughts in about four sentences (we shall revisit this issue in Chapter 3).

2.15. The set Odd of odd numbers is a proper subset of Nat. It is true that Odd "appears to be smaller" than Nat - yet, both sets contain an infinite number of elements. How can this be? Is the 'infinity' that measures the size of Odd a 'smaller infinity' than that which measures the size of Nat? Again, express your thoughts in about four sentences.

2.16. Let E be the set of Even natural numbers. Express the set $E \times 2^E$ using set comprehension.

2.17. An *undirected graph* G is a pair $\langle V, E \rangle$, where V is the set of vertices and E is the set of edges. For example, a triangular graph over $V = \{0, 1, 2\}$ is

$$\langle \{0, 1, 2\}, \{\langle 0, 1 \rangle, \langle 1, 2 \rangle, \langle 0, 2 \rangle\} \rangle.$$

We follow the convention of not listing symmetric variants of edges - such as $\langle 1, 0 \rangle$ for $\langle 0, 1 \rangle$.

Now, this question is about *cliques*. A triangle is a 3-clique. A *clique* is a graph where every pair of nodes has an edge between them. We showed you, above, how to present a 3-clique using set-theoretic notation.

Present the following n-cliques over the nodes $i \in n$ in the same set-theoretic notation. Also draw a picture of each resulting graph:

1. 1-clique, or a point.
2. 2-clique, or a straight-line.
3. 4-clique.
4. 5-clique.

2.18. Write a function signature for the sin and tan functions that accept inputs in degrees.

2.19. Decipher the signature given below by writing down four distinct members of the domain and the same number from the range of this function. Here, X stands for "don't care," which we add to $Bool$. For your examples, choose as wide a variety of domain and range elements as possible to reveal your detailed understanding of the signature:

$$(Int \cup \{-1\}) \times 2^{Int} \times Bool \rightarrow 2^{2^{Bool \cup \{X\}}} \times Int.$$

2.20. Write a function signature for the function $1/(1-x)$ for $x \in Nat$.

2.21. Express the successor function over Nat using the Lambda notation.

2.22. Express the function that sums 1 through N using the Lambda notation.

2.23. Simplify $(\lambda zy.(\lambda x.(z + ((\lambda v.(v + x))5))))$ 2 3 4

2.24. A half-wave rectifier receives a waveform at its input and produces output voltage as follows. When fed a positive voltage on the input, it does not conduct below 0.7 volts (effectively producing 0 volts). When fed a positive voltage above 0.7 volts, it conducts, but diminishes the output by 0.7 volts. When fed a negative voltage, it produces 0 volts, except when fed a voltage below -100 volts, when it blows up in a cloud of smoke (causing the output to be undefined). View the functionality of this rectifier as a function that maps input voltages to output voltages. Describe this function using the Lambda notation. You can assume that *ifthenelse* and *numbers* are primitives in Lambda calculus.

2.25. Provide one example of a bijection from Nat to Int.

2.26. Point out which of the following functions can exist and which cannot. Provide reasons for functions that cannot exist, and examples for functions that can exist.

1. A bijection from $\emptyset$ to $\emptyset$.
2. A bijection from $\{\varepsilon\}$ to $\{\emptyset\}$.
3. A partial 1-1 and onto function from Nat to Nat.
4. A partial 1-1 and onto function from Int to Nat.
5. A 1-1 into function from Nat to Nat.
6. A 1-1 into, but not onto, function from Nat to Nat.
7. A bijection from Int to $Real$.
8. A bijection from a set to its powerset. (Recall that we cannot have $\emptyset$ as either the domain or range of a function.)
9. A many-to-one function from the powerset of a set to the set.
10. An into map from a set to its powerset.

2.27. Describe a bijection from the set $\{\varepsilon\}$ to the set $\{\emptyset\}$. Here, ε is the empty string.

2.28. Think about the following question, writing your thoughts in a few sentences, in case you cannot definitely answer the question (these will be addressed in Chapter 3).

- Can there be a bijection between Int and 2^{Int}?
- How about a finite subset, F, of Int, and 2^F?
- How about an infinite subset, I, of Int, and 2^I?
- Which other kinds of functions than bijections may exist?

2.29. Do you see any problems calling a function g "computable" if g were to accept a subset of Nat, output 4 if given $\{1, 2\}$, and output 5 given any other subset of Nat? How about a variant of this problem with "any other subset" replaced by "any other *proper* subset?"

2.30. Which of the following functions are computable?:

1. A function that inverts every bit of an infinite string of bits.
2. A function that inverts every bit of a finite (but arbitrarily long) string of bits.
3. A function that outputs a 1 when given π, and 0 when given any other Real number. (Recall that π is not 22/7, 3.14, or even 3.1415926. In fact, π is not a Rational number.)

2.31. Does there exist a procedure that, given a C program P and its input x, answers whether P halts on x? Does there exist an algorithm for this purpose?

2.32. Can there be an algorithm that, given two C programs, checks that they have identical functionality (over *all* their inputs)?

2.33. Can there be an algorithm that, given two `Yacc` grammar files (capturing context-free grammar productions) checks whether the grammars encode the same language or not? (`Yacc` is a tool to generate parsers). Write your 'best guess' answer for now; this problem will be formally addressed in Chapter 17.

2.34. What is the symmetric difference between '$\leq$' and '$\geq$'? How about the symmetric difference between '$<$' and '$\leq$'?

2.35. Consider the binary relation *relprime* over $Nat \times Nat$ such that $relprime(x, y)$ exactly when x and y are relatively prime (the greatest common divisor of x and y is 1). Is *relprime* a functional relation? What is its *inverse*? What is its *complement*?

2.36. Consider the 3-ary relation "unequal3," which consists of triples $\langle a, b, c \rangle$ such that $a \neq b$, $b \neq c$, and $a \neq c$. Is this relation a functional relation? Provide reasons.

2.37. How many functions with signature $Bool^k \to Bool$ exist as a function of k? Here $Bool^k$ is $Bool \times Bool \times \ldots \times Bool$ (k times). Think carefully about *all possible distinct functions* that can have this signature. Each k-ary Boolean function can be presented using a $k + 1$-column truth table with each row corresponding to one input and its corresponding output.

2.38. Repeat Exercise 2.37 for partial functions with signature $Bool^k \rightarrow Bool$. View each partial function as a table with the output field being either a Boolean or the special symbol $\perp$, standing for *undefined*.

3

Cardinalities and Diagonalization

In this chapter, we discuss the important idea of measuring sizes of infinite sets. In addition to helping reinforce many mathematical concepts, we obtain a better appreciation of the work of pioneers, notably George Cantor, who originated many of the fundamental ideas in this area. We will employ many of the ideas found in this chapter in later chapters to argue the existence of *non-computable* functions and certain languages called *non Turing-recognizable*—languages for which the membership test (testing whether an arbitrary string is a member of the language)—*cannot* be performed by any machine.

3.1 Cardinality Basics

The *cardinality* of a set is its size. The cardinality of a finite set is measured using natural numbers; for example, the size of $\{1,4\}$ is 2. How do we "measure" the size of infinite sets? The answer is that we use "funny numbers," called *cardinal numbers*. The smallest cardinal number is $\aleph_0$, the next larger cardinal number is $\aleph_1$, and so on. If one infinite set has size $\aleph_0$, while a second has size $\aleph_1$, we will say that the second is larger than the first, even though both sets are infinite. For now, $\aleph_0$ is the number of elements of Nat, while $\aleph_1$ is the number of elements of $Real$. All these ideas will be made clear in this section.

To understand that there could be "smaller" infinities and "bigger" infinities, think of two infinitely sized dogs, Fifi and Howard. While Fifi is infinitely sized, every finite patch of her skin has a finite amount of hair. This means that if one tries to push apart the hair on Fifi's back, they will eventually find two adjacent hairs between which there is no other hair. Howard is not only huge - every finite patch of his skin has *an infinite* amount of hair! This means that if one tries to push apart the hair on Howard's back, they will never find two hairs that are truly adjacent. In other words, *there will be a hair between every pair of hairs!* This can happen if Fifi has $\aleph_0$ amount of hair on her entire body while Howard has $\aleph_1$ amount of hair on his body.[1] Real numbers are akin to hair on Howard's body; there is a real number that lies properly between any two given real numbers. Natural numbers are akin to hair on Fifi's body; there is no natural number between adjacent natural numbers.

We begin with the question of how one "counts" the number of elements in an infinite set. For example, are there the same "number" of natural numbers as there are real numbers? Since we cannot count infinite sets, let us adopt a method that our ancestors sometimes used when they could not count certain *finite* sets.[2] Our ancestors used to conduct trade successfully through the *barter* system without actually counting the number of objects; say, a cabbage for an elephant, and so on.[3] The real idea behind barter is to establish a *bijection* or *correspondence* between two sets of elements without actually counting them. The same technique works quite well when we have to count the contents of infinite sets; in fact, that is the *only technique* that works! But what does *'counting'*, *'countable'*, etc., mean?

3.1.1 Countable sets

A set S is said to be *countable* if there is a 1-1 total mapping from it to natural numbers. (This mapping *need not* be onto). Clearly, finite sets are countable. Consider the infinite set *Odd*, the set of odd numbers. Since there is a total 1-1 mapping $\lambda x.(x-1)/2$ from *Odd* numbers to *Nat*, *Odd* is countable. The set *Real* is not countable, as we shall show in this chapter.

3.1.2 Cardinal numbers

We now discuss the use of cardinal numbers more precisely. The cardinality of *Nat* is *defined* to be $\aleph_0$, written $|Nat| = \aleph_0$. Two sets A and B *have the same cardinality* if *there is a bijection from A to B*. Function $\lambda x.(x-1)/2$ actually serves as a bijection from *Odd* numbers to *Nat*. To sum up,

---

[1] Hope this wouldn't be viewed as splitting hairs...

[2] It is good that Romans didn't discover the concept of Avogadro's number - how could they have carved it out on stone tablets?

[3] Cabbages with magical powers, perhaps.

- *Odd* is countable,
- $|Odd| = |Even| = |Nat| = \aleph_0$,
- But, note that $Odd \subset Nat$ and $Even \subset Nat$;
- Therefore, it is entirely possible that for two sets A and B, $A \subset B$, and yet $|A| = |B|$.

The above example demonstrates that one cannot determine the cardinality of sets purely based on subset relationships. One correct (and handy) method for using subset relationships to determine the cardinality of sets is using *cardinality traps*.

3.1.3 Cardinality "trap"

To motivate the notion of cardinality trap, consider the question, "how many points are there in the map of mainland USA?" Let us treat this map as a region of $Real \times Real$. The theorem which we call *cardinality trap* says:

> If, for three sets A, B, and C, we have $|A| = |C|$ and $A \subset B \subset C$, then $|A| = |B| = |C|$.

Specifically, cardinality trap allows one to "trap" the cardinality of a set B to be between those of two sets A and C. Exercise 3.12 asks you to prove that cardinality trap is a simple corollary of the famous *Schröder-Bernstein theorem*. For the question at hand,

- Any given map of the USA (set B of points) can be properly inscribed within a (larger) square (set C of points).
- Within the given map of the USA, one can properly inscribe a (smaller) square (set A of points).
- All squares in $Real \times Real$ have the same number of points, $\aleph_1$ (this is a result we shall prove later ($|A| = |C|$).
- Therefore, $|A| = |B| = |C|$, or the map of the USA, however drawn, has the same number of points as in a square, namely $\aleph_1$.

Now, we present one of George Cantor's central results, which allows us to prove that two sets have different cardinalities. Known as *the diagonalization method*, it is basically a specific application of the principle of *proof by contradiction*.

3.2 The Diagonalization Method

Let us return to our original question, "is there a bijection from *Nat* to *Real*?" The answer is *no* and we proceed to show how. We follow

the *powerful* approach, developed by Cantor, called *diagonalization.*
Diagonalization is a particular application of the principle of *proof by
contradiction* or *reductio ad absurdum* in which the solution-space is
portrayed as a square matrix, and the contradiction is observed along
the diagonal of this matrix. We now walk you through the proof, providing section headings to the specific steps to be performed along the
way.

Most textbooks prove this result using numbers represented in decimal, which is much easier than what we are going to present in this
section - namely, prove it in binary. We leave the proof in decimal as an
exercise for you. In addition to being a 'fresh,' as well as illuminating
proof, a proof for the binary case also allows us to easily relate cardinality of *Real*s to that of languages over some alphabet. Here, then,
are the steps in this proof.

3.2.1 Simplify the set in question

We first simplify our problem as follows. Note that $(\lambda x.1/(1 + x))$ is
a bijection from $[0, \infty] \subset Real$ to $[0, 1] \subset Real$. Given this, it suffices
to show that there is *no* bijection from *Nat* to $[0, 1] \subset Real$, since
bijections are closed under composition. We do this because the interval
$[0, 1]$ is "easier to work with." We can use *binary fractions* to capture
each number in this range, and this will make our proof convenient to
present.

3.2.2 Avoid dual representations for numbers

The next difficulty we face is that certain numbers have *two* fractional
representations. As a simple example, if the manufacturer of Ivory soap
claims that their soap is 99.99% pure, it is not the same as saying it is
99.999% pure.[4] However, if they claim it is 99.9$\overline{9}$% pure (meaning an infinite number of 9s following the fractional point), then it is equivalent
to saying it is 100% pure. Therefore, in the decimal system, infinitely
repeating 9s can be represented without infinitely repeating 9s. As another example, 5.123$\overline{9}$ = 5.124. The same 'dual representations' exist in
the binary system also. For example, in the binary system, the fraction
0.010$\overline{0}$ (meaning, 0.010 followed by an infinite number of 0s) represents
0.25 in decimal. However, the fraction 0.010$\overline{1}$ (0.010 followed by an infinite number of 1s) represents 0.011$\overline{0}$ in binary, or 0.375 in decimal.
Since we would like to avoid dual representations, we will avoid dealing

---

[4] Such Ivory soap may still float.

with number 1.0 (which has the dual representation of $0.\bar{1}$). Hence, we
will perform our proof by showing that there is no bijection from Nat
to $[0, 1) \subset Real$. This would be an even stronger result.

Let us represent each real number in the set $[0, 1) \subset Real$ in binary.
For example, 0.5 would be $0.100\ldots$, 0.375 would be $0.01100\ldots$. We
shall continue to adhere to our convention that *we shall never use any
bit-representation involving* $\bar{1}$. Fortunately, every number in $[0, 1)$ can
be represented without ever using $\bar{1}$. (This, again, is the reason for
leaving out 1.0, as we don't wish to represent it as $0.\bar{1}$, or 1.0).

3.2.3 Claiming a bijection, and refuting it

For the simplicity of exposition, we first present a proof that is "nearly
right," and much simpler than the actual proof. In the next section, we
repair this proof, giving us the actual proof. Suppose there is a bijection
f that puts Nat and $[0, 1)$ in correspondence C1 as follows:

$$0 \rightarrow .b_{00}b_{01}b_{02}b_{03}\ldots$$
$$1 \rightarrow .b_{10}b_{11}b_{12}b_{13}\ldots$$
$$\ldots$$
$$n \rightarrow .b_{n0}b_{n1}b_{n2}b_{n3}\ldots$$
$$\ldots$$

where each b_{ij} is 0 or 1.

Now, consider the real number

$$D = 0.\neg b_{00} \ \neg b_{11} \ \neg b_{22} \ \neg b_{33}\ldots.$$

This number is *not* in the above listing, because it differs from the i-
th number in bit-position b_{ii} **for every** i. Since this number D is not
represented, f cannot be a bijection as claimed. Hence such an f does
not exist.

3.2.4 'Fixing' the proof a little bit

Actually the above proof needs a small "fix"; *what if the complement
of the diagonal happens to involve a* $\bar{1}$? The danger then is that we
cannot claim that a number equal to the complemented diagonal does
not appear in our listing. It might then end up existing in our listing
of Reals in a "non $\bar{1}$ form."

We overcome this problem through a simple correction.[5] This correction ensures that the complemented diagonal will never contain a

---

[5] Exercise 3.6 asks you to propose an alternative correction.

$\overline{1}$. In fact, we arrange things so that the complemented diagonal will contain zeros infinitely often. This is achieved by placing a 1 in the un-complemented diagonal every so often; we choose to do so for all *even* positions, by listing the *Real* number $.1^{2n+1}\overline{0}\ldots$ ($2n+1$ 1s followed by $\overline{0}$) at position $2n$, for all n. Consider the following correspondence, for example:

$$0 \to .1\overline{0}$$
$$1 \to .c_{00}c_{01}c_{02}c_{03}\ldots$$
$$2 \to .111\overline{0}$$
$$3 \to .c_{10}c_{11}c_{12}c_{13}\ldots$$
$$4 \to .11111\overline{0}$$
$$5 \to .c_{20}c_{21}c_{22}c_{23}\ldots$$
$$6 \to .1111111\overline{0}$$
$$\ldots$$
$$2n \to .1^{2n+1}\overline{0}\ldots$$
$$2n+1 \to .c_{n0}c_{n1}c_{n2}c_{n3}\ldots$$
$$\ldots$$

Call this correspondence C2. We obtain C2 as follows. We know that the numbers $.1\overline{0}$, $.111\overline{0}$, $.11111\overline{0}$, etc., exist in the original correspondence C1. C2 is obtained from C1 by first permuting it so that the above elements are moved to the *even positions* within C2 (they may exist arbitrarily scattered or grouped, within C1). We then go through C1, strike out the above-listed elements, and list its remaining elements in the odd positions within C2. We represent C2 using rows of $.c_{ij}$, as above.

We can now finish our argument as follows. The complemented diagonal doesn't contain a $\overline{1}$, because it contains 0 occurring in it infinitely often. Now, this complemented diagonal cannot exist anywhere in our $.c_{ij}$ listing. The complemented diagonal is certainly a Real number missed by the original correspondence C1 (and hence, also missed by C2). Hence, we arrive at a contradiction that we have a *correspondence*, and therefore, we cannot assign the same cardinal number to the set $[0,1) \subseteq Real$. It is therefore of higher cardinality.

The conclusion we draw from the above proof is that *Real* and *Nat* have different cardinalities. Are there any cardinalities "in between" that of *Real* and *Nat*? Loosely speaking, "is there a $\aleph_{0.5}$?!" The hypothesis that states "no there isn't a cardinality between $\aleph_0$ and $\aleph_1$," or in other words, "there isn't a $\aleph_{0.5}$," is known as the *Continuum Hypothesis*. It has been a problem of intense study over the last 120 years, and in fact is the *first* of Hilbert's 23 challenges to computer science [54].

These challenges helped spur considerable amounts of research in Computer Science, and contributed to much of the foundational knowledge of the subject area (e.g., as covered in this book). For further details, please see [26]. We shall use cardinality arguments when comparing the set of all functions and the set of all *computable* functions.

3.2.5 Cardinality of 2^{Nat} and $Nat \rightarrow Bool$

In this section, we argue that the sets 2^{Nat} (the powerset of Nat) and $Nat \rightarrow Bool$ (the set of functions from Nat to $Bool$) have the same cardinality as *Real*. Notice that each set within 2^{Nat} can be represented by an infinitely long characteristic sequence. For instance, the sequence $10010100\overline{0}$ represents the set $\{0, 3, 5\}$; the sequence $101010\ldots$ represents the set *Even*; the sequence $010101\ldots$ represents the set *Odd*; and so on. Notice that the very same characteristic sequences also represent functions from Nat to $Bool$. For instance, the sequence $10010100\overline{0}$ represents the function that maps 0, 3, and 5 to *true*, and the rest of Nat to *false*; the sequence $101010\ldots$ represents the function $\lambda x.even(x)$; and the sequence $010101\ldots$ represents the function $\lambda x.odd(x)$. Hence, the above two sets have the same cardinality as the set of all infinitely long bit-sequences. How many such sequences are there? By putting a "0." before each such sequence, it appears that we can define the *Reals* in the range $[0, 1]$. However, we face the difficulty caused by infinite 1s, i.e., we will end up having $\overline{1}$ occurring within an infinite number of infinite sequences. Therefore, we cannot directly use the arguments in Section 3.2.2, which rely on such numbers being absent from the listing under consideration.

We now present the Schröder-Bernstein Theorem which allows us to handle this, and other "hard-to-count" sets, very cleanly. We present the theorem and its applications in the next section.

3.3 The Schröder-Bernstein Theorem

Theorem 3.1. (Schröder-Bernstein Theorem): *For any two sets A and B, if there is a 1-1, total, and into map f going from A to B, and another 1-1, total, and into map g going from B to A, then these sets have the same cardinality.*

Section 3.3.3 discusses a proof of this theorem.

3.3.1 Application: cardinality of all C Programs

As our first application of the Schröder-Bernstein Theorem, let us arrive at the cardinality of the set of all C programs, CP. We show that this is $\aleph_0$ by finding 1-1, total, and *into* maps from Nat to CP and vice versa. The real beauty of this theorem is that we can find such maps *completely arbitrary*. For instance, we consider the class of C programs beginning with `main(){}`. This is, believe it or not, a *legal C program*! The next longer, such "weird but legal" C program, is `main(){;}`. The next ones are `main(){;;}`, `main(){;;;}`, `main(){;;;;}`, and so on! Now,

- A function $f : Nat \to CP$ that is 1-1, total, and into is the following:
  - Map 0 into the legal C program, `main(){}`
  - Map 1 into another legal C program `main(){;}`
  - Map 2 into another legal C program `main(){;;}`
  - ..., map i into the C program `main(){;`^{i}`}`—*i.e.*, one that contains i occurrences of `;`.
- A function $g : CP \to Nat$ that is 1-1, total, and into is the following: view each C program as a string of bits, and obtain the value of this bit-stream viewed as an unsigned binary number.

By virtue of the existence of the above functions f and g, from the Schröder-Bernstein Theorem, it follows that $|CP| = |Nat|$.

3.3.2 Application: functions in $Nat \to Bool$

We have already shown that such functions can be viewed as infinite bit-sequences, IBS. As already pointed out, we cannot interpret such sequences straightforwardly as Real numbers in $[0, 1)$ because of the presence of $\bar{1}$ that gives rise to multiple representations. We use the following alternative approach:

- We map every member of IBS that *does not contain* an occurrence of $\bar{1}$ into the range $[0, 1)$ by putting a "0." before it, and interpreting it as a Real number.
- We map every member of IBS that *contains* an occurrence of $\bar{1}$ into the range $[1, 2]$ by putting a "1." before it, interpreting it as a Real number in $[1, 2]$, and converting it into an equivalent form without $\bar{1}$. For example, $0.01010\bar{1}$ is mapped to $1.01011\bar{0}$.
- This is a 1-1, total, and *into* map from IBS to $[0, 2]$. Composing this map with the scale-factor $\lambda x.(x/2)$, we obtain the desired function f that goes from IBS into $[0, 1]$.

- Now, we obtain function g that is 1-1, total, and into, going from $[0, 1]$ to IBS as follows:
  - map every number except 1 in this range to the corresponding number in IBS (with a "0." before it) that does not contain $\bar{1}$.
  - map 1 to $0.\bar{1}$.
  This mapping hits every member of IBS except those containing $\bar{1}$ (the only exception being for 1). This is 1-1, total, and into.

From the Schröder-Bernstein Theorem, it follows that $|IBS| = |Real|$.

Illustration 3.3.1 *Using the Schröder-Bernstein Theorem, define a bijection between $Nat \times Int$ and Int.*
Solution: Using the SB theorem, we just need to find a 1-1 total into maps going from $Nat \times Int$ to Int and one going from Int to $Nat \times Int$. Here is the first map:

$$\lambda\langle x, y\rangle.sign(y) \times (2^x \times 3^{|y|}).$$

That this map is total is obvious because $\times$ is defined everywhere. Why is this 1-1? That's because of the unique prime decomposition of any number (note that we are not ignoring the sign). It is into because we can't generate numbers that are a multiple of 5, 7, etc.

The reverse map is much easier: just pair the Int with some arbitrary Nat:

$$\lambda x.\langle 0, x\rangle.$$

How about finding a bijection directly? It can be done as follows (but it will become apparent *how much harder* this is, compared to using the Schröder-Bernstein Theorem):

List all $\langle Nat, Int \rangle$ pairs systematically, then list all Ints systematically against it.

Listing all of the former proceeds systematically as follows:

- All that "add up to 0 ignoring signs," with signs later attached in every possible way. $\langle 0, 0 \rangle$ is the only pair that adds up to 0. Additionally, -0 is 0.
- All that "add up to 1 ignoring signs," with signs later attached in every possible way. $\langle 0, 1 \rangle$ and $\langle 1, 0 \rangle$ add up to 1. List them as follows:
  $$\langle 0, 1 \rangle, \langle 0, -1 \rangle, \langle 1, 0 \rangle.$$
- All that "add up to 2 ignoring signs," with signs later attached in every possible way. $\langle 0, 2 \rangle, \langle 1, 1 \rangle$, and $\langle 2, 0 \rangle$ add up to 2. List them as follows:

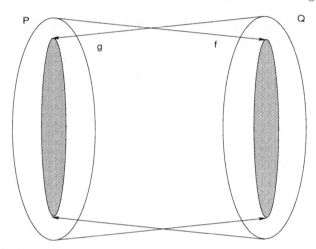

Fig. 3.1. Proof of the Schröder-Bernstein Theorem

$$\langle 0,2 \rangle, \langle 0,-2 \rangle, \langle 1,1 \rangle, \langle 1,-1 \rangle, \langle 2,0 \rangle.$$

The full bijection looks like the following:

```
<0,0>  -> 0    <0,1>  -> 1      <0,2>  -> -2
               <0,-1> -> -1     <0,-2> -> 3
               <1,0>  -> 2      <1,1>  -> -3
                                <1,-1> -> 4
                                <2,0>  -> -4


<0,3>  -> 5    <0,4>  -> -8
<0,-3> -> -5   <0,-4> -> 9
<1,2>  -> 6    <1,3>  -> -9
<1,-2> -> -6   <1,-3> -> 10
<2,1>  -> 7    <2,2>  -> -10
<2,-1> -> -7   <2,-2> -> 11
<3,0>  -> 8    <3,1>  -> -11
               <3,-1> -> 12
               <4,0>  -> -12
```

3.3.3 Proof of the Schröder-Bernstein Theorem

We offer a proof of this theorem due to Hurd [62]. Page 49 provides an alternative proof similar to that in [50].

Our goal is to show that if there are two sets P and Q, with injections $f : P \rightarrow Q$ and $g : Q \rightarrow P$, there exists a bijection $h : P \rightarrow Q$ as in Figure 3.1.

Imagine P and Q being mirrors facing each other. Rays emanating from P fall within Q, and vice versa. It stands to reason that these sets

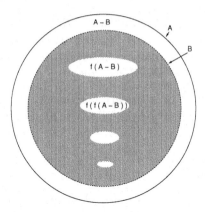

Fig. 3.2. Proof approach

then have the same cardinality. Our proof will exploit this analogy. We will first prove Lemma 3.2 and then finish our proof.

Lemma 3.2. *For two sets A and B, if $B \subset A$ and there exists an injection $f : A \to B$, then there exists a bijection $j : A \to B$.*

Proof: Consider A and B as in Figure 3.2. The ring-shaped region $A - B$ can be regarded as $f^0(A - B)$. The diagram also shows $f(A - B)$, $f(f(A - B))$, etc. (We will call $f^i(A - B)$ for $i \geq 1$ as "thought bubbles" in what follows). Since f is an injection from A to B, $f(A - B)$ is embedded within B, and likewise are the other $f^i(A - B)$ regions. Let $X = \cup_{i \geq 0} f^i(A - B)$, i.e., the union of the ring and all the "thought bubbles" embedded within B. Also, observe that $f(X)$ are the "thought bubbles" alone. So A can be written as $X \cup (B - f(X))$. We will exploit this fact in creating a bijection from A to B, as follows:

- Suppose we propose the mapping $j : A \to B$ as $\lambda a.$ *if* $a \in X$ *then* $f(a)$ *else* a. In other words, everything within the ring and the "thought bubbles" maps to the "thought bubbles" $(f(X))$, and everything in the remaining region (which is the shaded region $B - f(X)$) maps to itself.
- Suppose we show that f is a bijection from X to $f(X)$ (Claim 1 below); then j will be a bijection from A to B (j is defined partly by the identity map and partly by f, both of which are injective; furthermore, j's domain covers A and range covers B).

Now, going back to the original problem, the desired bijection from P to Q is found as follows:

- $g(Q)$ is a subset of A. Furthermore, the composition of f and g, $g \circ f$, is an injection from A to $g(B)$. Hence, there is a bijection j from A to $g(B)$. Now, since g is injective, g^{-1} exists. Therefore, the desired bijection from A to B is $inv(g) \circ j$.

Proof of Claim 1: f is a bijection from X to $f(X)$ because f is surjective, and f is injective from A to B; hence, it must be so when restricted to domain X.

Illustration of the Proof

Let us choose $P = O$ (O standing for *Odd*, the set of positive odd numbers) and $Q = E$ (standing for *Even*, the set of positive even numbers). Clearly, $\lambda x.x - 1$ is a bijection that can be found manually. Let's see which bijection is found by the method in the proof (which, of course, works automatically for any situation).

According to the Schröder-Bernstein theorem, if there is an injection from O into E and vice versa, there exists a bijection between O and E. Two such injections are easy to find. For example, $f = \lambda x.2x$ and $g = \lambda x.(2x + 1)$. The proof gives a way to construct another bijection automatically from these.

- The set A of the lemma is O, the *Odd* numbers.
- Applying f to O, we get $f(O) = 2, 6, 10, 14, \ldots$.
- Applying g to E, we get $g(E) = 1, 5, 9, 13, \ldots$. Call this set O_1.
- Now, $g \circ f = \lambda x.(4x + 1)$. Applying this to O, we get $g \circ f(O) = 5, 13, 21, 29, \ldots$. Call this set O_2. Notice that $O_2 \subset O_1 \subset O$.
- Now, as far as the lemma goes, the set "A" is O and "B" is O_1. There is a bijection (namely $g \circ f$) from O to O_2. Therefore, the same function is an injection from O to O_1.
- Now, we can build function j as suggested in the lemma as follows:
  - The set $A - B$ is $O - O_1 = 3, 7, 11, 15, 19, \ldots$. This is the "outer ring."
  - The "f" of the lemma is $g \circ f = \lambda x.(4x + 1)$. Applying this to $O - O_1$, we get $13, 29, 45, 61, 77, \ldots$, which is the first "thought bubble."
  - Similarly, $f^2(O - O_1) = 53, 117, 181, 145, \ldots$.
  - X is the union of all the above sets.
- Now, the j function is $\lambda a.if\ a \in (O_1 - \cup_{i \geq 1} f(O - O_1))\ then\ a\ else\ f(a)$
- Equivalently, j is $\lambda a.if\ a \in (O - X)\ then\ a\ else\ f(a)$
- Hence, the j function is given by the map
  $1 \to 1$, $3 \to 13$, $5 \to 5$, $7 \to 29$, $9 \to 9$, $11 \to 45, \ldots$ This is because $1, 5, 9, \ldots$ is what's missed by X.

- Finally, the bijection from O to E is given by $g^{-1} \circ j$, which is illustrated by the following table where the first map is the application of j and the second is the application of g^{-1}:

```
 1 ->  1 ->  0
 3 -> 13 ->  6
 5 ->  5 ->  2
 7 -> 29 -> 14
 9 ->  9 ->  4
11 -> 45 -> 22
...
```

Alternate Proof—a sketch

The classical proof of the Schröder-Bernstein Theorem goes as follows (the reader may draw a diagram to better understand the proof). Consider the injections $f : P \to Q$ and $g : Q \to P$. Let $P_0 = P - g(Q)$ and $Q_0 = Q - f(P)$. Let $Q_i = f(P_{i-1})$ and $P_i = g(Q_{i-1})$. Now, $f : P_{i-1} \to Q_i$ and $g : Q_{i-1} \to P_i$ are bijections.

Define $P_{even} = \cup_{even\ n} P_n$, and likewise define P_{odd}, Q_{odd}, and Q_{even}. Define $P_\infty = P - (P_{even} \cup P_{odd})$, and similarly for Q. Now define the desired bijection $h : P \to Q$ to be f on P_{even}, g^{-1} on P_{odd}, and either on P_∞.

Cardinal Numbers

The topic of cardinalities is quite detailed; we have barely scratched the surface. One of the more important take away messages is that sets such as natural numbers are *denumerable* or *countable*, while sets such as Reals are not *denumerable* or are *uncountable*. Hence, we cannot list real numbers as "the first real," "the second real," etc., because they are not denumerable. How about the powerset of Reals? Are they of the same cardinality as Reals? We conclude with a theorem that shows that there are more than a finite set of cardinalities (we omit delving into sharper details).

Theorem 3.3. *For a set S and its powerset 2^S, we have $|S| \prec |2^S|$, where $\prec$ says "of lower cardinality."*

The notion of cardinal numbers basically stems from this theorem. The cardinal numbers form a sequence $\aleph_0$, $\aleph_1$, $\aleph_2$, ..., beginning with the cardinality of the set of natural numbers $\aleph_0$ and going up in cardinality each time a powerset operation is invoked on the earlier set ($\aleph_1$ for Reals, $\aleph_2$ for the powerset of Reals, etc.).

The above would be a generalization of the diagonalization proof. The style of the proof is precisely what will be used to prove the undecidability of the Halting problem.

Chapter Summary

This chapter introduced the concept of cardinality and illustrated how various infinite sets might have incomparable cardinalities. The main technique for demonstrating that two sets have *different* cardinalities is a proof technique by contradiction known as *diagonalization*. A useful theorem for showing that two sets have the same cardinality is the Schröder-Bernstein theorem. A thorough description of these concepts in this early of a chapter has been found to be helpful to many students when they study later chapters of this book.

Exercises

3.1. What is the cardinality of the set of all possible C programs?

3.2. What is the cardinality of the set of all binary files created by C compilers (`a.out` files)?

3.3. What is the cardinality of the set of all fundamental particles in the universe?

3.4. Show that $|Real| = |Real \times Real|$. *Hint:* First map the numbers into $[0, 1]$. Thereafter, interleave the bits of pairs of reals to get a unique mapping to a single real. For example, given the pair

$$\langle .b_{00}b_{01}b_{02}b_{03}\ldots, .c_{00}c_{01}c_{02}c_{03}\ldots\rangle,$$

we can uniquely obtain the single Real number $.b_{00}c_{00}b_{01}c_{01}b_{02}c_{02}\ldots$.

3.5. Show that there are more points on the map of the USA (viewed as a set of points $\subset |Real \times Real|$) than hair on an infinitely sized dog such as Fifi (even though Fifi is infinitely sized, there isn't a hair between every pair of her hairs – unlike with Howard). *Hint:* Use the idea of *cardinality trap*.

3.6. Propose another trick to deal with Reals of a "non $\overline{1}$ form" discussed in Section 3.2.4.

3.7. What is the cardinality of the set of all binary search routines one can write in C to search an array of characters (a through z) of size 4?

3.8. Compare the cardinality of the powerset of *Odd* and the powerset of the powerset of *Even*. Argue whether these cardinalities are the same, giving a semi-formal proof.

3.9. What is the cardinality of the set of all finite staircase patterns in 2D space? Finite staircase patterns are curves that go a finite distance along the positive x axis ('right'), then a finite distance along the positive y axis ('upwards'), and repeats this pattern a finite number of times. The corners of the staircase may have coordinates that are pairs of real numbers.

3.10. Repeat Exercise 3.9 if the corners of the staircase are restricted to integer coordinate pairs.

3.11. Consider the directed graphs in Figure 22.6. For each graph, answer the following questions:

1. What is the cardinality of the set of all finite paths?
2. What is the cardinality of the set of all infinite paths?

3.12. Prove the theorem of *cardinality trap* presented on page 39.

3.13. Illustrate this proof of the Schröder-Bernstein Theorem on the above example involving *Odd* and *Even*.

3.14. Prove Theorem 3.3 by assuming there exists a total bijection f from S to 2^S, and arguing whether there can exist an element of S which maps to the following set:

$$D = \{x \mid x \in S \wedge x \notin f(x)\}.$$

If such an element d exists, first ask whether $d \in D$, and then ask if $d \notin D$. Derive a contradiction.

3.15. (The first part of the following problem was narrated to me by Prof. Riesenfeld, my faculty colleague.) (i) Someone starts somewhere on earth, walks 10 miles south, then 10 miles west, and finally 10 miles north, and returns to where he started from. Where could he have started such a journey from (provide two examples of where he could have started the journey from). (ii) Find out the size of the *set* of solutions to this problem.

4

Binary Relations

Binary relations help impart structure to sets of related elements. They help form various meaningful *orders* as well as *equivalences*, and hence are central to mathematical reasoning. Our definitions in this chapter follow [50, 77, 63] closely.

4.1 Binary Relation Basics

A binary relation R on S is a subset of $S \times S$. It is a relation that can be expressed by a 2-place predicate. Examples: (i) x loves y, (ii) $x > y$.

Set S is the *domain* of the relation. Theoretically, it is possible that the domain S is empty (in which case R will be empty). In all instances that we consider, the domain S will be non-empty. However, it is quite possible that S is non-empty and R is empty.

We now proceed to examine various types of binary relations. In all these definitions, we assume that the binary relation R in question is *on S*, *i.e.*, a subset of $S \times S$. For a relation R, two standard prefixes are employed: *irr-* and *non-*. Their usages will be clarified in the sequel.

Relations can be depicted as graphs. Here are conventions attributed to Andrew Hodges in [63]. The domain is represented by a closed curve (e.g., circle, square, etc) and the individuals in the domain by dots labeled, perhaps, a, b, c, and so on. The fact that $\langle a, b \rangle \in R$ will be depicted by drawing a *single arrow* (or equivalently *one-way arrow*) from dot a to dot b. We represent the fact that both $\langle a, b \rangle \in R$ and $\langle b, a \rangle \in R$ by drawing a double arrow between a and b. We represent the fact that $\langle a, a \rangle \in R$ by drawing a double arrow from a back to itself (this is called a *loop*). We shall present examples of these drawings in the sequel.

4.1.1 Types of binary relations

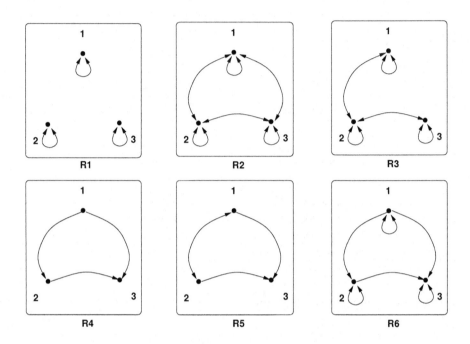

Fig. 4.1. Some example binary relations

We shall use the following examples. Let $S = \{1, 2, 3\}$, $R_1 = \{\langle x, x \rangle \mid x \in S\}$, $R_2 = S \times S$, and

$$R_3 = \{\langle 1, 1 \rangle, \langle 2, 2 \rangle, \langle 3, 3 \rangle, \langle 1, 2 \rangle, \langle 2, 1 \rangle, \langle 2, 3 \rangle, \langle 3, 2 \rangle\}.$$

All these (and three more) relations are depicted in Figure 4.1.

Reflexive, and Related Notions

R is **reflexive**, if for all $x \in S$, $\langle x, x \rangle \in R$. Equivalently,

    In R's graph, there is no dot without a loop.

Informally, "every element is related to itself."

A relation R is **irreflexive** if there are *no* reflexive elements; i.e., for no $x \in S$ is it the case that $\langle x, x \rangle \in R$. Equivalently,

    In R's graph, no dot has a loop.

Note that irreflexive is not the negation (complement) of reflexive. This is because the logical negation of the definition of reflexive would be, "there exists $x \in S$ such that $\langle x, x \rangle \notin R$. This is not the same as irreflexive because *all* such pairs must be absent in an irreflexive relation.

A relation R is **non-reflexive** if it is neither reflexive nor irreflexive. Equivalently,

> In R's graph, at least one dot has a loop and at least one dot does not.

Examples:

- R_1, R_2, R_3 are all reflexive.
- $R = \emptyset$ is reflexive and irreflexive. It is not non-reflexive.
- For $x, y \in Nat$, $x = y^2$ is non-reflexive (true for $x = y = 1$, false for $x = y = 2$).

Symmetric, and Related Notions

R is *symmetric* if for all $x, y \in S$, $\langle x, y \rangle \in R \Rightarrow \langle y, x \rangle \in R$. Here, x and y need not be distinct. Equivalently,

> In R's graph, there are no single arrows. If the relation holds one way, it also holds the other way.

Examples: R_1, R_2, and R_3 are symmetric relations. Also note that $\emptyset$ is a symmetric relation.

R is **asymmetric** if there exists no two distinct $x, y \in S$ such that $\langle x, y \rangle \in R \wedge \langle y, x \rangle \in R$. In other words, **if** $\langle x, y \rangle \in R$, **then** $\langle y, x \rangle \notin R$. Example: "elder brother" is an asymmetric relation, and so is $<$ over Nat. Equivalently,

> There are no double arrows in its graph; if the relation holds one way, it does not hold the other.

Again, note that *asymmetric* is *not* the same as the negation of (the definition of) symmetric. The negation of the definition of symmetric would be that *there exists* distinct x and y such that $\langle x, y \rangle \in R$, but $\langle y, x \rangle \notin R$.

R is non-symmetric if it is neither symmetric nor asymmetric (there is at least one single arrow and at least one double arrow).

Example: $\emptyset$ is symmetric and asymmetric, but not non-symmetric.

R is **antisymmetric** if for all $x, y \in S$, $\langle x, y \rangle \in R \wedge \langle y, x \rangle \in R \Rightarrow x = y$ (they are the same element). Equivalently,

There is no double arrow unless it is a loop.

Antisymmetry is a powerful notion that, unfortunately, is too strong for many purposes. Consider the elements of 2^S, the powerset of S, as an example. If, for any two elements x and y in S, we have $x \subseteq y$ and $y \subseteq x$, then we can conclude that $x = y$. Therefore, the set containment relation $\subseteq$ is *antisymmetric*; and hence, antisymmetry is appropriate for comparing two sets in the "less than or equals" sense.

Consider, on the other hand, two basketball players, A and B. Suppose the coach of their team defines the relation $\preceq_{BB}$ as follows: $A \preceq_{BB} B$ if and only if B has more abilities or has the same abilities as A. Now, if we have two players x and y such that $x \preceq_{BB} y$ and $y \preceq_{BB} x$, we can conclude that they have identical abilities - they don't end up becoming the very same person, however! Hence, $\preceq_{BB}$ must not be antisymmetric. Therefore, depending on what we are comparing, antisymmetry may or may not be appropriate.

Transitive, and Related Notions

To define transitivity in terms of graphs, we need the notions of a *broken journey* and a *short cut*. There is a broken journey from dot x to dot z via dot y, if there is an arrow from x to y and an arrow from y to z. Note that dot x might be the same as dot y, and dot y might be the same as dot z. Therefore if $\langle a, a \rangle \in R$ and $\langle a, b \rangle \in R$, there is a broken journey from a to b via a. Example: there is a broken journey from Utah to Nevada via Arizona. There is also a broken journey from Utah to Nevada via Utah.

There is a short cut just if there is an arrow direct from x to z. So if $\langle a, b \rangle \in R$ and $\langle b, c \rangle \in R$ and also $\langle a, c \rangle \in R$, we have a broken journey from a to c via b, together with a short cut. Also if $\langle a, a \rangle \in R$ and $\langle a, b \rangle \in R$, there is a broken journey from a to b via a, together with a short cut.

Example: There is a broken journey from Utah to Nevada via Arizona, and a short cut from Utah to Nevada.

R is **transitive** if for all $x, y, z \in S$, $\langle x, y \rangle \in R \ \wedge \ \langle y, z \rangle \in R \ \Rightarrow \ \langle x, z \rangle \in R$. Equivalently,

There is no broken journey without a short cut.

R is **intransitive** if, for all $x, y, z \in S$, $\langle x, y \rangle \in R \ \wedge \ \langle y, z \rangle \in R \ \Rightarrow \ \langle x, z \rangle \notin R$. Equivalently,

There is no broken journey *with* a short cut.

R is **non-transitive** if and only if it is neither transitive nor intransitive. Equivalently,

> There is at least one broken journey with a short cut and at least one without.

Examples:

- Relations R_1 and R_2 above are transitive.
- R_3 is non-transitive, since it is lacking the pair $\langle 1, 3 \rangle$.
- Another non-transitive relation is $\neq$ over Nat, because from $a \neq b$ and $b \neq c$, we cannot always conclude that $a \neq c$.
- R_4 is irreflexive, transitive, and asymmetric.
- R_5 is still irreflexive. It is not transitive, as there is no loop at 1. It is not intransitive because there *is* a broken journey (2 to 3 via 1) with a short cut (2 to 1). It is non-transitive because there is one broken journey without a short cut and one without.
- R_5 is not symmetric because there *are* single arrows.
- R_5 is not asymmetric because there *are* double arrows.
- From the above, it follows that R_5 is non-symmetric.
- R_5 is not antisymmetric because there is a double arrow that is not a loop.

4.1.2 Preorder (reflexive plus transitive)

If R is reflexive and transitive, then it is known as a *preorder*. Continuing with the example of basketball players, let the $\preceq_{BB}$ relation for three members A, B, and C of the team be

$$\{\langle A, A \rangle, \langle A, B \rangle, \langle B, A \rangle, \langle B, B \rangle, \langle A, C \rangle, \langle B, C \rangle, \langle C, C \rangle\}.$$

This relation is a *preorder* because it is reflexive and transitive. It helps compare three players A, B, and C, treating A and B to be equivalent in abilities, and C to be superior in abilities to both.

In Section 4.3, we present a more elaborate example of a preorder.

4.1.3 Partial order (preorder plus antisymmetric)

If R is reflexive, antisymmetric, and transitive, then it is known as a *partial order*. As shown in Section 4.1.1 under the heading of *antisymmetry*, the subset or equals relation $\subseteq$ is a partial order.

4.1.4 Total order, and related notions

A *total order* is a special case of a partial order. R is a total order if for all $x, y \in S$, either $\langle x, y \rangle \in R$ or $\langle y, x \rangle \in R$. Here, x and y need not be distinct (this is consistent with the fact that total orders are reflexive).

The $\leq$ relation on Nat is a total order. Note that '$<$' is not a total order, because it is not reflexive.[1] However, '$<$' is transitive. Curiously, '$<$' is antisymmetric.

A relation R is said to be **total** if for all $x \in S$, there exists $y \in S$ such that $\langle x, y \rangle \in R$. In other words, a "total" relation is one in which every element x is related to at least one other element y. If we consider y to be the image (mapping) of x under R, this definition is akin to the definition of a *total* function.

Note again that R being a *total order* is **not** the same as R being a partial order and a total relation. For example, consider the following relation R over set $S = \{a, b, c, d\}$:

$$R = \{\langle a, a \rangle, \langle b, b \rangle, \langle c, c \rangle, \langle d, d \rangle, \langle a, b \rangle, \langle c, d \rangle\}$$

R is a partial order. R is also a total relation. However, R is *not* a total order, because there is *no* relationship between b and c (neither $\langle b, c \rangle$ nor $\langle c, b \rangle$ is in R).

4.2 Equivalence (Preorder plus Symmetry)

An equivalence relation is reflexive, symmetric, and transitive. Consider the $\preceq_{BB}$ relation for three basketball players A, B, and C. Now, consider a "specialization" of this relation obtained by leaving out certain edges:

$$\equiv_{BB} = \{\langle A, A \rangle, \langle A, B \rangle, \langle B, A \rangle, \langle B, B \rangle, \langle C, C \rangle\}.$$

This relation is an equivalence relation, as can be easily verified.

Note that $\equiv_{BB} = \preceq_{BB} \cap \preceq_{BB}^{-1}$. In other words, this equivalence relation is obtained by taking the preorder $\preceq_{BB}$ and intersecting it with its inverse. The fact that $\preceq_{BB} \cap \preceq_{BB}^{-1}$ is an equivalence relation is not an accident. The following section demonstrates a general result in this regard.

---

[1] Some authors are known to abuse these definitions, and consider $<$ to be a total order. It is better referred to as *strict* total order or *irreflexive* total order.

4.2.1 Intersecting a preorder and its inverse

Theorem 4.1. *The relation obtained by intersecting a preorder r with its inverse r^{-1} is an equivalence relation.*

Proof: First we show that $R \cap R^{-1}$ is reflexive. Let R be a preorder over set S. Therefore, R is reflexive, *i.e.*, it contains $\langle x, x \rangle$ pairs for every $x \in S$. From the definition of R^{-1}, it also contains these pairs. Hence, the intersection contains these pairs. Therefore, $R \cap R^{-1}$ is reflexive.

Next we show that $R \cap R^{-1}$ is symmetric. That is, to show that for every $x, y \in S$, $\langle x, y \rangle \in R \cap R^{-1}$ implies $\langle y, x \rangle \in R \cap R^{-1}$. If the antecedent, *i.e.*, $\langle x, y \rangle \in R \cap R^{-1}$ is false, the assertion is vacuously true. Consider when the antecedent is true for a certain $\langle x, y \rangle$. These x and y must be such that $\langle x, y \rangle \in R$ and $\langle x, y \rangle \in R^{-1}$. The former implies that $\langle y, x \rangle \in R^{-1}$. The latter implies that $\langle y, x \rangle \in R$. Hence, $\langle y, x \rangle \in R \cap R^{-1}$. Hence, $R \cap R^{-1}$ is symmetric.

Next we prove that $R \cap R^{-1}$ is transitive. Since R is a preorder, it is transitive. We now argue that the inverse of any transitive relation is transitive. From the definition of transitivity, for every $x, y, z \in S$, from the antecedents $\langle x, y \rangle \in R$ and $\langle y, z \rangle \in R$, the consequent $\langle x, z \rangle \in R$ follows. From these antecedents, we have that $\langle y, x \rangle \in R^{-1}$ and $\langle z, y \rangle \in R^{-1}$ respectively. From the above conclusion $\langle x, z \rangle \in R$, we can infer that $\langle z, x \rangle \in R^{-1}$. Hence, R^{-1} is transitive and so is the conjunction of R and R^{-1}.

4.2.2 Identity relation

Given a set S, *the* identity relation R over S is $\{\langle x, x \rangle \mid x \in S\}$. An identity relation is one extreme (special case) of an equivalence relation. This relation is commonly denoted by the equality symbol, $=$, and relates equals with equals. Please note the contrast with Theorem 4.1.

4.2.3 Universal relation

The *universal relation* R over S is $S \times S$, and represents the other extreme of relating everything to everything else. This is often an uninteresting binary relation.[2]

---

[2] This is sort of what would happen if one were to give everyone in a theory class an 'A' grade.

4.2.4 Equivalence class

An equivalence relation R over S partitions $elements(R) = domain(R) \cup codomain(R)$ into *equivalence classes*. Intuitively, the equivalence classes E_i are those subsets of $elements(R)$ such that every pair of elements in E_i is related by R, and E_is are the maximal such subsets. More formally, given an equivalence relation R, there are two cases:

1. *R is a universal relation.* In this case, there is a single equivalence class E_1 associated with R, which is $elements(R)$ itself.
2. *R is not a universal equivalence relation.* In this case, an equivalence class E_i is a maximal proper subset of $elements(R)$ such that the restriction of R on E_i, $R\mid_{E_i}$, is universal (meaning that every pair of elements inside each of the E_is is related by R).

Putting it all together, the set of all equivalence classes of an equivalence relation R is written "$elements(R)_{/R}$." It can be read, "the elements of R partitioned according to R." In general, we will write $S/\equiv$, meaning "set S partitioned according to the equivalence relation $\equiv$."

in Section 4.3.1, we will demonstrate Theorem 4.1 on the *Power* relation that relates machine types.

4.2.5 Reflexive and transitive closure

The *reflexive closure* of R, denoted by R^0, is

$$R^0 = R \cup \{\langle x, x \rangle \mid x \in S\}.$$

This results in a relation that is reflexive.
The *transitive closure* of R, denoted by R^+, is

$$R^+ = R \cup \{\langle x, z \rangle \mid \exists y \in S : \langle x, y \rangle \in R \wedge \langle y, z \rangle \in R^+\}.$$

The use of '+' highlights the fact that transitive closure relates items that are "one or more steps away."
The reflexive *and* transitive closure of a relation R, denoted by by R^*, is

$$R^* = R^0 \cup R^+.$$

The use of '*' highlights the fact that reflexive and transitive closure relates items that are "zero or more steps away."
Example: Consider a directed graph G with nodes a, b, c, d, e, and f. Suppose it is necessary to define the *reachability* relation among the

nodes of G. Oftentimes, it is much easier to instead define the one-step reachability relation

$$Reach = \{\langle a, b \rangle, \langle b, c \rangle, \langle c, d \rangle, \langle e, f \rangle\}$$

and let the users[3] perform the reflexive and transitive closure of *Reach*. Doing so results in $Reach_{RTclosed}$, that has all the missing reflexive and transitive pairs of nodes in it:

$$Reach_{RTclosed} = \{\langle a, b \rangle, \langle b, c \rangle, \langle c, d \rangle, \langle e, f \rangle, \langle a, a \rangle, \langle b, b \rangle, \langle c, c \rangle, \langle d, d \rangle,$$
$$\langle e, e \rangle, \langle f, f \rangle, \langle a, c \rangle, \langle a, d \rangle, \langle b, d \rangle\}.$$

4.3 The *Power* Relation between Machines

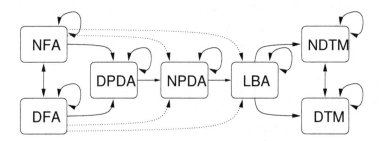

Fig. 4.2. The binary relation *Power* is shown. The dotted edges are *some* of the edges implied by transitivity. Undotted and dotted means the same in this diagram. Therefore, *Power* actually contains: (i) the pairs corresponding to the solid edges, (ii) the pairs indicated by the dotted edges, (iii) and those pairs indicated by those dotted transitive edges not shown.

An example that nicely demonstrates the versatility of preorders is the one that defines the "power" of computing machines. Let

$$MT = \{dfa, nfa, dpda, npda, lba, dtm, ndtm\}$$

represent the set of machine types studied in this book. These acronyms stand for deterministic finite automata, nondeterministic finite automata, deterministic push-down automata, nondeterministic push-down automata, linear bounded automata, deterministic Turing machines, and nondeterministic Turing machines, respectively. A binary

---

[3] "Why sweat? Let the end users do all the work."

relation called *Power* that situates various machine types into a dominance relation is shown in Figure 4.2. Each ordered pair in the relation shows up as an arrow ($\rightarrow$). We draw an arrow from machine type m_1 to m_2 if for every task that a machine of type m_1 can perform, we can find a machine of type m_2 to do the same task. Spelled out as a set, the relation *Power* is

$Power = \{\langle dfa, dfa \rangle, \langle nfa, nfa \rangle,$
$\quad\quad \langle dpda, dpda \rangle, \langle npda, npda \rangle,$
$\quad\quad \langle lba, lba \rangle,$
$\quad\quad \langle dtm, dtm \rangle, \langle ndtm, ndtm \rangle,$

$\quad\quad \langle dfa, nfa \rangle, \langle nfa, dfa \rangle, \langle dtm, ndtm \rangle,$
$\quad\quad \langle ndtm, dtm \rangle, \langle dfa, dpda \rangle, \langle nfa, dpda \rangle,$
$\quad\quad \langle dfa, lba \rangle, \langle nfa, lba \rangle,$
$\quad\quad \langle dpda, npda \rangle, \langle npda, dtm \rangle, \langle npda, ndtm \rangle,$
$\quad\quad \langle dpda, lba \rangle, \langle npda, lba \rangle,$
$\quad\quad \langle lba, dtm \rangle, \langle lba, ndtm \rangle,$

$\quad\quad \langle dfa, npda \rangle, \langle nfa, npda \rangle,$
$\quad\quad \langle dpda, dtm \rangle, \langle dpda, ndtm \rangle,$

$\quad\quad \langle dfa, dtm \rangle, \langle dfa, ndtm \rangle,$
$\quad\quad \langle nfa, dtm \rangle, \langle nfa, ndtm \rangle$
$\quad\}.$

We will now study this dominance relation step by step.

Any machine is as powerful as itself; hence, *Power* is reflexive, as shown by the 'self-loops.' *Power* is *transitive* relation because for every $m_1, m_2, m_3 \in MT$, if $\langle m_1, m_2 \rangle \in Power$ and $\langle m_2, m_3 \rangle \in Power$, then certainly $\langle m_1, m_3 \rangle \in Power$. (We do not show the transitive edges in the drawing). *Power* is *not* antisymmetric because even though *dfa* and *nfa* dominate each other, they have *distinct* existence in the set of machine types *MT*. *Power* is a preorder, and in our minds captures exactly how the space of machines must be subdivided. It is not a partial order.

4.3.1 The equivalence relation over machine types

Applying Theorem 4.1 to *Power*, we obtain the equivalence relation $\equiv_{MT}$ over machine types:

$Power \cap Power^{-1} = \{\langle dfa, dfa \rangle, \langle nfa, nfa \rangle,$
$\quad\quad\quad\quad\quad\quad \langle dpda, dpda \rangle, \langle npda, npda \rangle,$

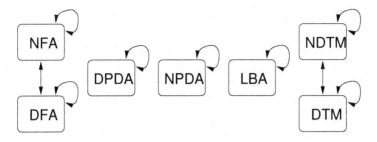

Fig. 4.3. The equivalence relation $Power \cap Power^{-1}$

$$\langle dtm, dtm \rangle, \langle ndtm, ndtm \rangle,$$
$$\langle lba, lba \rangle,$$

$$\langle dfa, nfa \rangle, \langle nfa, dfa \rangle, \langle dtm, ndtm \rangle, \langle ndtm, dtm \rangle$$
$$\}.$$

Figure 4.3 illustrates $\equiv_{MT}$. We can see that $\equiv_{MT}$ subdivides MT into *five equivalence classes*: $\{dfa, nfa\}$, $\{dpda\}$, $\{npda\}$, $\{lba\}$, and $\{dtm, ndtm\}$. Let us now look at this equivalence relation formally. As pointed out earlier, an equivalence relation partitions the underlying set into equivalence classes. The set of equivalence classes is denoted by $elements(R)_{/R}$ below. The mathematical definition below elaborates on equivalence relations and equivalence classes:

$$elements(R)_{/R} = \{ \ \rho \ | \ \rho \subseteq elements(R)$$
$$\wedge \ \rho \times \rho \subseteq R$$
$$\wedge \neg \exists Y : \ \rho \subset Y \wedge Y \subseteq elements(R)$$
$$\wedge \ Y \times Y \subseteq R \ \}$$

This definition says the following:

- Each equivalence class E_i is a subset of the elements of the underlying set
- For each E_i, $E_i \times E_i$ is a subset of the equivalence relation.
- Each such set E_i is maximal (no bigger Y containing each E_i exists)

Coming to our specific example, there are four equivalence classes over $Power \cap Power^{-1}$: $\{dfa, nfa\}$, $\{dpda\}$, $\{npda\}$, and $\{dtm, ndtm\}$. As can be seen from Figure 4.3, they are four maximal universal relations. Given an equivalence relation R, the equivalence class of an element $x \in elements(R)$ is denoted by $[x]$.

4.4 Lattice of All Binary Relations over S

A lattice $(Q, \preceq)$ is a partially ordered set such that for any two elements $x, y \in Q$, the *greatest lower-bound* $glb(x, y)$ and the *least upper bound* $lub(x, y)$ exist in Q, and are unique. An element x' is lower than x if $x' \preceq x$. An element x' is a lower-bound of x and y if x' is lower than x and y. $glb(x, y)$ is then a lower-bound that is below no other lower bound. Similarly, we can define $lub(x, y)$ to be an upper-bound that is above no other upper-bound. As an example, the powerset of a set S, 2^S, forms a lattice under the partial order $\subseteq$. In this lattice, the *glb* among *all* elements of S is $\emptyset$, and the *lub* among all elements is S itself.

As another illustration, the set of *all equivalence relations over S* forms a *lattice* under the normal set inclusion operator. The *glb* of this lattice is the *universal relation* $S \times S$, and the *lub* of this lattice is the *identity relation* $\{\langle x, x \rangle \mid x \in S\}$.

4.5 Equality, Equivalence, and Congruence

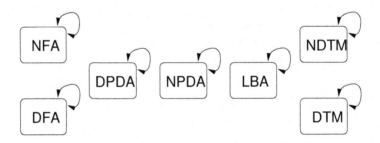

Fig. 4.4. The identity relation over MT

By now, we have seen the notion of the identity relation, which is what the $=$ symbol really corresponds to. Our use of the term "equality" will, without further qualifications, refer to $=$. Figure 4.4 illustrates identity. Our use of the word *equivalence*, denoted by $\equiv$, has already been explained.

4.5.1 Congruence relation

A *congruence relation* captures the notion of substitutability. For example, not only are 2 and $(3-1)$ equivalent, but in most contexts, one can use 2 instead of $(3-1)$. As discussed in Section 4.1.2, a resistor may be substituted with another having the same Ohmic value. Let us take a generic example. Consider something "big", say B, and let x be a "small part of it." In symbols, we denote B as $B[x]$, or "B that contains x." Let $x \equiv y$ be true under some equivalence relation $\equiv$. If $\equiv$ is a congruence relation (or simply "congruence"), then $B[x] \equiv B[y]$. As an example of *big*, think of a radio receiver, with x being a 10Ω resistor inside the receiver, and y being two 5Ω resistors in series. Most radio receivers will continue to behave the same if one were to pull out a 10Ω resistor and replace it by two 5Ω resistors in series. This ability to replace "equals for equals" in various contexts is really the idea behind *congruence*.

A context (or "hole") inside a "big" expression is elegantly captured by a *lambda* expression. For example, $\lambda z.z + 1$ gives the context "$. + 1$." Now, 2 and $200 - 198$ are congruent modulo operator $\lambda z.z + 1$ because $(\lambda z.z + 1)2 = (\lambda z.z + 1)(200 - 198)$. Such congruence can be *lost* in systems where addition is implemented in a resource-bounded fashion, causing overflows or loss of precision. Therefore, while arithmetic substitutability is a congruence, in real hardware implementations, this substitutability may be lost! In other words, congruence *respects* the operators in the underlying context. In the examples we discussed, these operators are addition $(+)$ and resistor parallel composition.

Formally, an equivalence relation R modulo operator

$$op : \times^n \; elements(R) \rightarrow elements(R)$$

for some $n \geq 0$ is called a *congruence relation*, denoted $R_{/\equiv}$, if

$op(\langle x_1, \ldots, x_n \rangle) = op(\langle y_1, \ldots, y_n \rangle), \; for \; all \; i \in n,$ and
$\forall i : y_i \in [x_i]$ (here, y_i in the equivalence class of x_i).

Illustration 4.5.1 *(Binary relations:) Let a relation $\prec$ over intervals of Reals be defined as follows. Let $[a, b]$ for $a, b \in Real$ be an interval. For intervals $[a, b]$ and $[c, d]$, $[a, b] \prec [c, d]$ iff $a \leq c \vee d \leq b$. Find out the exact nature of $\prec$: i.e., is it reflexive, symmetric, antisymmetric, transitive, etc.?*
Solution: We will show the following. First of all, $\prec$ is reflexive. It is not symmetric. If it were symmetric, then we must have

$$((a \leq c) \vee (d \leq b)) \Rightarrow ((c \leq a) \vee (b \leq d))$$

This would mean that

$$((c > a) \wedge (b > d)) \Rightarrow ((a > c) \wedge (d > b))$$

But this is not so; consider $(a, b) = (0, 10)$ and $(c, d) = (2, 9)$.

It is not transitive. If it were transitive, it would mean that

$$((a \leq c) \vee (d \leq b) \wedge (c \leq e) \vee (f \leq d)) \Rightarrow ((a \leq e) \vee (f \leq b))$$

But this is not so; consider $(a, b) = (2, 10), (c, d) = (3, 15)$ and $(e, f) = (1, 11)$.

Illustration 4.5.2 We will need this result in Illustration 4.5.3. Let us prove that there are exactly 2^{2^N} Boolean functions over N Boolean variables. Any N-ary Boolean function can be portrayed as a *truth table* (see Table 18.1 if you have never seen a truth table) of 2^N rows, giving a value of the function for each of the 2^N input combinations. For each way of filling these output value columns, we obtain one distinct function (*e.g.*, the 2-input nand function in Table 18.1 has outputs 1, 1, 1, and 0). Since there are 2^{2^N} ways of filling the value column, there are that many functions of N inputs. For example, there are 16 2-input Boolean functions.

Illustration 4.5.3 (Equivalence over Boolean formulas): Let $V = \{x, y, z\}$ be a set of Boolean variables. Let F be the set of all Boolean formulas formed over V using $\wedge, \vee, \neg$. Let $\Rightarrow$ stand for the implication symbol. Define the relation R to be the set of all formula pairs $\langle f_1, f_2 \rangle$ such that $f_1 \Rightarrow f_2$ is true, where f_1 and f_2 come from F. It is easy to see that R is a preorder. Also, $R \cap R^{-1}$ is an equivalence relation where each equivalence class contains formula pairs that are logically equivalent. Since there are only 3 Boolean variables used, there are 2^{2^3} equivalence classes (see Illustration 4.5.2). Each equivalence class is a set with cardinality $\aleph_0$ (e.g., f, $\neg\neg f$, $\neg\neg\neg\neg f$, etc., are in the equivalence class of formula f).

Also, the set F with the operator $\Rightarrow$ forms a lattice, with $f_1 \vee f_2$ being the least upper-bound and $f_1 \wedge f_2$ being the greatest lower-bound of two formulas f_1 and f_2.

Chapter Summary

This chapter provided a thorough self-contained introduction to binary relations. It went through many familiar definitions, such as *reflexive*, *symmetric*, *transitive*, etc. It also introduced the "*irr-*" and "*non-*" variants of most relations. For instance, it is clearly shown why an irreflexive relation is *not* the negation (complement) of a reflexive relation. Preorders—very important in comparing machines with more or equivalent behaviors—are introduced. It was shown that given a pre-order, one can define an equivalence relation by intersecting it with its inverse. This was illustrated by taking the "power" of various machines we are going to study in this book into account. We then introduced universal as well as identity relations, defined congruence, and briefly looked at the "power" of machines.

Exercises

4.1. Argue that $\emptyset$ is both symmetric and asymmetric.

4.2. Find a, b, and c for which transitivity holds, and those for which transitivity does not hold.

4.3. Refereeing to Figure 4.1,

1. Which relations are preorders?
2. Which are partial orders?
3. Which are equivalence relations?
4. Which are universal equivalence relations?
5. Which are identity equivalence relations?

4.4. Characterize these human relationships in as many ways as you can (ascribe as many attributes as you can think of):

1. x is the spouse of y
2. x is a sibling of y
3. x is an ancestor of y
4. x likes y
5. x does not know y

4.5. Explain the notions of a *total relation* and a *total order* in terms of the required graph properties (in terms of dots and arrows).

4.6.
1. Give two examples (each) of a binary relation and a ternary relation over Nat (choose practical examples of these relations).
2. Draw a directed graph G with nodes being the subsets of set $\{1, 2\}$, and with an edge from node n_i to node n_j if either $n_i \subseteq n_j$ or $|n_i| = |n_j|$. $|S|$ stands for the *size* or *cardinality* of set S. Now, view the above graph G as a binary relation R_G. Write down the pairs in this relation. A relation is written as

$$\{\langle a_0, b_0 \rangle, \langle a_1, b_1 \rangle, \ldots\}$$

where $\langle a_i, b_i \rangle$ are the pairs in the relation.

3. Is R_G a functional relation? Why, or why not? A functional relation is one for which for one 'input', at most one 'output' is produced. As far as an n-tuple $\langle a_1, a_2, \ldots, a_n \rangle$ goes, the last position a_n is regarded as the 'output' and the remaining $n - 1$ positions are regarded as the 'input.' Therefore, for a functional binary relation, if $\langle a_i, b_j \rangle$ and $\langle a_i, b_k \rangle$ are in the relation, then $b_j = b_k$.

4.7. Consider the *Power* relation defined in Section 4.3. Depict a simplified power relation ("*Spower*") that retains only DFA, NFA, DTM, and NDTM. Argue that *Spower* is a preorder. Compute the equivalence relation obtained by intersecting *Spower* and its inverse.

4.8. Consider electrical *resistors*. Let the set of resistors be defined as follows. If x is a *Real* number, then the *term res(x)* is a resistor. If r_1 and r_2 are in *resistor*, then $series(r_1, r_2)$ is a resistor. (Again, $series(r_1, r_2)$ is a term). Some of the elements in set *Resistor* are:

- $res(1)$, $res(2)$, $res(3)$, $series(res(1), res(2))$,
- $series(res(1), series(res(2), res(3)))$, and
- $series(series(res(2), res(3)), res(1))$.

In any[4] circuit, we must be able to substitute $res(3)$ by $series(res(1), res(2))$, as series connected resistors add up the resistivity. However, we *cannot* regard $res(3)$ and $series(res(1), res(2))$ as *identical* because they have distinct representations in the set. To compare two resistors "properly," we define the relation *Resistor_leq*:

$$Resistor\_leq = \{\langle x, y \rangle \mid sumR(x) \leq sumR(y)\}$$

where $sumR$ is defined as follows:

$sumR(res(i)) = i,$
$sumR(series(x, y)) = sumR(x) + sumR(y).$

Given the above, show that *Resistor_leq* is reflexive and transitive, and hence a preorder. Also show that *Resistor_leq* is not antisymmetric.

4.9. Show that the 'less than' relation, '$<$', defined over *Nat*, is antisymmetric.

4.10. Prove that the intersection of a partial order R over set S and its inverse R^{-1} is the *identity* relation over S.

---

[4] We are ignoring aspects such as size, weight, tolerance, etc.

4.11. Define the following relation $\prec$ over intervals of *Reals* as follows. Let $[a, b]$ for $a, b \in Real$ be an interval. For intervals $[a, b]$ and $[c, d]$, $[a, b] \prec [c, d]$ iff $a \leq c \wedge d \leq b$. Now, is $\prec$ a preorder? Is it a partial order? Show that $\prec \cap \prec^{-1}$ is an identity relation.

4.12. Continuing with the above question, let's change $\prec$ as follows. For intervals $[a, b]$ and $[c, d]$, define

$$[a, b] \prec [c, d] \Leftrightarrow \exists x : (a \leq x \leq b) \wedge (c \leq x \leq d).$$

Now, is $\prec$ a preorder or a partial order? Is $\prec \cap \prec^{-1}$ an identity equivalence relation or simply an equivalence relation?

4.13. Find out which are true and which are false assertions:

1. The intersection of two preorders is a preorder.
2. The intersection of a preorder order and its inverse can never be the identity relation.
3. The universal relation is a preorder.
4. The empty relation is a preorder.

4.14. Choose all correct answers. The intersection of a partial order and its inverse is:

1. The universal relation.
2. The identity relation.
3. An equivalence relation.
4. A preorder.

4.15. Choose all correct answers. The intersection of a preorder and its inverse is:

1. a reflexive relation.
2. a congruence relation, modulo any given function f.
3. a transitive relation.
4. an antisymmetric relation.

4.16. Find a relation that is symmetric, asymmetric, antisymmetric, irreflexive, and transitive (all at the same time).

4.17. List some of the equivalence classes of the equivalence relation $\prec \cap \prec^{-1}$ of Exercise 4.12.

4.18. Compare the equivalence classes defined by $\prec \cap \prec^{-1}$ as defined by Exercise 4.12 and Exercise 4.11. It is the case that each equivalence class of one of these examples contains many equivalence classes of the other example. Explain this statement. Specifically, consider the interval $[x, x]$ for some Real number x. Consider its equivalence class $[[x, x]]$. In which example does this equivalence class contain more members besides $[x, x]$?

4.19. Define the following family of equivalence relations over Nat: equivalence relation E_i, for $2 \leq i \leq 4$, contains all pairs $\langle x, y \rangle$ such that $(x = y) \ mod \ i$. For example, for $i = 2$, one equivalence class is Odd and another equivalence class is $Even$. Sketch the equivalence classes generated by $i = 2$, $i = 3$, and $i = 4$. Show the inclusion relationships between these equivalence classes. Also show the inclusion relationships between these equivalence classes and the universal and the identity relations over Nat.

4.20. The two DFAs in Figure 10.11 are language-equivalent deterministic finite-state automaton (DFA) which will be discussed in Chapter 8.

A string s is in the language of a DFA if there is an s-labeled path from the start state (state labeled by the arrow) to one of the final states. Given two DFAs D_1 and D_2, define $D_1 \leq_{DFA} D_2$ exactly when for every string in D_1's language there exists the same string in D_2's language.

1. Show that $\leq_{DFA}$ is a preorder.
2. Describe the equivalence relation $\leq_{DFA} \cap \leq_{DFA}^{-1}$.
3. Show that the DFAs of Figure 10.11 (say DFA_1 and DFA_2) are such that $DFA_1 \leq_{DFA} DFA_2$ and $DFA_2 \leq_{DFA} DFA_1$.

4.21. Electrical engineers can realize a resistor of a certain Ohmic value by connecting resistors in series, parallel, or a combination of both. Two resistors of Ohmic value R_1 and R_2 in series are equivalent to a single resistor of Ohmic value $R_1 + R_2$, while two such resistors in parallel are equivalent to a resistor of $\frac{R_1 R_2}{R_1 + R_2}$ Ohms. This equivalence is a congruence relation modulo the series connection and parallel connection operators, as far as resistivity goes.

Some attributes of circuits *may not* be preserved by this substitution.

- Derive a formula for the *tolerance* of a parallel connection, and check whether tolerance is preserved by parallel composition.
  For example, if each resistor in a collection has a tolerance in Ohmic value of 5% (meaning that the actual Ohmic value may be $\pm 5\%$) regardless of its Ohmic value, then do two 10Ω resistors from this collection that are connected in parallel have the same overall tolerance as a single 5Ω resistor, also drawn from this collection?

- Is tolerance preserved by any assembly? (In other words, suppose we realize r Ohms through two different circuits c_1 and c_2 with different serial/parallel structures; do these circuits have the same tolerance overall?)
- Is the *planarity* of the circuit layout preserved by serial/parallel resistive combinations? How about the wattage rating?
- Think of three other attributes that are not preserved.

This exercise drives home the point that not everything of interest may be preserved in a substitution under a congruence.

5

Mathematical Logic, Induction, Proofs

In this chapter, we present a sampling of ideas from mathematical logic. We consider various operators such as *if* and *if and only if* that are crucial to the proper understanding of formal statements made while discussing automata theory. We then describe methods to write down formal definitions and formal proofs, including inductive definitions, and proofs by contradiction.

5.1 To Prove or Not to Prove!

Writing proofs is tedious work. Therefore, it goes without saying that this activity is rewarding only if the proofs that one embarks on succeed in establishing the proof goal. Unfortunately, there are no guaranteed safe methods that one can follow to ensure success in proving (or disproving) things. Here are some familiar (and some unfamiliar) situations in *slanted* fonts, and the safety net that the human society relies on to avoid them:

- *One makes mistakes in the deductive steps, and nobody else who proof-reads our proofs is able to spot mistakes in the proof.* While relatively infrequent, this situation is avoided in practice by one of two means: (i) having multiple people proof-check, perhaps spanning several years or even decades; (ii) having modern proof-checkers ("theorem provers") check the steps. The Flyspeck project of Thomas Hales [69] attempts to find such machine-checked proofs for several mathematical conjectures for which the problem statement is easy, but the proofs are so immensely hard that it requires experts *plus* reliable computer programs working in concert. Several impressive results have already been obtained, including a solution to the Kepler conjecture.[1]

- *We end-up proving something like 6 = 6, and then realize that we have perhaps applied a circular chain of reasoning steps that led us to a tautology. We crumple and discard our worksheets and try another attack.*

  This could happen in a reasoning tool that loops due to the presence of an unfortunate circular chain of reasoning steps. The solution is to build proof tools that avoid such direct circularities from ever manifesting. The ACL2 theorem prover [3] is an example of a tool that incorporates such precautions.

- *We attempt to prove a famous open conjecture, such as $P \neq NP$, and end up deducing, after a few hours, that 5 = 5. This allows us to conclude nothing! (Had we ended up proving 5 = 6 through a sound chain of reasoning steps, we would have ended up proving that $P = NP$). In Chapter 12 on Pumping lemmas, we will revisit this issue.*

- *We are asked to prove that 5 = 6. By negating the proof goal, we try to prove that $5 \neq 6$ and get nowhere. In this particular case, we might stop after a few hours and try instead to prove $5 \neq 6$. We then begin by assuming the negation of $5 \neq 6$, which is indeed 5 = 6, and easily finish the proof by obtaining a contradiction (end up proving false). At this point, we would have proved $5 \neq 6$ (see Section 5.2.3).*

  This situation shows that one has to know what proof goal is likely to be true before wasting much time.

- *We are asked to prove that $2^{67} - 1$ is a prime number. We might attempt to search for all factors in a certain range, but find such brute-force techniques to be utterly infeasible. We sit utterly hopeless until someone gives us one of the factors (761, 838, 257, 287), and then we easily find the other factor.*

  This situation demonstrates that some proofs are hard because of the sheer computational complexity involved. In fact, finding the prime factors of an arbitrary natural number is a very hard problem. Cryptography systems rely on this fact for the secure transmission of data. Rivest, Shamir, and Adleman invented an algorithm that goes by the acronym RSA, and depends on this computational complexity [102].

---

[1] The Kepler conjecture is about the most compact 3-dimensional arrangement for cannon balls – or oranges in a fruit-stand for that matter. For details, see [69].

In summary, before embarking on any proof, one must attempt to assess the likelihood of the proof goal being true as well as the inherent difficulty of finding the proof. Unfortunately, as discussed in Chapters 17 and 18, these are, in general, impossible to do. Fortunately, in practice, such proofs are routinely carried out in computer companies such as AMD and Intel (e.g., [52]), with respect to systems such as floating-point hardware.

5.2 Proof Methods

A proof is a sequence of *steps* justifying why something is *true*. More formally, a proof of a statement s_n in formal logic is a *sequence* $s_0, s_1, \ldots, s_n$, such that each s_i is either an *axiom* or follows from some of the earlier s_j ($j < i$) through a *rule of inference*. In Chapter 18, we shall introduce the formal logical machinery necessary to define what a *step* can be, and what *true* means. In this section, we list a few styles of proofs and their uses throughout this book. We begin with a review of some of the basic operators.

5.2.1 The implication operation

Implication, or $\Rightarrow$, stands for "if then." It is used to assert the truth of its *consequent* conditional on the truth of its *antecedent*. For example,

$$\textbf{if } (x > 2 \land \mathit{even}(x)) \textbf{ then } \mathit{composite}(x)$$

is a true assertion. Written using $\Rightarrow$, the above statement reads

$$(x > 2 \land \mathit{even}(x)) \Rightarrow \mathit{composite}(x).$$

This assertion is true because every even number is a composite number, and for odd numbers, the above assertion is vacuously true (its antecedent is false). What about the statement

$$(x > 2 \land x < 2) \Rightarrow (x = x + 1)?$$

The antecedent that asserts that x is both > 2 and < 2 is *false*. In this case also, the assertion is vacuously true. Note that we are not drawing the conclusion $x = x + 1$ - the moment we find that the antecedent is false, we leave the consequent alone. Another sentence of this form is, "*if the moon is made of blue cheese, then horses can fly,*" in that it is a true assertion.[2] The assertion

---
[2] Notice that we are *not* concluding that horses can fly.

$$(x > 2 \ \wedge \ even(x)) \Rightarrow prime(x)$$

is, on the other hand, neither true nor false (it is both falsifiable and satisfiable). **Note that $a \Rightarrow b$ is equivalent in meaning to $\neg a \vee b$.**

An exercise in reading *iff*

We now turn to the *iff* operator. Basically, *a iff b*, also written $a \Leftrightarrow b$ or $a = b$, means $a \Rightarrow b$ *and* $b \Rightarrow a$. In other words, a has the same truth value as b; if a is true, so is b; if a is false, so is b.

Definitions involving *iff* are sometimes tricky to read. Consider the following definition that talks about *ultimately periodic sets* (see also Section 10.5):

Definition 5.1. *(Ultimately Periodic)* A set $S \subseteq N$ is said to be *ultimately periodic* (UP) if there exists a number $n \geq 0$ and another number (called 'period') $p > 0$ such that for all $m \geq n$, $m \in S$ iff $m + p \in S$.

See if you understand the above definition by doing the following exercise. The fact that the set in Question 3 is UP is clear, because $n = 26$ and $p = 15$ works. Think how you get the other two answers.

5.2.2 'If,' or 'Definitional Equality'

When one presents mathematical ideas, one often adopts a certain systematic style consisting of first introducing a bunch of definitions, followed by the statement and proof of various theorems. While introducing definitions, you may have noticed that people use various kinds of equality or arrow symbols, some of the typically used symbols being $\doteq$, $\overset{\triangle}{=}$, or sometimes even $\Leftarrow$. Also while defining Boolean quantities, we might even employ *iff* or $\Leftrightarrow$ instead of these other variants. We will prefer $\doteq$ over all these various forms.

This is known as *definitional equality*. Definitional equalities are used to introduce *new* definitions into the domain of discourse. Note that such a definition, by itself, causes no contradictions. For example, let there be a definition

$$foo(x) \doteq odd(x) \ \wedge \ prime(x)$$

in some domain of discourse, where *foo* has never before been introduced, while *odd* and *prime* have already been defined. However, if one continues defining functions 'willy nilly,' and later introduces the definition $foo(x) \doteq even(x)$, one would have introduced a contradiction that asserts that $even(x) = odd(x) \wedge prime(x)$! In many frameworks, a *second* definition of something that is already defined is flagged as an error. Consequently, some protection (through checking) is available while using definitional equalities.

Considering all this, a real conundrum seems to arise when one uses a definitional equality symbol to introduce a new symbol which also appears on the right-hand side of the definition! How can we be introducing something "afresh" when we are also using it on the right-hand side, as if it were already existing? This is, however, how recursive definitions are! Do such recursive definitions always 'make sense?' Can they cause contradictions of the kind alluded to above? How do we understand recursive definitions that, on one hand, are so very convenient, but on the other hand prone to circularity? Chapter 6 examines these questions in greater detail, and helps set the stage for using recursive definitions with more assurance. If you are planning to skip Chapter 6, try to at least study the following recursive definitions to see which ones define a function (any function at all), which ones define a function *uniquely*. Also, try to understand carefully why uniqueness is achieved.

> **LHS and RHS:** We define two abbreviations that will often be used: LHS stands for left-hand side and RHS for right-hand side. These abbreviations may refer to equations, production rules, etc., all of which have two "sides" to them.

The *as* Notation

Sometimes, when we define a function f, we might be interested not only in the *entire* argument fed to f but also the *substructure* of the arguments. In this case, the use of the '*as*' notation comes in handy.
Example: $f(x \text{ as } \langle y, z \rangle, w) = $ **if** $w > 0$ **then** $f(x, w - 2)$ **else** z.

In this function, if $w > 0$ then we recurse, passing x "as is" in the first argument. If $w = 0$, we return the second component of x, which is z. The *as* notation allows both x and its components to be referred to on the right-hand side.

5.2.3 Proof by contradiction

Proof by contradiction, or *reductio ad absurdum*, is a reasoning principle one uses frequently in day-to-day life (for example in playing games, in detecting contradictions in arguments, etc.). This principle is captured by the *contrapositive* rule

$$(A \Rightarrow B) \Leftrightarrow (\neg B \Rightarrow \neg A),$$

as can easily be seen by converting $A \Rightarrow B$ to $\neg A \vee B$ and also similarly converting $(\neg B \Rightarrow \neg A)$.[3] If we assume something ("A") and derive a contradiction, we have essentially proved $A \Rightarrow false$, which, by the contrapositive, allows us to conclude $true \Rightarrow \neg A$, or $\neg A$.

We have already seen how proof by diagonalization uses proof by contradiction. We shall also use proof by contradiction in the context of the *Pumping lemma* in Chapter 12, and in the context of *mapping reducibility arguments* in Chapter 17.

5.2.4 Quantification operators $\forall$ and $\exists$

We now present some facts about the $\forall$ and $\exists$ operators. These operators allow you to perform *iteration* over a certain range, similar to what Σ (summation) and Π (product) do in traditional mathematics. The universal quantification operator $\forall$ is used to make an assertion about all the objects in the domain of discourse. (In mathematical logic, these domains are assumed to be non-empty). For example,

$$\forall x : Nat : \ P(x)$$

is equivalent to an infinite conjunction

$$P(0) \ \wedge \ P(1) \ \wedge \ P(2) \ \dots$$

or equivalently $\wedge_{x \in Nat} : \ P(x)$. Note that we are saying "$x : Nat$." This is simply saying "x belongs to type Nat." It is entirely equivalent to $x \in Nat$—we show both uses here, as both usages may be found in the literature; therefore, it pays to be familiar with both usages.
Here is an example of how $\forall$ works for a finite range:

$$\forall x \in \{0, 1, 2\} : \ P(x)$$

The above formula is equivalent to $P(0) \ \wedge \ P(1) \ \wedge \ P(2)$.
The above definitions allow us to establish numerous identities, for instance:

$$(\forall \ x : Nat : \ P(x)) \wedge (\forall \ x : Nat : \ Q(x)) \equiv (\forall \ x : Nat : \ (P(x) \wedge Q(x))),$$

as the explicit $\wedge$ gets absorbed into the iterated conjunction denoted by $\forall$.

---

[3] The above theorem can also be written $(A \Rightarrow B) \equiv (\neg B \Rightarrow \neg A)$, as the equality symbol $\equiv$ takes the meaning of $\Leftrightarrow$ for Booleans.

Similarly,

$$\exists x : Nat : \ P(x)$$

is equivalent to an infinite disjunction

$$P(0) \ \lor \ P(1) \ \lor \ P(2) \ \ldots$$

or equivalently $\lor_{x \in Nat} : \ P(x)$.

In [75], Lamport introduces a notational style for writing long mathematical formulas in a readable manner. His insight is that by using suitable indentations, and by employing the operators $\land$ and $\lor$ in a manner similar to bullets, one can avoid many parentheses that would otherwise be needed. For instance, the above formula would be written as

$\lor \ P(0)$
$\lor \ P(1)$
$\lor \ P(2)$
$\lor \ \ldots$

We will *often* employ this style in the following illustration.

Illustration 5.2.1 *Fully expand out the $\forall$ and $\exists$ below, and simplify the resulting Boolean formula. "and" and "or" are equivalent to $\land$ and $\lor$, except they are written in the prefix syntax (not infix). Your final answer should be T or F (true or false), with justifications.*

$$(\forall x \in Bool : \ \forall y \in Bool : \ (or(x,y) \Rightarrow \exists z \in Bool : \ and(y,z)))$$

Solution: The given formula is

$\forall x \in Bool : \ \forall y \in Bool :$
$\qquad or(x,y) \Rightarrow \exists z \in Bool : \ and(y,z)$

Expanding x, we get

$\qquad \land \ \forall y : or(0,y) \Rightarrow \exists z : and(y,z)$
$\qquad \land \ \forall y : or(1,y) \Rightarrow \exists z : and(y,z)$

Now expand y:

$\qquad \land \ \land \ or(0,0) \Rightarrow \exists z : and(0,z)$
$\qquad\qquad \land \ or(0,1) \Rightarrow \exists z : and(1,z)$
$\qquad \land \ \land \ or(1,0) \Rightarrow \exists z : and(0,z)$
$\qquad\qquad \land \ or(1,1) \Rightarrow \exists z : and(1,z)$

Now, expand z:

$$\wedge \ \wedge \, or(0,0) \Rightarrow \vee \ and(0,0)$$
$$\vee \ and(0,1)$$
$$\wedge \, or(0,1) \Rightarrow \vee \ and(1,0)$$
$$\vee \ and(1,1)$$
$$\wedge \ \wedge \, or(1,0) \Rightarrow \vee \ and(0,0)$$
$$\vee \ and(0,1)$$
$$\wedge \, or(1,1) \Rightarrow \vee and \ (1,0)$$
$$\vee and \ (1,1)$$

This simplifies to 0, or *false*.

In *first-order* logic, quantification is allowed over the domain of *individuals* (numbers, strings, etc.) while in the *second-order* logic, quantification over functions and relations is allowed. Further facts about quantification will be introduced as necessary.

5.2.5 Generalized DeMorgan's Law Relating ∀ And ∃

DeMorgan's law relates $\wedge$ and $\vee$. For two Boolean variables a and b, the DeMorgan's laws are:

$$\neg(a \wedge b) = (\neg a \vee \neg b), \ \ and \ \ \neg(a \vee b) = (\neg a \wedge \neg b).$$

Extending the above to infinite conjunctions (∀) and infinite disjunctions (∨), we have the following generalized DeMorgan's laws, otherwise known as the *duality* between ∀ and ∃:

$$\forall \, x : Nat : \ P(x) \equiv \neg \exists \, x : Nat : \ \neg P(x)$$

and

$$\exists \, x : Nat : \ P(x) \equiv \neg \forall \, x : Nat : \ \neg P(x).$$

5.2.6 Inductive definitions of sets and functions

The term *inductive definition* finds widespread use throughout computer science, and tends to mean a whole bunch of (related) things. We introduce two specific usages of this term now.

Inductively Defined Sets and Functions

A set that is inductively defined is constructed in a principled way, by first introducing the *basis elements*, and then "pumping" new elements out of existing elements through *constructors*. For example, the set of all lists over N, denoted $list[N]$, is defined as follows:

$nil \in list[N]$
if $l \in list[N]$ and $n \in N$, then $cons(n, l) \in list[N]$.

The intent is to say that $list[N]$ is the smallest set closed under the above two rules. In a general setting, the process of inductively defining a set S consists of first including the *basis elements* $B = \{b_i \mid i \in m\}$ into S. Thereafter, for every element x in the set S, one of a (usually finite) set of constructors $C = \{c_i \mid i \in n\}$ are applied to obtain elements of the form $c_i(x)$. The set S being defined is the *least set* that includes the basis-case elements and is closed under constructor application. Said more formally:

Rule 1: For every $i \in m, b_i \in S$.
Rule 2: If $x \in S$, and c_i is a constructor for $i \in n$, then $c_i(x) \in S$.
Rule 3: S is the least such set.

There are actually two other (and somewhat non-obvious) ways to express the intent of *Rule 3*. They are explained below, as they tend to occur in other books as well as papers.

1. S is the *intersection of all such sets*.
2. Express S directly as follows:

$$S = \{x \mid (x \in B) \vee (\exists c_i \in C : x = c_i(y) \wedge y \in S)\}.$$

The constructors of an inductively defined set S have the status of *primitive* functions that operate over the set. In most cases, a whole list of other *inductively* (or, equivalently, *recursively*) defined functions are also defined over an inductively defined set. One can then establish properties of these functions by *structural induction*; for instance, if we define the *length* of a list, and the *append* function on lists, *inductively* as follows:

$length(nil) = 0$
$length(cons(n, L)) = 1 + length(L)$

$append(nil, L) = L$
$append(cons(x, L1), L2) = cons(x, append(L1, L2))$.

One can then easily prove properties such as

$$length(append(L1, L2)) = length(L1) + length(L2)$$

by inducting on the number of applications of the `cons` operator (more detailed discussions of structural induction are provided in Section 5.3.3). We shall use proof by structural induction when we deal with data structures such as trees and lists.

Free inductive definitions: The inductive definition of a set S is called a *free inductive definition* if, for every object x in S, either $x \in B$, or $x = c_i(y)$ for exactly one constructor c_i and exactly one element $y \in S$. Only one of these conditions must hold. Such sets are called free*ly* generated, from a set of basis elements B and a set of constructors C.

To understand the notion of freely generated sets better (and to see why it matters), suppose we have a set S such that for some element y in it, $y \in B$ *and* $y = c_i(z)$. Or suppose we have a set S such that for some element y in it, $y = c_i(z_1)$ *and also* $y = c_j(z_2)$ for $z_1 \neq z_2$ or $c_i \neq c_j$. Then S is not freely generated. For example, in $list[N]$ if some element $y = nil$, as well as $y = cons(x_1, cons(x_2, cons(x_3, nil)))$ for some arbitrary x_1, x_2, and x_3, what can go wrong? (Exercise 5.18.) With freely generated sets, we can be assured that functions that are inductively defined over them exist and are *unique*. With non-freely generated sets, such functions may not have a unique definition.

5.3 Induction Principles

We now provide a brief tour through some of the induction principles that we shall use repeatedly. We begin with a recap of the two familiar principles of induction over N, namely *arithmetic* induction and *complete* induction. We shall demonstrate that these principles are *equivalent*.

5.3.1 Induction over natural numbers

Generally, the induction principle over natural numbers is stated in two forms: *arithmetic* induction and *complete* induction. We now state these forms and prove them equivalent.

Arithmetic: The principle of arithmetic induction states that for all P, if we can show $P(0)$ and for all x we can show that $P(x-1)$ implies $P(x)$, then for all x, $P(x)$. In symbols:

$$\forall P : \wedge \ P(0)$$
$$\wedge \ \forall x > 0: \ P(x-1) \Rightarrow P(x)$$
$$\Rightarrow (\forall x : P(x)).$$

Complete: The principle of complete induction states that for all P, if we can show that for all y, for all $x < y$, $P(x)$ implies $P(y)$, then for all x, $P(x)$. In symbols:

$\forall P : \forall y \ : \ (\forall x < y : P(x)) \Rightarrow P(y)$
$\qquad \Rightarrow (\forall x : P(x)).$

Proof of equivalence: To prove that these induction principles are equivalent, we proceed as follows. First pick an arbitrary predicate P for which the proof is being done. Then, these induction principles are of the form $A \Rightarrow B$ and $C \Rightarrow B$ (A standing for *arithmetic* and C for *complete*). Therefore, we merely need to prove $A \Leftrightarrow C$, i.e., that the antecedents of these induction theorems are equivalent.

1. We need to prove $A \Leftrightarrow C$, where

$$A = P(0) \wedge \forall x > 0 : \ P(x-1) \Rightarrow P(x)$$

   and

$$C = \forall y \ : \ (\forall x < y : P(x)) \Rightarrow P(y).$$

2. $(A \Leftrightarrow C) \equiv (A \Rightarrow C) \wedge (C \Rightarrow A).$
3. Let's prove $(C \Rightarrow A)$. I'll leave $(A \Rightarrow C)$ as an exercise for you (Exercise 5.21).
4. Here there are two approaches possible: one is to prove $(C \Rightarrow A)$ itself, and the other is to prove the contrapositive form $(\neg A \Rightarrow \neg C)$. We now detail the proof using the contrapositive form (the other form is quite similar).
5. $\neg A = \neg P(0) \vee \neg(\forall x > 0 : \ P(x-1) \Rightarrow P(x)).$
6. Let's assume $\neg P(0)$.
7. Now, $\neg C = \exists y \ : \ (\forall x < y : P(x)) \wedge \neg P(y).$
8. Consider $y = 0$. For this y, there is no $x < y$. Hence, the $(\forall x < y : P(x))$ part is vacuously true.
9. Therefore, by the "infinite disjunction" property of $\exists$,

$$\exists y \ : \ (\forall x < y : P(x)) \wedge \neg P(y)$$

   reduces to

$$\neg P(0) \vee \ldots,$$

   which obviously follows from the assumption $\neg P(0)$.
10. This finishes one of the cases of $\neg A$. Let us move on to the other case, by now assuming

$$\neg(\forall x > 0 : \ P(x-1) \Rightarrow P(x)).$$

   This means $(\exists x > 0 : \ P(x-1) \wedge \neg P(x)).$
11. *(I know how to "pull" a magic assumption here, but won't do it, as I want to illustrate this proof getting stuck and having to come back to discover the magic).*

12. Since $(\exists x > 0 : P(x-1) \wedge \neg P(x))$, assume a witness for x.[4] Call it x_0.
13. Therefore, we have $P(x_0 - 1) \wedge \neg P(x_0)$ as our assumption.
14. We can instantiate the y in C to x_0, to get $(\forall x < x_0 : P(x)) \Rightarrow P(x_0)$.
15. Here, we are stuck. We cannot instantiate the x in the above formula, as it lies under an "implicit negation." (Anything belonging to the antecedent of an implication is under an implicit negation). Said another way, if $\forall$, an infinite conjunction, implies something, we cannot say that one of the conjuncts implies the same thing).
16. Revisit Step 12. In there, assert that the witness x_0 is *the first such* x such that $P(x-1) \wedge \neg P(x)$.
17. Now, consider $(\forall x < x_0 : P(x)) \Rightarrow P(x_0)$ again. Clearly, because of the way we selected x_0, for all $x < x_0$ we *do* have $P(x)$ holding true. So we are left with $P(x_0)$. Since this is false, we again have $\neg C$.
18. Consequently, for both cases resulting from $\neg A$ we have $\neg C$, or in other words, $\neg A \Rightarrow \neg C$, or $C \Rightarrow A$.

5.3.2 Noetherian induction

Both arithmetic and complete induction are special cases of the *noetherian* induction rule stated in terms of *well-founded partial orders*. A well-founded partial order $\preceq$ is one that has no infinite descending chains of the form $\ldots \preceq a_{-3} \preceq a_{-2} \preceq a_{-1} \preceq a_0$. The total order $(N, \leq)$ is a special case of a well-founded partial order. Other well-founded partial orders include the lexicographic ordering of words, subset ordering, etc. In a general setting, well-founded partial orders can have multiple minimal elements (giving rise to multiple basis cases). For example, consider the ordering between closed intervals over natural numbers:

$$\preceq = \{\langle [a, b], [A, B] \rangle \mid a \geq A \wedge b \leq B\}.$$

We have $[2, 2] \preceq [1, 3]$, but neither $[2, 2] \preceq [3, 3]$ nor $[3, 3] \preceq [2, 2]$. In this case, we have an infinite number of minimal elements.

The principle of noetherian induction: Suppose $(W, \preceq)$ is a well-founded partial order. Suppose property P holds for all minimal elements in W. Also, for all non-minimal elements x, if for all $y \preceq x$ $P(y)$ implies $P(x)$, then $\forall x. P(x)$. In symbols:

---

[4] A *witness* is a value that, when substituted in place of a variable that is existentially quantified in a formula, makes the formula true. For example, a witness for y in $\exists y. y < 5$ is 4.

$\forall P : \wedge_{i \in min(W)} \; P(i)$
$\quad \wedge \; \forall x \notin min(W) : (\forall y \preceq x : P(y)) \Rightarrow P(x)$
$\quad \Rightarrow (\forall x : P(x)).$

5.3.3 Structural

The principle of structural induction can easily be seen to be a special case of the principle of noetherian induction by selecting $\preceq$ to keep track of the number of "layers" of application of constructors. Specifically, $min(W) = B = \{b_i \mid i \in m\}$, and

$$preceq = \{\langle x, y \rangle \mid x \in B \vee x \longrightarrow y\},$$

where $x \longrightarrow y$ means there exists a sequence of constructors that can be applied to x to obtain y.

5.4 Putting it All Together: the Pigeon-hole Principle

The colloquial version of the pigeon-hole principle states that if there are n pigeons and $n - 1$ holes, then surely there is a hole with more than one pigeon. Stated in general, we can say that there is no *total injection* from a **finite** set S *into* any of its *proper subsets*. This is not true of infinite sets, as Exercise 5.25 shows.

Illustration 5.4.1 We can prove the pigeon-hole principle using induction, as well as proof by contradiction, as follows (adapted from [106]). Consider finite sets B_i from some universe, where i is the size of the set. We denote the size of a set (e.g., S) by the $|\,|$ operator (e.g., $|S|$). Now, in order to show that for all k, there is no total injection f from any B_k set to one of its proper subsets, we proceed as follows:

- *Basis case:* Consider sets B_2 with two elements. There is no total injection from B_2 into any of its proper subsets.
- *Induction hypothesis:* Now assume by induction hypothesis that for $2 < i \leq (k - 1)$, there is no total injection from any of the B_is to any of their proper subsets. Suppose a particular set B'_k of size k has a total bijection into its proper subset $A_{B'_k}$. But now, we can remove any element x from B'_k and its image under f, namely $f(x)$, from $A_{B'_k}$, and get a total injection from a set in B_{k-1} to one of its proper subsets. This is a contradiction. Therefore, we conclude that for any k, there does not exist a total injection f from any set B_k to one of its proper subsets.

Chapter Summary

This chapter began with some cautionary remarks pertaining to conducting formal proofs: how one has to spend some time assessing whether the proof goal is achievable, and then how to carry out the proof in a manner that is easily verifiable. It then discussed the operators 'if' ($\Rightarrow$) and 'iff' ($\Leftrightarrow$). Proof by contradiction was introduced through the game of Mastermind. After discussing quantifications, inductively defined sets and functions, and induction principles, a proof of equivalence between arithmetic and complete induction was given. Various other induction principles were also discussed. Many of these concepts were illustrated by the pigeon-hole principle.

Exercises

5.1. Read about the following man/machine produced proofs that have made headlines (all references obtainable from
`http://en.wikipedia.org/wiki` or other sources):

1. The four-color theorem including the classic proof of Appel and Haken and the recent proof of Georges Gonthier.
2. Andrew Wiles's proof of Fermat's Last Theorem.

5.2. With respect to the discussions concerning $2^{67} - 1$ on Page 74, determine the other factor of $2^{67} - 1$ by long division.

5.3. Indicate which of the following implicational formulas are true, and which are false:

1. $(1 = 2) \Rightarrow (2 = 3)$.
2. $(gcd(x, y) = z) \Rightarrow (gcd(x + y, y) = z)$. Here, gcd stands for the greatest common divisor of its arguments.
3. $(k > 1) \land hasclique(G, k) \Rightarrow hasclique(G, k - 1)$. Here, $hasclique(G, k)$ means that graph G has a clique of size k.
4. $(k > 1) \land hasclique(G, k) \Rightarrow hasclique(G, k + 1)$.
5. $((a \Rightarrow (b \Rightarrow c)) \Rightarrow (a \Rightarrow b)) \Rightarrow (a \Rightarrow c)$.

5.4.
1. Justify that $\emptyset$ is ultimately periodic (UP).
2. Justify that $\{0, 1, 27\}$ is UP.
3. Justify that $\{0, 1, 5, 23, 24, 26, 39, 41, 54, 56, 69, 71, 84, 86, 99, 101, \ldots\}$ is UP.

5.5. Study the following "recursive definitions:" Assume that x ranges over Int. The operator $|x|$ takes the absolute value of x:

1. $f(x) \doteq f(x) + 1$
2. $f(x) \doteq f(x + 1)$
3. $f(x) \doteq f(x)$
4. $f(x) \doteq if\ (x = 0)\ then\ 1\ else\ f(x + 1)$
5. $f(x) \doteq if\ (x = 0)\ then\ 1\ else\ f(x - 1)$
6. $f(x) \doteq if\ (x = 0)\ then\ 1\ else\ f(|x| - 1)$
7. $f(x) \doteq if\ (x = 0)\ then\ 1\ else\ x + f(|x| - 1)$

Answer the following questions:

(a) Which definitions are malformed (are not definitions but are contradictions)?

(b) Which definitions are well-formed but end up defining an everywhere undefined function?

(c) For all other cases, describe the mapping effected by the function.

5.6. Prove that $\sqrt{2}$ is irrational.

5.7. Consider the game of Mastermind played using three colors (to simplify the exposition), y, g, and b, standing for yellow, green, and blue.

```
CODE y g b b  Move number
--------------------------

rrr   g g b b   3
www   b b y y   2
rw    y y g g   1
```

The game is played between two players A and B. A selects a secret code hidden from view of the opponent B. In the example below, the code is y g b b. Player B must try and guess the code, improving his/her guesses based on scores assigned by A. In our example, B's moves are shown below, numbered as shown. B's first move is y y g g, which fetches a r (for red) and w (for white). A red means that one of the pegs is in the right place and of the right color, while a white means that there is a peg of the right color, but in the wrong place. In our example first move, one y peg fetches a red, and one of the g pegs fetches a white. The next two moves of B are as shown. With the game poised at this point, show that B can, by using the principle of proof by contradiction, finish the game in two more moves. (Hint: B views move 2 with interest. This move reveals that if there are two ys in the code, there must only be one b. Taking this assumption, to move 1 leads to an immediate contradiction. Continue to argue through the details).

5.8. How does one prove the statement, "There exist horses with a white tail?" (For this, and the following 'horse proofs,' conserve your horse power - simply offer proofs, in a few sentences, of the form: "go find a horse satisfying criterion x, then look for criterion y in them; hence, proved/disproved").

5.9. How does one *disprove* the statement, "There exist horses with a white tail?"

5.10. How does one prove the statement, "For all horses that have black ears there are some that have a white tail?"

5.11. How does one disprove the statement, "For all horses that have black ears there are some that have a white tail?"

5.12. Propose a quantified statement in first-order logic that expresses the fact that set $S \subseteq Nat$ is infinite. Construct a statement that goes like, "for all elements of S, ..."

5.13. Propose a quantified statement in first-order logic that expresses the fact that set $S \subseteq Nat$ is finite. Construct a statement that goes like, "there exists...."

5.14. Eliminate the negation operator entirely from the following formula:

$$\neg(\forall x : \exists y : \forall z : (p(x,y) \Rightarrow \neg q(y,z))).$$

5.15.

1. Prove by induction that the sum of 1 through N is $N(N+1)/2$.
2. Obtain an expression for the total number of nodes in a balanced k-ary tree. Prove this result by induction.
3. Suppose we have the following kind of "almost binary" finite trees. All nodes of such a tree have a branching factor of 2, except for the father node of a leaf node, which has a branching factor of only 1 (these nodes "father" exactly one child).

   Such trees arise in the study of context-free grammars whose productions are of the A->BC or A->a. We shall study these grammars, called the *Chomsky Normal Form* grammars, in Chapter 15.

   Obtain an expression for the length of the frontier of such a tree as a function of the number of interior nodes. Verify your answer through an inductive proof.

5.16. Inductively define the set of all binary trees with nodes labeled by elements from Nat. Employ the constructor $Node(n)$ to introduce a tree of height zero containing just a node $n \in Nat$. Employ the constructor $Tree(n, T_1, T_2)$ to introduce a tree rooted at node n with left sub-tree and right sub-tree being T_1 and T_2, respectively. Write down the mathematical equations that achieve such a definition.

5.17. Given a tree such as defined in Exercise 5.16, inductively define a function that sums the values labeling all the nodes.

5.18. Pertaining to the definition of freely generated sets on Page 5.2.6, explain what can go wrong with a non-freely generated set. *Hint:* Consider *length(y)*. Can we get two possible answers when *length* is applied to the same y?

5.19. Find the smallest N such that using only 4-cent postage stamps and 7-cent postage stamps, it is possible to make postage for any denomination $k \geq N$ cents. Then prove your result by induction.

5.20. Prove using induction that using a 6-liter water jug and a 3-liter water jug, one *cannot* measure out exactly 4 liters of water. Here are further instructions.

1. What is wrong if you directly attack this induction proof, taking the given property as the proof-goal ("cannot measure out") (one sentence)?
2. How would you choose the new proof goal (one sentence)?
3. Show the new proof goal by induction.

5.21. Prove the remaining case of $A \Rightarrow C$ in the derivation on Page 84.

5.22. Provide an example of a well-founded partial order with more than one minimal element.

5.23. If you have not already, then first read Section 5.3.2. Now consider the relation $[c_1, c_2] \preceq_c [d_1, d_2]$ where c_1, c_2, d_1, d_2 are characters (belong to $\{a \ldots z\}$) and $d_1 \leq c_1$ and $c_2 \leq d_2$, with the character ordinal positions compared by $\leq$ in the usual way.

1. Show that $\leq_c$ is a well-founded partial order.
2. Show that any subset S of $\leq_c$ is also a well-founded partial order.
3. How many minimal elements does $\leq_c$ have?

4. Define the *height* of any non-minimal element $[c_1, c_2]$ of $\leq_c$ to be *one more than* the maximum number of elements that could *properly* exist between $[c_1, c_2]$ and one of the minimal elements of $\leq_c$. The height of minimal elements are defined to be 0. For example, the height of $[b, d]$ is 2, as either $[b, c]$ properly lies between $[b, d]$ and $[b, b]$, *or* $[c, d]$ properly lies between $[b, d]$ and $[d, d]$. Write down a general formula for the height of an element of $\leq_c$ and prove it by Noetherian induction.

5.24. Develop a formula for the number of leaf nodes in a k-ary tree. Prove the formula by structural induction.

5.25. Show that there exist infinite sets S such that there is a total injection from S into one of its proper subsets.

5.26. How many different ways can n identical pigeons occupy $n - 1$ pigeon holes? (by "identical," we mean that only the count of the pigeons in various holes matters; also, each hole can accommodate any number of pigeons).

5.27. I heard this one on NPR radio in the "Car Talk" show. A man and his wife went to a party where a total of 52 people attended. The party went well, with everyone mingling and shaking hands (the exact number of such handshakes, nor who shook whose hands is not known; also, a person shaking his/her own hands does not count as a handshake). However, while returning from the party, the man told his wife, "Gee, what a nice party!" The wife replied, "Yes! And, I know for a fact that there were at least two people (let's call them A and B) who shook the same number of hands." Prove this statement for 52, and then prove it for any N, stating all steps clearly.

5.28. Consider the set of characteristic formulas, $CF(x, y)$, over two variables x and y. This set can be inductively defined by the following rules:

1. *true* $\in CF(x, y)$ and *false* $\in CF(x, y)$.
2. If $f_1 \in CF(x, y)$ and $f_2 \in CF(x, y)$, then
   a) $f_1 \wedge f_2 \in CF(x, y)$
   b) $f_1 \vee f_2 \in CF(x, y)$
   c) $\neg f_1 \in CF(x, y)$
   d) $(f_1) \in CF(x, y)$.
3. $CF(x, y)$ is the intersection of all such sets.

While we allow the use of parentheses (and), they may be left out with the convention that $\neg$ has the highest precedence, followed by $\wedge$ and finally $\vee$.

Examples of formulas in $CF(x,y)$ are $(x \vee y)$, $x \vee y$, $false$, $(true)$, $(x \wedge \neg x)$, $(x \wedge y \vee x \wedge y \vee \neg x)$, etc. Clearly many of these characteristic formulas are equivalent, in the sense that they denote the same relation.

1. Show that $(CF, \Rightarrow)$ forms a preorder. $\Rightarrow$ is the implication operator. $\langle x, y \rangle \in \Rightarrow$ precisely when the formula $x \Rightarrow y$ is *valid* (or 'is a *tautology*,' or 'is true for all x, y').
2. Show that $\equiv = (\Rightarrow \cap \Rightarrow^{-1})x$ is an equivalence relation. Recall that $\Rightarrow^{-1}$ can be written as $\Leftarrow$.
3. How many equivalence classes does the equivalence relation $(CF, \equiv)$ have? Show by going through the definition of equivalence classes given earlier (specifically, $elements(R)_{/R}$).
4. Argue that two formulas, $f_1 \equiv f_2$, denote the same relation, and hence $[f_1] = [f_2]$.
5. Arrange the relations denoted by the equivalence classes of $CF(x,y)$ into a lattice, clearly pointing out the *glb* and *lub*.

Two Puzzles

The following two puzzles are due to Lewis Carroll (the author of *Alice in Wonderland* and a mathematician at Christ Church, Oxford).

5.29. From the premises

1. Babies are illogical;
2. Nobody is despised who can manage a crocodile;
3. Illogical persons are despised.

Conclude that *Babies cannot manage crocodiles.*

5.30. From the premises

1. All who neither dance on tight ropes nor eat penny-buns are old.
2. Pigs, that are liable to giddiness, are treated with respect.
3. A wise balloonist takes an umbrella with him.
4. No one ought to lunch in public who looks ridiculous and eats penny-buns.
5. Young creatures, who go up in balloons, are liable to giddiness.
6. Fat creatures, who look ridiculous, may lunch in public, provided that they do not dance on tight ropes.
7. No wise creatures dance on tight ropes, if liable to giddiness.

8. A pig looks ridiculous carrying an umbrella.
9. All who do not dance on tight ropes and who are treated with respect are fat.

Show that *no wise young pigs go up in balloons.*

6

Dealing with Recursion

Recursion is a topic central to computer science. Lambda calculus offers us a very elegant (and fundamental) way to model and study recursion. In a book on computability and automata, such a study also serves another purpose; to concretely demonstrate that Lambda calculus provides a universal mechanism to model computations, similar to the role played by Turing machines. While this chapter can be skimmed, or even skipped, we have taken sufficient pains to make this chapter as "friendly" and intuitive as possible, permitting it to be covered without spending too much time. Covering this chapter will permit fixed-point theory to be used as a conceptual "glue" in covering much important material, including studying context-free grammars, state-space reachability methods, etc.

6.1 Recursive Definitions

Let's get back to the discussion of function 'Fred' introduced in Chapter 2, Section 2.6. Now consider a recursively defined function, also called 'Fred.' We fear that this will make it impossible to 'de-Fred:'

```
function Fred x = if (x=0) then 0 else x + Fred(x-1)
```

It is quite tempting to arrive at the following *de-Fred*ed form:

```
function(x) = if (x=0) then 0 else x + self(x-1)
```

However, it would really be nice if we can avoid using such *ad hoc* conventions, and stay within the confines of the Lambda notation as introduced earlier. We shall soon demonstrate how to achieve this; it is, however, instructive to examine one more recursive definition, now involving a function called Bob:

```
function Bob(x) = Bob(x+1).
```

One would take very little time to conclude that this recursive definition is "plain nonsense," from the point of view of recursive programming. For example, a function call `Bob(3)` would never terminate. However, unless we can pin down exactly why the above definition is "nonsense," we will have a very ad hoc and non-automatable method for detecting nonsensical recursive definitions - relying on human visual inspection alone. Certainly we do not expect humans to be poring over millions of lines of code manually detecting nonsensical recursion. Here, then, is how we will proceed to "de-Fred" recursive definitions (*i.e.*, create irredundant names for them):

- We will come up with a 'clever' Lambda expression, called Y, that will allow recursion to be expressed purely within the framework of Lambda calculus. No arbitrary strings such as "Fred" will be stuck inside the definition.
- Then we will be able to understand recursion in terms of solving an equation involving a function variable F.
- We will then demonstrate that recursive definitions 'make sense' when we can demonstrate that such equations have a *unique* solution. Or if there are multiple solutions, we can select (using a sound criterion) which of these solutions 'makes sense' as far as computers go.

6.1.1 Recursion viewed as solving for a function

Let us write our Fred function as an equation:

```
Fred=lambda x. if (x=0) then 0 else x + Fred(x-1).
```

The above equation can be rewritten as

```
Fred=(lambda Fred' .lambda x. if(x=0) then 0 else x + Fred'(x-1)) Fred
```

The fact that this equation is equivalent to the previous one can easily be verified. We can apply the *Beta* rule, plugging in `Fred` in place of `Fred'`, to get back the original form.

We can now apply the *Alpha* rule, and change `Fred'` to y, to obtain the following:

```
Fred = (lambda y . lambda x . if(x=0) then 0 else x + y(x-1)) Fred.
```

Well, we are almost done eliminating the redundant name "Fred." What we have achieved is that we have expressed `Fred` using an equation of the form

```
Fred = H Fred
```

where H is the Lambda expression
`(lambda y . lambda x . if(x=0) then 0 else y(x-1))`. Note that H contains no trace of `Fred`.

6.1.2 Fixed-point equations

We now make a crucial observation about the nature of the equation
Fred = H Fred. Recall that juxtaposing H and Fred is, in Lambda
calculus, equivalent to applying H to Fred. If we wish, we can rewrite the
above equation as Fred = H(Fred) to express it in the more familiar
syntax of "$f(x)$"—for f applied to x. Note that there is something
peculiar going on; when we feed Fred to H, it spits out Fred. The
application of H seems to be "stuck" at Fred.

In mathematics, an equation of the form $x = f(x)$ is called a *fixed-
point equation*. Think of a *fixed-point* as a *fixed point*, i.e., an "immov-
able point." It appears that things are "stuck" at x; even if we apply
f to x, the result is stuck at x.

Can we find such "stuck function applications?" Surely!

- Take a calculator, and clear its display to get 0 on the display. Then
  hit the *cos* (cosine) key to compute $cos(0) = 1$.
- Hit *cos* again, to get 0.9998477.
- Hit *cos* again, to get 0.99984774.
- Hit *cos* again, to get 0.99984774 - we are now "stuck" at *the* fixed-
  point of *cos*. (The number of steps taken to achieve the fixed-point
  will, of course, depend on the precision of your calculator).
- Now (assuming you have the factorial function on your calculator),
  compute 0, 0!, 0!!, etc., then 2, 2!, 2!!, etc., and finally 3, 3!, 3!!, etc.
  How many fixed-points did you discover for factorial?

Here is another way to get a "fixed-point out of a photocopying
machine." More specifically, we can get the fixed-point of the image
transformation function of a photocopying machine. Go photocopy your
face; then photocopy the photocopy, photocopy the photocopy of the
photocopy, etc. In most photocopying machines, the image stabilizes,
by turning all gray regions to black or (sometimes) white. Such a stable
image is then one fixed-point of the *image transformation function* of
the photocopying machine.

Coming back to our example, we want a fixed-point, namely a *function* Fred, that solves the fixed-point equation Fred = H(Fred). Ideally, we would like this fixed-point to be unique, but if that's not possible, we would like some way to pick out the fixed-point that corresponds
to what computers would compute should they handle the recursive
program.

The beauty of Lambda calculus is that it does not fundamentally
distinguish between functions and "values," and so the principles of
obtaining fixed-points remain the same, independent of whether the

fixed-points in question are functions, numbers, or images. For the example at hand, if **Fred** is a fixed-point of H, we must

- either ensure that there is only one fixed-point, thus giving a unique solution to **Fred**, or
- in case multiple fixed-points exist, find a canonical way of picking the desired fixed-point (solution).

We shall resolve these issues in what follows.

6.1.3 The Y operator

There exists a class of (almost "magical") functions, the most popular one being function Y below, that can *find* the unique fixed-points of functions such as H!

```
Y = (lambda x. (lambda h. x(h h)) (lambda h. x(h h)))
```

In other words, we claim that (Y f) is the fixed-point of any arbitrary function **f**. Things seem almost too good to be true: if we desire the fixed-point of *any* function f expressed in Lambda calculus, simply "send Y after it." In other words, $Y(f)$ would be the fixed-point. For the above to be true, the following fixed-point equation must hold:

```
(Y f) = f(Y f).
```

Here is how the proof goes:

```
Y f     = (lambda x. (lambda h. x(h h)) (lambda h. x(h h))) f

        = (lambda h. f(h h)) (lambda h. f(h h)) <-- look --|
                                                           |
                                                           |
        = f( (lambda h. f(h h)) (lambda h. f(h h)) ) -------

        = f( Y f ).
```

In the above derivation steps, in the penultimate step, we have **f** applied to a big, long Lambda form that was obtained in the second step as an expansion of Y f. Therefore, in the last step, we can obtain the simplification indicated. Okay, now, finally, we have successfully "de-Freded" our original recursive definition for **Fred**. We can write **Fred** as

```
Fred = Y (lambda y . lambda x . if(x=0) then 0 else x + y(x-1))
```

where the right-hand side contains *no trace* of **Fred**. The right-hand side is now an irredundant name defining what was originally cast as an explicit recursion.

6.1.4 Illustration of reduction

While we have shown how to turn the equation `Fred = H Fred` into `Fred = (Y H)`, does the new irredundant name really capture the reduction semantics of functional evaluation? Let us demonstrate that indeed it does, by applying `Fred` to an argument, say 5:

```
Fred 5
  = (Y H) 5
  = H (Y H) 5
  = (lambda y . lambda x . if(x=0) then 0 else x + y(x-1)) (Y H) 5
  = (lambda x . if(x=0) then 0 else x + (Y H)(x-1)) 5
  = (if(5=0) then 0 else 5 + (Y H)(5-1))
  = 5 + (Y H) 4
  = ...
  = 5 + 4 + (Y H) 3
  = ...
  = 5 + 4 + 3 + 2 + 1 + 0
  = 15
```

From the above, one can observe that the main characteristic of Y is that it has the ability to "self-replicate" a Lambda expression. Notice how a copy of `(Y H)` is "stashed away" "just in case" there would be another recursive call. Self-replication, unfortunately, is also the basis on which many malicious programs such as computer viruses operate. In advanced computability theory, the deep connections between "self-replication" and computability are captured in the so-called *recursion theorem*. The interested reader is encouraged to read up on this topic, including Ken Thompson's widely cited article "*Reflections on trusting trust*," to be fully informed of the true potentials, as well as societal consequences, of computers. On one hand, computers are mankind's most impressive invention to date; on the other hand, they are prone to abuse, stemming either from innocent oversight or malicious intent - in both cases demanding the eternal vigil of the computing community to guard against, detect outbreaks, and restore normal operations if bad things do happen.

6.2 Recursive Definitions as Solutions of Equations

The reduction behavior using `Y` indeed tracks the normal function evaluation method.

```
Fred 5 = (lambda x . if(x=0) then 0 else x + Fred(x-1)) 5
       = if(5=0) then 0 else 5 + Fred(5-1)
       = 5 + (Fred 4)
```

```
= (lambda x . if(x=0) then 0 else x + Fred(x-1)) 4
= 5 + (if(4=0) then 0 else 4 + Fred(4-1))
= 5 + 4 + (Fred 3)
= ...
= 5 + 4 + 3 + 2 + 1 + 0
= 15.
```

What really is the advantage of a fixed-point formulation? To better understand this, let us study the connection between recursion and solving for functions more deeply.

The fact that we have equated **Fred** to a Lambda term (**lambda x. if (x=0) then 0 else x + Fred(x-1)**) containing **Fred** suggests that **Fred** is a solution for f in an equation of the form

$$f = (\lambda x.if\ (x = 0)\ then\ 0\ else\ x + f(x - 1)).$$

How many solutions exist for f? In other words, how many different functions can be substituted in place of f and satisfy the equation? Also, if there are multiple possible solutions, then which of these solutions did (Y H) correspond to? Might the function f_0 below, which is undefined over its entire domain N, for instance, be a solution?

$$f_0 = (\lambda x.\ \bot)$$

Here, $\bot$ stands for *"undefined"* or *"bottom"* value.

Substituting f_0 for f, the right-hand simplifies to the function

$$(\lambda x.if\ (x = 0)\ then\ 0\ else\ x + \bot),$$

or

$$(\lambda x.if\ (x = 0)\ then\ 0\ else\ \bot).$$

This function is different from f_0 in that it is defined for one input, namely 0. Hence the above equation is not satisfied. Calling this function f_1, let us see whether it would serve as a solution. Substituting f_1 on the right-hand side, we get

$$(\lambda x.if\ (x = 0)\ then\ 0\ else\ x + f_1(x - 1)).$$

Simplifying $f_1(x - 1)$, we get

$$(\lambda x.if\ (x = 0)\ then\ 0\ else\ \bot)(x - 1)$$
$$= if\ ((x - 1) = 0)\ then\ 0\ else\ \bot$$
$$= if\ (x = 1)\ then\ 0\ else\ \bot.$$

The right-hand side now becomes

$$(\lambda x.if\ (x = 0)\ then\ 0\ else\ x + (if\ (x = 1)\ then\ 0\ else\ \bot))$$

$$= (\lambda x.if\ (x = 0)\ then\ 0\ else\ if\ (x = 1)\ then\ x + 0\ else\ x + \bot)$$
$$= (\lambda x.if\ (x = 0)\ then\ 0\ else\ if\ (x = 1)\ then\ 1\ else\ \bot)$$

which (calling it f_2) is yet another function that defined for one more input value. In summary, substituting f_0 for f, one gets f_1, and substituting f_1, one gets f_2. Continuing this way, each f_i turns into a function f_{i+1} that is defined for one more input value. While *none* of these functions satisfies the equation, *in the limit* of these functions is a *total* function that satisfies the equation, and hence is a fixed-point (compared with earlier examples, such as the *cos* function, which seemed to "stabilize" to a fixed-point in a few steps on a finite-precision calculator; in case of the f_i series, we achieve the fixed-point only in the limit). This limit element happens to be the *least fixed-point* (in a sense precisely defined in the next section), and is written

$$\mu x.(if\ (x = 0)\ then\ 0\ else\ x + f(x - 1)).$$

Let this least fixed-point function be called h. It is easy to see that h is the following function:

$$h(n) = \Sigma_{i=0}^{n}\ i.$$

It is reassuring to see that the least fixed-point is indeed the function that computes the same "answers" as function **Fred** would compute if compiled and run on a machine. It turns out that in recursive programming, the "desired solution" is always the least fixed-point, while in other contexts (*e.g.*, in reachability analysis of finite-state machines demonstrated in Chapter 9), that need not be true.

The 'solution' point of view for recursion also explains recursive definitions of the form

```
function Bob(x) = Bob(x+1).
```

The only solution for function **Bob** is the everywhere undefined function $\lambda x.\ \bot$. To see this more vividly, one can try to **de-Bob** the recursion to get

```
Bob = Y (lambda y . lambda x . y(x+1)).
```

Suppose H = (lambda y . lambda x . y(x+1)). Now, supplying a value such as 5 to **Bob**, and continuing as with function **Fred**, one obtains the following reduction sequence:

```
Bob 5  =  (Y H) 5
       =  H (Y H) 5
       =  (lambda y . lambda x . y(x+1)) (Y H) 5
       =  (lambda x . (Y H)(x+1)) 5
       =  (Y H)(5+1)
```

```
=  (Y H) 6
=  ...
=  (Y H) 7
=  ...
=  (non-convergent).
```

In other words, Bob turns out to be the totally undefined function, or "bottom function."

6.2.1 The *least* fixed-point

In the above example, we obtained a fixed-point h which we asserted to be the "least" in a sense that will now be made clear. In general, given a recursive program, if there are multiple fixed-points,

- The desired meaning of recursive programs corresponds to the *least* fixed-point, and
- The fixed-point finding combinator Y is guaranteed to compute the least fixed-point [114].

To better understand these assertions, consider the following four definitions of functions of type $Z \times Z \to Z$, where Z is the integers (this example comes from [78]):

$$f_1 = \lambda(x,y) \, . \, if \; x = y \; then \; y + 1 \; else \; x + 1$$
$$f_2 = \lambda(x,y) \, . \, if \; x \geq y \; then \; x + 1 \; else \; y - 1$$
$$f_3 = \lambda(x,y) \, . \, if \; x \geq y \; and \; x - y \; is \; even \; then \; x + 1 \; else \; \perp$$

Now consider the recursive definition:

$$F(x,y) = if \; x = y \; then \; y + 1 \; else \; F(x, F(x - 1, y + 1)).$$

We can substitute f_1, f_2, or f_3 in place of F and get a true equation! Exercise 6.8 asks you to demonstrate this. However, of these functions, f_3 is the least defined function in the sense that

- whenever $f_3(x)$ is defined, $f_i(x)$ is defined for $i = 1, 2$, but not vice versa, and
- there is no other function (say f_4) that is less defined than f_3 and also serves as a solution to F.

To visualize concepts such as "less defined," imagine as if we were plotting the graphs of these functions. Now, when a function is undefined for an x value, we introduce a "gap" in the graph. In this sense, the graph of the least fixed-point function is the "gappiest" of all; it is a solution, and is the most undefined of all solutions – it has the most number of "gaps." These notions can be captured precisely in terms of *least upper-bounds* of *pointed complete partial orders* [107]. In our example, it can be shown that f_3 is the least fixed-point for the recursion.

6.3 Fixed-points in Automata Theory

We will have many occasions to appeal to the least fixed-point idea in this book. Here are some examples:

- Cross-coupled combinational gates attain stable states defined by fixed-points. For instance, two inverters in a loop become a flip-flop that stores a 0 or a 1. These are the two solutions of the recursive circuit-node equation $x = not(not(x))$.[1]
- Three inverters in a loop form a ring oscillator. The reason, to a first approximation, is that they are all trying, in vain, to solve the recursive node equation $x = not(not(not(x)))$; in addition, there is a 3-inverter delay around the loop.
- Context-free grammar productions are recursive definitions of languages. The language defined by such recursive definitions can be determined through a least fixed-point iteration similar to what we did by starting with $\perp$; for context-free grammars, the empty language $\emptyset$ serves as the "seed" from which to begin such iterations.
- In the section pertaining to Binary Decision Diagrams, we will formulate a process of computing the set of reachable states as a fixed-point computation.
- In the section on nondeterministic automata, we will write recursive definitions for *Eclosure* and justify that such recursive definitions are well-defined (i.e., they are not nonsensical, similar to the recursion for function Bob, for which no solution exists).

It is also worth noting that while least fixed-points are most often of interest, in many domains such as *temporal logic*, the greatest fixed-points are also of interest.

The functional (higher order function) H from Section 6.1.1, namely

$$H = (\lambda y.\lambda x.if\ (x = 0)\ then\ 0\ else\ x + y(x - 1))$$

has an interesting property: it works as a "bottom refiner!" In other words, when fed the "everywhere undefined function" f_0 (called the

---

[1] Of course, this is a simplified explanation, as a golden wedding ring can also be thought of as setting up a recursive node equation $x = x$ or as $x = not(not(x))$, since the two *not*s cancel. The crucial differences are of course that there is no *amplification* around the wedding ring, and there is far less *loop delay* in a ring than in case of a flip-flop loop.

"bottom function"), it returns the next better approximation of the least fixed-point h in the form of f_1. When fed f_1, it returns the next better approximation f_2, and so on. It may sound strange that the meaning of recursive programs is determined starting from the most uninteresting of functions, namely the 'bottom' or $\perp$ function. However, this *is* the most natural approach to be taking, because

- $\perp$ contains *no* information at all,
- going through the fixed-point iteration starting from $\perp$, pulls into subsequent approximants f_1, f_2, etc., only *relevant information* stemming from the structure of the recursion.

Later when we study state-space reachability analysis techniques in Section 11.3.2, and context-free grammars in Section 13.2.2, we will appeal to this "bottom-refining" property. We will be identifying a suitable "bottom" object in each such domain.

Chapter Summary

Fixed-point theory is often a dreaded topic. We attempted to present the "'friendliest foray into fixed-point theory" that we could muster. We start with simple recursive function definitions and motivate the need to have irredundant forms of these definitions. We arrive at such irredundant forms using Lambda calculus, and the Y operator. We study fixed-point equations and how to solve them using fixed-point iteration, starting from the totally undefined function 'bottom' ($\perp$). We shall employ fixed-points in numerous chapters of this book, especially in two contexts: (i) viewing context-free grammars as recursive equations (Section 13.2.2), and (ii) computing the reachable set of states in state transition systems (Section 11.3.2).

Exercises

6.1. Present a recursive algorithm to mirror a binary tree "along the y axis." Mirroring an empty tree gives back an empty tree. Mirroring a non-empty tree rooted at node n with left sub-tree L and right sub-tree R results in a tree rooted at n with left sub-tree ... (complete this definition). Write the pseudocode.

6.2.
1. Write a recursive program (pseudocode will do) to traverse a binary tree in postorder.
2. Write a recursive program to solve the Towers of Hanoi problem.
3. Describe a recursive descent parser for arithmetic expressions.
4. Describe a recursive descent parser for Reverse Polish expressions.

6.3. Define the Fibonacci function `fib` that determines the nth Fibonacci number using tree recursion (non-linear recursion) using the recipe "the first and second Fibonacci numbers are both 1, and the nth Fibonacci number for $n > 2$ is obtained by adding the previous two Fibonacci numbers." Then compute `fib(3)` by first obtaining the function H and using the Y combinator – as illustrated in Section 6.1.4 for function **Fred**. Show all the reduction steps.

6.4. With respect to the function f_3 defined in Section 6.2.1, try to arrive at different solutions by returning something non-$\bot$ instead of $\bot$ in f_3, and see if such variants serve as solutions. If they do, explain why; also explain if they don't.

6.5. Perform other fixed-point experiments (similar to the ones described in Section 6.1.2) using a calculator. Try to find a fixed-point in each case. In each case, answer whether the fixed-point is unique (are there multiple fixed-points for any of the functions you choose?). Try to find out at least one function that has multiple fixed-points.

6.6. This is Exercise 63, page 89 of [48], where it is attributed to Barendregt: show that

$$(\lambda xy.y(xxy))(\lambda xy.y(xxy))$$

is a fixed-point operator.

6.7. What is one fixed point of Y itself? (Hint: Straightforward from definition of Y as $Yf = f(Yf)$.)

6.8. Verify the assertion on page 100 (that f_1, f_2, and f_3 are solutions to F).

"Oh, never mind him -- he's just looking for fixed-points again."

7

Strings and Languages

Now we embark on studying *strings* of symbols over a given alphabet, and *languages*, which are sets of strings. An *alphabet* is a finite collection of *symbols*. Usually, an alphabet is denoted by Σ. **Alphabets are always non-empty.** In most cases, alphabets are finite, although that is strictly not necessary.

We said that an alphabet Σ is a finite collection of *symbols*. What can symbols be? They can be letters, such as $\{a, b, c\}$, bits $\{0, 1\}$, musical notes, Morse code characters, a collection of smoke-signal puffs, hieroglyphics, kicks you might administer to an errant vending machine, or just about anything we wish to regard as an *indivisible unit* of communication or activity. We *do not* worry about the inner structure of a symbol, even if it has one. We emphasize this point early on, lest you be totally surprised when we start regarding such objects as *regular expressions* (expressions that denote languages - they are discussed in Chapter 8) as *symbols*.

Strings are zero or more symbols in juxtaposition. Each *symbol* from an alphabet can also be regarded as a string of length one. *Languages* are *sets of strings*. We will also use the term **sequence** instead of "string," and employ operations on strings on sequences also.

Each string or sequence can model sequences of key strokes you might type on a computer, a piano, a telegraph machine, etc. We call each such sequence or string a *computation*, a *run* of a machine, etc. A vending machine has a computation consisting of coin plinks, candy ejects, and kicks. A formal machine such as a Turing machine has computations depending on what program is loaded into it, and what input is submitted on its tape. In most chapters, we consider only finite strings. Beginning with Chapter 21, we will consider automata on infinite strings. Infinite strings are useful to model certain properties of *reactive* systems - systems, such as operating systems that are supposed to run forever (at least at a conceptual mathematical level). Such behaviors cannot be adequately modeled at a high level using only finite strings.[1]

Strings and languages are *fundamental* to computer science theory. By measuring run times or storage requirements of programs in terms of the lengths of their input strings, we are able to study time and space complexity. By studying *patterns* that sets of strings possess, we are able to taxonomize degrees of problem solving abilities (computational "power"). In fact, life seems to be full of strings (pun intended[2]). We now proceed to systematically examine this central role played by strings and languages in computer science theory.

7.1 Strings

7.1.1 The *empty* string ε

A string may also be empty, meaning it has zero symbols in it. We denote empty strings by ε. When the available type fonts (mostly in mechanically-generated illustrations) do not permit, we also write ε as `e`, or `epsilon`. Empty strings are the *identity element* for concatenation, meaning the concatenation of ε at the beginning or at the end of a string s yields s itself. Many programming languages denote empty strings by `""`. Many books on automata employ λ instead of ε (we *totally* avoid this usage, as for us λ is reserved for use in Lambda expressions). Examples of non-empty strings from programming languages are `"abcd"`, `"aaa"`, etc. In automata theory, we shall model these strings using *abcd* and *aaa*, or in ordinary fonts sometimes quoted—as in 'abcd'.

To avoid some confusion early on, we stress upon a few basic conventions. In a context where we are talking about strings, '123' is a string of length three, consisting of the symbols 1, 2, and 3, and not a single integer. As an example, when asked to design a DFA that has

---

[1] Of course, by taking timing into account, one can model everything in terms of computations occurring over a finite number of steps—or finite strings. The purpose of employing infinite strings is to reason about propositions such as "event *a* happens, but following that event *b* *never* happens" with due precision. For this, the concept of '*never*' must be modeled in a manner that does not depend on absolute timings. The only way to achieve this is to allow for an infinite string after *a* wherein *b* is never found to occur.

[2] DNA sequences are strings over the symbol-set *A*, *C*, *T*, and *G* – see "Cartoon Guide to Genetics," by Larry Gonick and Mark Wheelis.

alphabet $\{0, \ldots, 9\}$ and accepts all numeral sequences divisible without remainder by 3, the DFA reads the symbols one by one, and checks for the familiar condition for divisibility by 3 (the digits add up to a number divisible by 3). Also, when given the input 126, it reads three symbols—not the entire number 126 in one fell swoop. These conventions mirror the reality that physical machines may never consume an unbounded amount of information in a finite amount of time. The notion of a *symbol* connotes this reality.

7.1.2 Length, character at index, and substring of a string

Given string s, its length is denoted by $length(s)$. We have, for every string s, $length(s) \geq 0$. We of course have $length(\varepsilon) = 0$.

For each position $i \in length(s)$ (*i.e.*, $0 \leq i \leq (length(s) - 1)$) of a string s, $s[i]$ denotes the character at position i. Notice the following: since $length(\varepsilon) = 0$, the $s[.]$ notation is undefined, as there are no i such that $i \in length(s)$, if s were to be ε.

If i and j are positions within a string s, and $j \geq i$, $substr(s, i, j)$ is the substring of s beginning at i and ending at j, both inclusive. If $j < i$, $substr(s, i, j) = \varepsilon$. For instance, $substr(apple, 0, 0) = a$, $substr(apple, 1, 1) = substr(apple, 2, 2) = p$, $substr(apple, 0, 2) = app$, $substr(apple, 2, 4) = ple$, and $substr(apple, 4, 3) = \varepsilon$. For every string s, $s = substr(s, 0, length(s) - 1)$.

7.1.3 Concatenation of strings

Given two strings s_1 and s_2, their concatenation, written as $s_1 s_2$, yields a string of length $s_1 + s_2$ in the obvious way. Sometimes we write string concatenation as $s_1 \circ s_2$ to enhance readability. Examples: *apple worm* results in *appleworm*, as does *apple $\circ$ worm*.

7.2 Languages

A language is a set of strings. Like strings, languages are a central notion in computer science theory. Questions pertaining to undecidability as well as complexity can be studied in terms of languages. The smallest language is $\emptyset$, since $|\emptyset| = 0$. The next larger language has one string in it. Of course there are many such languages containing exactly one string. One example is the language $\{\varepsilon\}$, which contains one string that happens to be empty. Given the alphabet $\Sigma = \{0, 1\}$, a few languages of size 1 are $\{1\}$, $\{0\}$, $\{0101\}$; a few languages of size 2 are $\{0, 00\}$,

{0, 0101}, etc. Many (most?) of the languages we study will be infinite in size.

7.2.1 How many languages are there?

Are there countably many languages, or uncountably many languages? The answer is that there are *uncountably many* languages over any alphabet. In other words, there are as many languages over $\Sigma = \{0,1\}$ as there are Real numbers. This is because each language L can be represented by an *infinite* characteristic sequence which tells which (among the infinite strings over $\Sigma = \{0,1\}$) are present in L and which are absent. Since each such characteristic sequence can be read as a real number in the range $[0,1]$, the set of all such characteristic sequences is uncountable, as explained in Section 3.2.5.

Why does it matter that there are uncountably many languages? To answer this question, let us briefly consider how we shall model computations. We shall study computations in terms of *languages accepted by Turing machines*. Each Turing machine M is a very simple computer program written for a very simple model of a computer. M is started with an input string x on its tape, and "allowed to run." Each such run has three outcomes: M halts in the *accept* state, it halts in the *reject* state, or it loops (much like a computer program can go into an infinite loop). The set of all strings x that cause M to halt in the *accept* state constitute the *language* of M. The language of a Turing machine is, in a sense, the *essence* or *meaning* of a Turing machine. Now, each Turing machine M itself can be represented by a single *finite* string over $\Sigma = \{0,1\}$ (this is akin to viewing an entire `a.out` file obtained by compiling a C program as a single string). Since the set of all *finite* strings is countable, there can only be countably many Turing machines, and hence countably many Turing machine languages. Hence, there are languages that are the language of *no* Turing machine at all. In a sense, these languages (called *non-Turing recognizable*) carry "patterns" that are "beyond the reach of" (or "defy analysis by") any man-made computational device. Their existence is confirmed by the above cardinality arguments.

7.2.2 Orders for Strings

There are two systematic methods available for listing the contents of sets of words. The first is known as the *lexicographic* order, or "order as in a dictionary" while the second is called the *numeric* order which lists strings by length-groups: all strings of a lower length are listed before a string of a higher length is listed. Within a length-group, the strings are listed in lexicographic order. We now present some details as well as examples.

Lexicographic order for strings

In this order, given two strings x and y, we compare the characters comprising x and y position by position. We do this for each position $i \in min(length(x), length(y))$ (here, min finds the minimum of two numbers). For example, if we compare *apple* and *pig*, we compare for each position $i \in 3$, as *pig* is the shorter of the two and has length 3. While comparing characters at a position, we compare with respect to the ASCII code of the character at that position.[3]

Given all this, for strings x and y such that $x \neq y$, x said to be *strictly before* y in lexicographic order, written $x <_{lex} y$, under one of two circumstances:

1. If there exists a position $j \in min(length(x), length(y))$ such that $x[j] < y[j]$, and for all $i \in j$ (meaning, for positions $0 \ldots (j - 1)$), $x[i] = y[i]$. For example, *aaabb* $<_{lex}$ *aaaca*, *aaron* $<_{lex}$ *abate*, *apple* $<_{lex}$ *pig*, and *pig* $<_{lex}$ *putter*.
2. For all positions $j \in min(length(x), length(y))$, we have $x[j] = y[j]$, and $length(x) < length(y)$.

Definition: Define *string identity*, $x =_{lex} y$, to hold for two *identical* strings x and y. Also define $\leq_{lex}$ to be $<_{lex} \cup =_{lex}$, and $>_{lex}$ to be the complement of $\leq_{lex}$.

Definition: Given two strings x and y, x is *before* y in *lexicographic order* iff $x \leq_{lex} y$.

Let us now go through some examples of lexicographic order:

- $\varepsilon <_{lex} aa$ follows from condition 2 above. This is because there are no positions within a length of 0, and so the only condition to be satisfied is $length(x) < length(y)$, which is true for $x \neq y$ and $x = \varepsilon$. Since $\varepsilon <_{lex} aa$, we also have $\varepsilon \leq_{lex} aa$.
- $a <_{lex} aa <_{lex} aaa <_{lex} aaaa \ldots$. Hence, these are also in the $\leq_{lex}$ relation.

---

[3] The ASCII code totally orders all the characters available on modern keyboards. Refer to the web or a book on hardware design to know what this total order is. We are taking this more pragmatic route of relying on ASCII codes to keep our definitions concrete.

Numeric order for strings

The idea of *numeric order* is of importance while systematically enumerating sets. To motivate this idea, consider a lexicographic ordering of all strings over alphabet $\{a, b\}$. Such a listing will go as follows: ε, a, aaa, $aaaa$, ... In other words, *not a single string containing a b will get listed*. To avoid this problem, we can define the notion of a *numeric order* as follows. For two strings x and y:

1. If $length(x) = length(y)$, then $x <_{numeric} y$ exactly when $x <_{lex} y$.
2. Otherwise, if $length(x) < length(y)$, then $x <_{numeric} y$.
3. Otherwise $(length(x) > length(y))$ $x >_{numeric} y$.

As before, we define $<_{numeric}$ to be *strictly before* in numeric order, and

$$\leq_{numeric} \; = \; <_{numeric} \cup =_{lex}$$

to be *before* in numeric order, or simply "in numeric order."
The numeric order of listing the strings over alphabet $\{a, b\}$ will yield the sequence

$$\varepsilon, a, b, aa, ab, ba, bb, aaa, aab, aba, abb, baa, bab, bba, bbb, aaaa, \ldots.$$

7.2.3 Operations on languages

Given languages L, L_1, and L_2, new languages may be defined using various operations. First, consider the set operations of *union, intersection, set difference,* and *symmetric difference.* These familiar set operations help define new languages as defined below:

> *Union:* $L_1 \cup L_2 = \{x \mid x \in L_1 \lor x \in L_2\}$
> *intersection:* $L_1 \cap L_2 = \{x \mid x \in L_1 \land x \in L_2\}$
> *Set difference:* $L_1 \setminus L_2 = \{x \mid x \in L_1 \land x \notin L_2\}$
> *Symmetric difference:* $(A \setminus B) \cup (B \setminus A)$

7.2.4 Concatenation and exponentiation

The concatenation operation on two languages L_1 and L_2 performs the concatenation (juxtaposition) of strings from these languages. Exponentiation is *n*-ary concatenation. These ideas are basically quite simple. Suppose we have a finite-state machine M_1 that accepts strings containing either an odd number of a's or an even number of b's. Suppose we have another finite-state machine M_2 that accepts strings containing either an even number of a's or an odd number of b's. We can now build a new machine M whose runs are of the form xy such that x is accepted by M_1 and y by M_2. We can also build a new machine M_1^* whose runs are of the form $xx \ldots x$, where x is repeated some $k \geq 0$ times, and x is a run of M_1. The set of runs of these new machines can be captured using concatenation and exponentiation.

We now formally define the concatenation and exponentiation operations on languages:

Concatenation: $L_1 \circ L_2 = \{xy \mid x \in L_1 \wedge y \in L_2\}$.

Note: Often, we will omit the 'o' operator and write $L_1 L_2$ instead of $L_1 \circ L_2$.

Exponentiation: First of all, for **any** language L, define $L^0 = \{\varepsilon\}$. That is, the zero-th exponent of any language L is $\{\varepsilon\}$. Note that this means that *even for the empty language* $\emptyset$, $\emptyset^* = \{\varepsilon\}$. (The reason for this convention will be clear momentarily).

Exponentiation may now be defined in one of two *equivalent* ways:

Definition-1 of Exponentiation: For all $n \geq 1$,

$$L^n = \{x_1 \ldots x_n \mid x_i \in L\}.$$

Definition-2 of Exponentiation:

$$For\ n > 0,\ L^n = \{xy \mid x \in L \wedge y \in L^{n-1}\}.$$

In the second definition, note that we should define the *basis case* for the recursion, which is L^0. We must put into L^0 anything that serves as the *identity element* for string concatenation, which is ε. *Hence, we* define $L^0 = \{\varepsilon\}$ for any language, L.

Consider a simple example:

$$\{a, aba\} \circ \{ca, da\} = \{aca, ada, abaca, abada\},$$

while

$$\{ca, da\}^3 = \{ca, da\} \circ \{ca, da\} \circ \{ca, da\},$$

which is the set

$$\{cacaca, cacada, cadaca, cadada, dacaca, dacada, dadaca, dadada\}.$$

Consider another example where L_1 (which models the runs of M_1) is the set

$$L_1 = \{x \mid x \text{ is a sequence of odd number of a's or even number of b's}\}$$

and L_2 (which models the runs of M_2) is the set

$L_2 = \{x \mid x$ is a sequence of even number of a's or odd number of b's$\}$.

Then, $L_1 \circ L_2$ is a language in which each string consists of

- an odd number of a's,
- an odd number of b's,
- an odd number of a's followed by an odd number of b's, or
- an even number of b's followed by an even number of a's.

In addition, L_1^k is a language in which each string can be obtained as a k-fold repetition of a string from L_1.

7.2.5 Kleene Star, '$*$'

Star performs the union of repeated exponentiations:

$$L^* = \{x \mid \exists k \in N : x \in L^k\}.$$

This definition can also be written as

$$L^* = \cup_{k \geq 0} L^k,$$

or even as

$$L^* = \cup_{k \in Nat} L^k.$$

Notice that according to these definitions, $\emptyset^* = \{\varepsilon\}$ This is because for any k, $\emptyset^k$ is defined to be $\{\varepsilon\}$. In turn, this is so because we must have a basis element in the recursive definition of concatenation. In our case, ε is the identity for string concatenation.

Examples:

- $\{a, b, c\}^*$ is *all possible strings* over a, b, and c, including the empty string ε.
- $\{a, bc\}^*$ is *all possible strings* formed by concatenating a and bc in some order some number of times. In other words, all the strings in $\{a, bc\}^k$, for all k, are present in this set.

In a programming sense, *star* is iteration that is unable to keep track of the iteration count. The ability to "forget" the number of iterations is crucial to obtaining simpler languages, and ultimately, *decidability*.

7.2.6 Complementation

Using the above-defined notion of *star*, we can now specify a universe Σ^* of strings, and define complementation with respect to that universe:

$$\overline{L} = \{x \mid x \in \Sigma^* \setminus L\}.$$

7.2.7 Reversal

The notion of reversing a string is pretty straightforward - we simply put the first character last, the second character penultimate, and so on. The definition of $reverse(s)$ for s being a string in some alphabet, is as follows:

$reverse(\varepsilon) = \varepsilon$
$reverse(s \text{ as } ax) = reverse(x) \circ a.$

We use the as construct to elaborate a quantity. 's as ax' means 'view string s as a concatenation of symbol a and string x.' The term '$reverse(x) \circ a$' says 'reverse x and append a to its end.'

For language L, $reverse(L)$ is then $\{reverse(x) \mid x \in L\}$.

7.2.8 Homomorphism

A homomorphism is a function that maps strings to strings and *that respects string concatenation*. In other words, if $h : \Sigma^* \to \Gamma^*$ is a homomorphism that takes strings from some alphabet Σ to strings in an alphabet Γ (not necessarily distinct), then it must satisfy two conditions:

- $h(\varepsilon) = \varepsilon$
- For string $s = xy$ in Σ^*, $h(xy) = h(x)h(y)$. In other words, the result of applying h to s is the same as the result of *concatenating* the application of h to its pieces x and y.

To obtain a better appreciation for the fact that a homomorphism "respects string concatenation," let us consider something that is *not* a homomorphism—say g. Let $\Sigma = \{a, b, c, d\}$ and $\Gamma = \{0, 1, 2\}$.

- $g(abcd) = 0$
- $g(ab) = 1$
- $g(cd) = 1$
- $g(s) = 2$, for all other strings.

g is not a homomorphism because $g(abcd) = 0$ while $g(ab)g(cd)$ is 11.

7.2.9 Ensuring homomorphisms

How do we *ensure* that a given function on strings *is* a homomorphism? The most commonly used approach to ensure that something is a homomorphism is to specify a mapping that goes from *symbols* to *strings*, and then to "lift" it up to map from strings to strings. Let h' be the symbol to string map, with signature

$$h' : \Sigma \cup \{\varepsilon\} \to \Gamma^*,$$

and define h **using** h'. Here is an example. Let h' be as follows:

- $h'(\varepsilon) = \varepsilon$
- $h'(a) = 0$
- $h'(b) = 1$
- $h'(c) = h'(d) = 2$

Formally, $h(s)$ is defined in terms of h' as follows:

- If $length(s) > 1$, then let $x \neq \varepsilon$ and $y \neq \varepsilon$ be such that $s = xy$. Then, $h(s) = h(x)h(y)$.
- If $length(s) = 1$, then $h(s) = h'(s)$.

Now, **by definition**, h respects string concatenation and all works out well!

Homomorphisms can also be defined over languages in a straightforward manner. Given L,

$$h(L) = \{y \mid h(x) = y \; for \; some \; x \; in \; L\}.$$

7.2.10 Inverse homomorphism

Given a homomorphism h, an *inverse* homomorphism h^{-1} maps a string y to the *maximal set of strings* that y could have come from; formally:

$$h^{-1}(y) = \{x \mid h(x) = y\}.$$

Example: Consider h to be

$h'(\varepsilon) = \varepsilon$
$h'(a) = 0$
$h'(b) = 1$
$h'(c) = h'(d) = 2$

Then, $h^{-1}(012) = \{abc, abd\}$ because 2 could have come either from c or from d. Inverse homomorphisms can also be defined over languages in a straightforward manner. Given L,

$$h^{-1}(L) = \cup_{y \in L} \; h^{-1}(y).$$

In other words, for each $y \in L$, take $h^{-1}(y)$, and then take the union of all these sets.

7.2.11 An Illustration of homomorphisms

In parsing a programming language such as C, the parser appeals to a *tokenizer* that first recognizes the structure of *tokens* (sometimes called *lexemes*) such as integers, floating-point numbers, variable declarations, keywords, and such. The tokenizer pattern-matches according to regular expressions, while the parser analyzes the structure of the token stream using a context free grammar (context free grammars are discussed in Chapters 13 and 14; basically, they capture the essential lexical structure of most programming languages, such as nested brackets, begin/end blocks, etc.). Suppose one wants to get rid of the tokenizer and write a context free grammar up to the level of individual characters. Such a grammar can be obtained by substituting the character stream corresponding to each token in place of each token in a *modular* fashion, according to a homomorphism. For example, if begin were to be a keyword in the language, instead of treating it as a token *keyword*, one would introduce additional productions of the form

> begin -> b e g i n

Thanks to the *modular* nature of the substitutions, it can be shown that the resulting grammar would also be context-free.

7.2.12 Prefix-closure

A language L is said to be prefix-closed if for every string $x \in L$, every prefix of x is also in L. If we are interested in *every* run of a machine, then its language will be prefix-closed. This is because a physically realizable string processing machine must encounter substrings of a string before it encounters the whole string. *Prefix closure* is an operation that 'closes' a set by putting in it all the prefixes of the strings in the set. For instance, given the set of strings $\{a, aab, abb\}$, its *prefix closure* is $\{\varepsilon, a, aa, ab, aab, abb\}$.

Chapter Summary

This chapter introduced the motivations for studying computations in terms of *strings* and *languages*. It put our previous discussions about cardinality in Chapter 3 to very good use by quickly showing the fact that there are uncountably many languages. After defining various orderings between strings, we discuss operations on languages: how to make new languages given existing languages. These discussions will set the stage for virtually all of what the rest of this book involves.

Exercises

7.1. How many languages over an alphabet of size 1 exist? Express your answer using the $\aleph$ notation.

7.2. Using the *Schröder-Bernstein Theorem*, argue that there are as many languages over the alphabet $\{0\}$ as there are over the alphabet $\{0, 1\}$.

7.3. Name the main difference between the proof of the uncountability of languages versus the uncountability of the set of Reals.

7.4. Choose among the sets given below, those with cardinality $\aleph_2$ (recall that Nat has cardinality $\aleph_0$):

1. The set of all languages over alphabet $\Sigma = \{0, 1\}$.
2. The powerset of the set of all languages over alphabet $\Sigma = \{0, 1\}$.
3. The set of all languages over alphabet $\Sigma = \{0\}$.
4. The powerset of the set of all languages over alphabet $\Sigma = \{0\}$.

7.5. Prove that $\leq_{lex}$ is a total order, while $<_{lex}$ is not.

7.6. Viewing C programs as ASCII strings, obtain a good estimate of the number C programs of less than n bytes, for $n \leq 12$? Obtain at least five such C programs; compile them and demonstrate that they do not cause any compilation errors.

Hint: `main(){}` is one such program, with byte count 9. View each C program as a string (so you could write `"main(){;}"` for clarity, if you wish). In later discussions of strings, we will omit the quotes (`"`) which serve as string delimiters.

7.7. What is the set consisting of the first ten words in English occurring in lexicographic order? (Follow any dictionary you may have access to. Treat the upper and lower cases the same).

7.8. What is the set consisting of the first ten words in English occurring in numeric order? (Follow any dictionary you may have access to).

7.9. Suppose that in a strange language, 'Longspeak', typeset in characters **a** through **z**, there are words of all lengths beginning with any letter. Consider a lexicographic order listing of all words in Longspeak. At what rank (ordinal position) does a word beginning with letter d appear? How about for a numeric order listing?

Considering the answers to these questions, suppose you are charged with the task of producing a dictionary for Longspeak. Fortunately, you are told that you don't have to include words of lengths beyond 30. Would you produce your dictionary in lexicographic order or numeric order? Explain your reasoning.

7.10. Define languages L_1 and L_2 over alphabet $\Sigma = \{0, 1\}$ as follows:

$$L_1 = \{x \mid x \text{ is a string over } \Sigma \text{ of odd length}\}$$

and

$$L_2 = L_1 \cap \{x \mid x \text{ is a string of } 0's\}.$$

What is the symmetric difference of languages L_1 and L_2?

7.11. Among the following languages, choose the ones for which $LL = L$:

- $L = \emptyset$.
- $L = \{\varepsilon\}$.
- $L = \{x \mid x \text{ is an odd length string over} \{0, 1\}\}$.
- $L = \{x \mid x \text{ is an even length string over} \{0, 1\}\}$.
- $L = \{x \mid \#_0(x) = \#_1(x)\}$; here, $\#_0$ determines the number of 0s in a given string, and likewise for $\#_1$.
- $L = \{x \mid \#_0(x) \neq \#_1(x)\}$.

7.12. Repeat Exercise 7.11 to meet the condition $L^* = L$.

7.13. Repeat Exercise 7.11 to meet the condition $L^* = LL = L$.

7.14. Complement each of the languages defined in Exercise 7.11 and express your answers using the set comprehension notation.

7.15. Consider the languages

- $L_1 = \{x \mid x \text{ is an odd length string over} \{0, 1\}\}$, and
- $L_2 = \{x \mid x \text{ is an even length string over} \{0, 1\}\}$.

1. Argue that the following map is a homomorphism (call it h):
   $h(\varepsilon) = \varepsilon$
   $h(0) = 0$
   $h(1) = 00$.
2. Determine $h(L_1) \cap h(L_2)$.

7.16. In Exercise 7.11, find those languages in this list such that $L = pref(L)$.

Machines, Languages, DFA

This chapter begins our study of machine models. We will first briefly examine the general characteristics of various abstract machines that can serve as language recognizers for the languages depicted in Figure 4.2 of Section 4.3. We will refer to this diagram as the *"Power"* diagram. We will then study the Deterministic Finite-state Automaton (DFA) model in the rest of the chapter.

8.1 Machines

All the machines discussed in the *Power* diagram, including DFAs discussed in this chapter, are primarily *string processors*. They are started in their initial state and are fed an input string. The inputs of interest are *finitely long* ("finite") strings. The main question of interest is whether, at the end of the input, the machine is in an *accepting state* or not. The Turing machine has, in addition to a set of accepting states, a set of *rejecting* states.

There are several axes along which machines can be distinguished. Some of these distinctions are now pointed out. We will remind the reader of these details when specific machine types are discussed. The purpose of the current discussion is to portray the overall nature of the variations that exist among machines.

All machine types in the *Power* diagram that are of lower power than the LBA (namely the DFA, NFA, DPDA, and NPDA) must read their input strings completely before it is deemed that they accept the input string. Furthermore, they must read their input string in the process of a single left-to-right scan (meaning, they cannot go back and reread a previously read symbol). In each *step* of computation, these machines are allowed to read atmost one input symbol (they may read none, as well).

The remaining machine types in the *Power* diagram, namely the LBA, DTM, and the NDTM, are not required to read the whole input on their input "tape." In other words, it is possible for these machines to accept an input without having read all of it—or for that matter,[1] without having read *any of it*! Furthermore, they are allowed to reread a previously read input. This rereading may happen in any order (*i.e.*, not confined to the order in which inputs are laid out on the tape).

All the machines in the *Power* diagram possess a *finite-state control flow graph* describing their behavior. The NFA, NPDA, and NDTM allow nondeterminism in their control flow graph (nondeterminism is described in the next section). While the DFA and NFA are totally described in terms of this finite-state control, the remaining machine types carry an additional data structure in the form of a stack or one or more tapes. The DPDA and NPDA have access to a single unbounded stack from which they pop a single symbol at every step. Additionally, in each step, zero or more symbols may be pushed back onto the stack.

The LBA, the DTM, as well as the NDTM have, in addition to a control flow graph, a single read/write tape. There is the notion of a *current position* on this input tape, usually specified by the position of a read/write "head." Also, there is the notion of presenting the input string over a contiguous sequence of cells on the input tape. The LBA is allowed to write a tape cell only if this cell belongs to the contiguous sequence of cells over which the initial tape input was presented. A Turing machine has no such restriction.

Note that we avoid depicting the *Büchi* automaton in the *Power* diagram of Chapter 3, as this machine type is not comparable with these other machine types. For example, one can simulate an LBA on a Turing machine, or a DFA on a push-down automaton; no such simulation of the Büchi automata on any of the machines in the *Power* diagram is possible. In a sense, the presentation of Büchi automata is a way of making the reader aware of the existence of these machines and the practical uses that these machines have in modeling time in an abstract manner, as well as broaden the reader's perspective early on. Details on Büchi automata appear in Chapters 22 and 23. It is also important to point out that due to space/time limitations, we have left out many machine types on *finite strings* that *could* easily have been

---

[1] While this sounds odd, this freedom is required when modeling machines with power comparable to general-purpose computers, which may, as we all know, often ignore reading their input entirely.

depicted on the *Power* diagram, such as two-way automata [108]. Many
of these machine have been proved equivalent to machines in the *Power*
diagram, as far as the class of languages they recognize is concerned.

With this general introduction, we proceed to examine the various
machine types beginning with the DFA.

8.1.1 The DFA

A *deterministic finite-state automaton* (DFA) is specified through a
transition graph such as in Figure 8.1. It starts its operation in an initial
state (in this case I). When a DFA is situated in one of its reachable
states s and is fed an input string w, it advances to a *unique* state as per
its transition graph. Formally, a *deterministic finite-state automaton*
D is described by five quantities presented as a tuple, $(Q, \Sigma, \delta, q_0, F)$,
where:

- Q, a *finite nonempty* set of states;
- Σ, a *finite nonempty* alphabet;
- $\delta : Q \times \Sigma \to Q$, a *total* transition function;
- $q_0 \in Q$, an initial state; and
- $F \subseteq Q$, a *finite, possibly empty* set of final (or *accepting*) states.

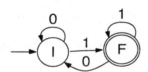

Fig. 8.1. A DFA that recognizes strings over $\{0, 1\}$ ending with 1

For example, in Figure 8.1, $Q = \{I, F\}$, $\Sigma = \{0, 1\}$, $q_0 = I$, $F = \{F\}$,
and $\delta = \{\langle I, 0, I \rangle, \langle I, 1, F \rangle, \langle F, 0, I \rangle, \langle F, 1, F \rangle\}$. A few useful conventions
pertaining to drawing DFAs as well as naming their states are now
described (these conventions will be followed in many of our initial
drawings, and possibly later also):

- The initial state will be called I. In drawings, it will be the state to
  which an arrow without a source node points to.
- If the initial state is also a final state, it will be called IF. It will
  then be a double-circled node.
- All final state names (other than the state named as IF, as explained
  above) will begin with letter F, and will be double-circled nodes.

```
/* Encoding of a DFA that accepts the language (0+1)*1 */
main()
{ bool state=0;
  char ch;
  while(1)
  { ch = getch();
    switch (state)
    { case I: switch-off the green light;
             switch (ch)
             { case 0 : break;
               case 1 : state=F;
             }
      case F: switch-on the green light;
             switch (ch)
             { case 0 : state=I;
               case 1 : break;
             }
    }
  }
}
```

Fig. 8.2. Pseudocode for the DFA of Figure 8.1

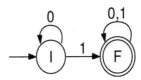

Fig. 8.3. Multiple symbols labeling a transition in lieu of multiple transitions

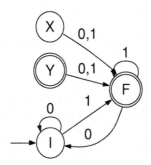

Fig. 8.4. A DFA with unreachable (redundant) states

- All other state names begin with letters other than I and F, and are
  best chosen with mnemonic significance. They are drawn as single
  circles.

- Generally one would expect a DFA to decode every move out of
  every state. If, for expediency, we leave out some of these moves, we
  recommend that the following comment accompany such diagrams:
  "we assume that all moves not shown lead the DFA to the black
  hole state." We recommend that the state name BH (for black hole)
  be used for that state from which no escape is possible.[2]
- One could have a DFA with multiple symbols labeling a single transition in lieu of separate transitions that bear these symbols, as
  shown in Figure 8.3.
- One could have a DFA with truly unreachable states, as shown in
  Figure 8.4. These states (states X and Y in this example) may be
  removed without any loss of meaning.

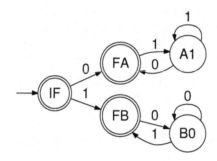

Fig. 8.5. Another example DFA

One can see the conventions explained above more clearly in Figure 8.5.

The DFA in Figure 8.1 is really encoding programs similar to the
one in Figure 8.2. All "DFA programs" are `while(1)` loop controlled
`switch` structures. While in a certain state, a DFA decodes the current
input symbol and decides to either update its state or keep its current
state. The next input symbol is decoded in this (possibly) updated
state. Whenever *any* final state (including the IF state) is entered, the
DFA turns on its *green light*. The green light stays on so long as the
DFA stays in one of the final state, and is turned off when it exits it. A
DFA program is not allowed to use *any* other variable than the single
variable called `state`. It cannot produce *any* other output than shine
the green light. It appears to be programming at its simplest - yet, such
humble DFAs are enormously powerful and versatile!

---

[2] This is similar to gravitationally hyper-dense stars from which nothing - including
light - escapes. We did vote down the alternate name RM (for 'Roach Motel')
proposed by some students.

The δ function of a DFA is also commonly presented as a table, with rows being states, columns being inputs, with the final state(s) starred. For our example, we have

```
        input
        -----
state|  0    1
-------------
  I   | I    F
 *F   | I    F
```

A DFA is said to accept a string s if its green-light is 'on' after processing that string; all other strings are *rejected* by the DFA. However, note that a DFA "never stops working"; after processing any string s, it will be in a position to process any one of the input symbols a, and may (depending on the DFA) accept or reject that string, sa. DFAs can be thought of as *string classifiers*, assigning a status (accepted/rejected) for *every* string within Σ^*. The set of strings accepted by a DFA is said to be its *language*. A DFA is said to *recognize* its language (not *accept* its language - the term 'accept' is used for individual strings).

Formally speaking, a DFA is said to *accept* an input $x \in \Sigma^*$ if x takes it from q_0 to one of its final states. Define a function $\hat{\delta}$ that maps a state and a string to a state, as follows:

For $a \in \Sigma$ and $x \in \Sigma^*$, $\hat{\delta}(q, ax) = \hat{\delta}(\delta(q, a), x)$
$$\hat{\delta}(q, \varepsilon) = q.$$

Then, a DFA accepts x exactly when $\hat{\delta}(q_0, x) \in F$.

An alternate definition of the language of a DFA is based on the notion of *instantaneous descriptions* (IDs). An ID is a "snapshot" of a DFA machine in operation. For a DFA, this snapshot contains the *current state* and the *unconsumed input string*. Knowing these, it is possible to predict the entire future course of a DFA.

Formally speaking, we define the *type* of the ID of a DFA to be $T_{ID\_DFA} = Q \times \Sigma^*$. In other words, we take the view that the *type* of the ID, T_{ID_DFA}, includes all possible $\langle state, string \rangle$ pairs, *i.e.*, $Q \times \Sigma^*$. We now define two relations, $\vdash$ and $\vdash^*$; the latter being the reflexive and transitive closure of the former. $\vdash$ is a one step move function that captures how a DFA moves from one ID to another.

Suppose id_1 is one snapshot, (q_1, σ_1), and id_2 is another snapshot, (q_2, σ_2), such that $\sigma_1 = a\sigma_2$. Now, if $\delta(q_1, a) = q_2$, we know that the DFA can go from snapshot (q_1, σ_1) to (q_2, σ_2). In other words, define the relation $\vdash \subseteq T_{ID\_DFA} \times T_{ID\_DFA}$. as follows:

$\vdash (id_1 \text{ as } (q_1, \sigma_1), id_2 \text{ as } (q_2, \sigma_2)) =$

$$\exists a \in \Sigma : \sigma_1 = a\sigma_2 \land \delta(q_1, a) = q_2.$$

In the infix syntax,

$$(q_1, \sigma_1) \vdash (q_2, \sigma_2) \text{ iff } \exists a \in \Sigma : \sigma_1 = a\sigma_2 \land \delta(q_1, a) = q_2.$$

As said before, define $\vdash^*$ is defined to be the reflexive transitive closure of $\vdash$. In other words, $\vdash^*$ is the smallest relation $\subseteq T_{ID\_DFA} \times T_{ID\_DFA}$ closed under the following rules:

- $\langle x, x \rangle \in \vdash^*$ for $x \in T_{ID\_DFA}$;
  In other words, $x \vdash^* x$ for $x \in T_{ID\_DFA}$,
- $\langle x, y \rangle \in \vdash \land \langle y, z \rangle \in \vdash^* \Rightarrow \langle x, z \rangle \in \vdash^*$;
  In other words, $x \vdash y \land y \vdash^* z \Rightarrow x \vdash^* z$,

Now, for DFA d, its language

$$\mathcal{L}(d) = \{x \mid \exists q \in F : (q_0, x) \vdash^* (q, \varepsilon)\}.$$

The above definition says that the language of a DFA consists of all strings x such that $\vdash^*$ can take the ID (q_0, x) to the ID (q, ε), where $q \in F$. Notice that the second ID contains ε, indicating that the DFA has consumed the entire string x.

Example: Consider the DFA in Figure 8.3. What is in the $\vdash$ relation with respect to snapshots that involve I and F? Since this is an infinite set, we show a few members below (we use $\langle$ and $\rangle$ as well as (and) to delimit tuples, to enhance readability):

$$\{\langle(I, 1), (F, \varepsilon)\rangle, \langle(I, 111), (F, 11)\rangle, \dots \langle(I, 1010100110), (F, 010100110)\rangle, \}.$$

Now we show a few members of $\vdash^*$:

$$\{\langle(I, 1111), (F, \varepsilon)\rangle, \dots \langle(I, 1010100110), (F, 0110)\rangle,$$
$$\langle(I, 1010100110), (F, 010)\rangle, \langle(I, 1010100110), (F, 01)\rangle,$$
$$\langle(I, 1010100110), (F, \varepsilon)\rangle, \dots\}.$$

Note that $\vdash^*$ can "consume" multiple symbols.

Some definitions, tips, and conventions:

- A language L is defined to be *regular* if there is a DFA D such that $L = \mathcal{L}(D)$.
- While it is helpful to leave out those transitions of a DFA that lead to the *black hole* - BH - state, *please remember to reinstate the black hole state and all the transitions going to this state before you perform DFA operations, such as complementation.*[3]

---

[3] Otherwise, you will find yourselves in a 'DH' - deep hole.

8.1.2 The "power" of DFAs

Despite their simple appearance, DFAs are incredibly powerful and versatile! Here are some examples of their power:

- The most touted applications of DFAs are in the area of compiler *scanner* generation, where a scanner is a program that recognizes the patterns hidden within strings representing keywords, floating-point numbers, etc.
- Exercises 8.16 and 8.17 demonstrate that using DFAs, one can perform division either going MSB-first (the 'high-school' method) or even LSB-first. Only a finite amount of information needs to be remembered across digits (essentially, the remainder after division; fully explained in these exercises).
- DFAs can be used to compactly represent Boolean functions (Chapter 11) or even help determine the *validity* of formulas in certain branches of mathematical logic, known as *Presburger arithmetic* (Chapter 20).

Most questions about DFAs are algorithmically decidable. In particular, the following questions about DFAs are decidable:

- Does a given DFA accept a given input?
- Does a given DFA have an empty language?
- Is the language of a given DFA all possible strings over Σ?
- Are two given DFAs equivalent?

DFAs have many desirable properties, some of which are the following:

- Given any DFA d, one can algorithmically obtain another DFA d' whose language is the reverse of the language of d.
- Given any DFA d, one can algorithmically obtain another DFA d' whose language is the complement of the language of d.

8.1.3 Limitations of DFAs

DFAs have two main limitations. First, despite being adequately expressive, they may require "too many" (an exponential number) states, which does adversely affect the space/time requirements of DFA-based algorithms. Second, for classifying strings based on many frequently occurring patterns, DFAs are simply inadequate. For instance, it is impossible to build a DFA that accepts *all* and *only* those strings containing an equal number of 0s and 1s. See Exercises 8.19, 8.20, and 8.21. We shall soon see that using the concept of nondeterminism, we can obtain exponentially more succinct finite-state representations, giving rise to the next machine type we shall study, namely, *nondeterministic finite automata* (NFA).

Considering Exercise 8.20, one can easily prove that there can exist no DFA. A proof sketch goes as follows:

- Suppose there exists an M-state DFA for this language. Consider the string $w = (^M \ )^M$.
- w causes $M + 1$ states to be visited even while processing $(^M$, a string of M left parentheses.
- Since only M of these states can be distinct states, one state repeats during the $(^M$ traversal.
- Therefore, there also exists a shorter, non state-repeating path leading to the *same* final state. However, taking this path causes the DFA to omit one or more (, thus causing it to accept $(^k)^M$ for some $k < M$ – a contradiction with the fact that this is the DFA for the language L.

It is obvious that to recognize a language such as L above, all we need to do is add a single stack to the DFA D. Doing so, we obtain a machine known as *deterministic push-down automaton* (DPDA). In general, by adding different kinds of data structures to the finite-state control, one can handle languages whose strings have more complicated structures. In Section 8.1.4, we shall now present a few examples that illustrate this point.

8.1.4 Machine types that accept non-regular languages

Consider the language

$$L_1 = \{a^i b^j c^k \mid i, j, k \geq 0 \ \wedge \ if \ i = 1 \ then \ j = k\}.$$

In all the strings in this language, the characters a, b, and c appear in that order, with b's and c's being equal in number if the number of a's is 1. This language can obviously be handled using a DPDA, using its stack. In fact,

$$reverse(L_1)$$

can also be handled using a DPDA, as the finite-state control can remember whether the number of b's and c's were equal by the time the a (or a's) appear.

Now consider

$$L_2 = \{a^i b^j c^k d^m \mid i, j, k, m \geq 0 \ \wedge \ if \ i = 1 \ then \ j = k \ else \ k = m\}.$$

This language can still be processed using only one stack, as the matching between b's and c's or c's and d's (whichever is to occur) can be decided by first seeing the a's. No such luck awaits

$$reverse(L_2),$$

which has to have two decisions, (whether the number of b's and c's match, or whether the number of c's and d's match), in hand by the time the a (or a's) arrive. Clearly, these two decisions cannot be generated using a single stack, thus showing that $reverse(L_2)$ cannot be processed by a DPDA. The same can be concluded about

$$L_3 = \{ww \mid w \in \{0,1\}^*\}$$

also, as the second w's head must be matched against the first w's head which, unfortunately, is at the bottom of the stack. It turns out (as we shall demonstrate later) that

$$L_4 = \overline{L_3}$$

indeed can be processed using a single stack push-down automaton.

One obvious solution to handling L_3, as well as $reverse(L_2)$, would be to employ two stacks instead of one. Unfortunately this gives more power than necessary (the machine becomes as powerful as a Turing machine). Another machine type, the linear bounded automaton (LBA), can be used. An LBA has finite control as well as a tape such that it can read/write only the region of the tape in which the input initially appeared. In addition, an LBA has a *finite tape alphabet* that may (and typically does) contain Σ, the input alphabet, *plus* (typically) many additional symbols. Exploiting the ability to repeatedly scan the input, an LBA can decide membership in all the languages listed above. By the same token, an LBA can decide membership in the language

$$\{a^n b^n c^n \mid n \geq 0\}$$

as well as

$$\{a^n b^{n^2} \mid n \geq 0\}.$$

We later present languages that cannot be decided using LBAs. To handle the full generality of languages, we remove the restriction on LBAs that they can write only where the input is presented, thus obtaining a *Turing machine*.

```
digraph G {
/* Defaults */
  fontsize = 12;
  ratio = compress;
  rankdir=LR;
/* Bounding box */
  size = "4,4";
/* Node definitions */
  I  [shape=circle, peripheries=1];
  F  [shape=circle, peripheries=2];
  "" [shape=plaintext];
/* Orientation */
  orientation = landscape;
/* The graph itself */
  ""  -> I;
  I ->   I [label="0"];
  I ->   F [label="1"];
  F ->   F [label="1"];
  F ->   I [label="0"];
/* Unix command: dot -Tps exdfa.dot >! exdfa.ps */
/* For further details, see the 'dot' manual    */
}
```

Fig. 8.6. Drawing DFAs using dot

```
% Include everything till %--- in a latex document, and run latex
\begin{figure}
{\hfill {\psfig{file=exdfa.ps,height=4.5cm,width=2cm,angle=-90}\hfill}}
\caption{Whatever caption you desire}
\label{fig:label-for-cross-referencing}
\end{figure}
%---
```

Fig. 8.7. Processing .ps files in Latex

8.1.5 Drawing DFAs neatly

We close this chapter off with some pragmatic tips for DFA drawing. Almost all drawings in this book are created using the dot package which is freely downloadable as part of the graphviz tools. After downloading and installing this package, you will see an executable file dot. You may then present your drawings in a file with extension .dot, and process the drawing into a postscript file for inclusion into your documents, including Latex documents, where you may further process the drawing. For example, Figures 8.6 and 8.7 provide all the commands needed to generate Figure 8.1.

Chapter Summary

This chapter began with a brief overview of various machine types as well as some key differences between them. It then introduced deterministic finite state automata (DFA). Various notations for representing DFA were discussed: tables, diagrams, and mathematical structures. After introducing the notion of acceptance of strings by DFA, the notion of a language L being *regular* was defined; there is at least one DFA that accepts all and only those strings in L. The chapter closed with a description of some of the limitations of DFAs in terms of being able to serve as recognizers for certain (non-regular) languages. We also pointed out how having one stack allows us to handle languages with more interesting structure such as balanced parentheses. We also presented a result that we shall see again much later; having two stacks gives a machine based on finite-state control as much power as a Turing machine.

Exercises

You are encouraged to use `grail` *and/or JFLAP to check your results in these exercises.*

8.1. Present the pseudocode of an algorithm to reverse a directed graph represented by an adjacency matrix. A directed graph is reversed by reversing all its edges.

8.2. What is the language of the DFA in Figure 8.1 if:

- the only accept state is the starting state?
- there are no accepting states (no double-circled states)?
- we reverse every arc, rename state I as state F, and rename state F as state I, making it the initial state?

8.3. What is the language recognized by the DFA in Figure 8.5? List ten strings in its language and ten not in its language in lexicographic order.

8.4. Draw a DFA that is different from (non-isomorphic to) the DFA shown in Figure 8.5, however with the same language.

8.5. What happens to the language of the DFA in Figure 8.5 if A1 jumps to B0 upon receiving a '1', and B0 jumps to A1 upon receiving a '0'?

8.6. Draw a DFA having language equal to the union of the languages of the DFAs in Figure 8.1 and Figure 8.5.

8.7. Draw a DFA having language equal to the symmetric difference of the languages of the DFAs in Figure 8.1 and Figure 8.5.

8.8. Apply the following homomorphism to the language of the DFA of Figure 8.1 and draw a DFA for the resulting language. Repeat for the DFA of Figure 8.5.

$$h(\varepsilon) = \varepsilon$$
$$h(0) = 11$$
$$h(1) = 00$$

8.9. Obtain a DFA for the prefix-closure of the language of the DFA of Figure 8.1. Repeat for the DFA of Figure 8.5.

8.10. Draw a DFA having language equal to the concatenation of the languages of the DFAs in Figure 8.1 and Figure 8.5.

8.11. Draw a DFA having language equal to the star of the language of the DFA in Figure 8.1. Repeat this exercise for the DFA of Figure 8.5.

8.12. Draw a DFA having language equal to the reverse of the language of the DFA in Figure 8.1. Repeat this exercise for the DFA of Figure 8.5.

8.13. Draw a DFA having language equal to the complement of the language of the DFA in Figure 8.1. Repeat this exercise for the DFA of Figure 8.5.

8.14. What happens if, in the DFA in Figure 8.1, every transition carries label '1'? Think of what the corresponding DFA program will do.

8.15. Give a DFA accepting the set of strings over $\Sigma = \{0,1\}$ such that each block of five consecutive symbols contains at least two 0s.

8.16. Develop a DFA that recognizes the following language:

$$L = \{x \mid x \in \{0,1\}^* \text{ and } (||x|| \bmod 5) = 0\}.$$

Here $||x||$ stands for the magnitude of x viewed in unsigned binary. Example: $|| 0101 ||= 5$, with the leftmost 0 coming first. *Hint:* Notice that the bits of x are being presented MSB-first. Act on each new bit and recalculate the remainder and remember it in the DFA states. Think of what happens to a number, say N, when it gets shifted left and a least significant bit b gets inserted. Basically N becomes $2 \times N + b$.

8.17. With respect to the language L of Exercise 8.16, design a DFA for $rev(L)$, the set of strings which are reverses of strings in L (this is tantamount to processing the bit stream LSB-first, as opposed to L which processes it MSB-first). Example: $\parallel 0101 \parallel = 5$, with the rightmost 1 coming first. *Hint:* N becomes $b \times 2^m + N$ where m is the position into which the new MSB b walks in. $N \bmod 5$ becomes $(b \times 2^m + N) \bmod 5$. We need a recurrence that keeps track of 2^m for various m. Each DFA state remembers a pair
("powers of 2 mod 5", "the mod of the number mod 5").
A few steps are:
$(1,0) - 1 \rightarrow (2,1) - 1 \rightarrow (4,3) - 1 \rightarrow (3,2)$.
   In general, the evolution of the state goes as follows:

State $N \bmod 5 \longrightarrow$
$(b \times (Weight \bmod 5) + N \bmod 5) \bmod 5$.

Now, $Weight \bmod 5$ is maintained as follows:

$2^m \bmod 5 \longrightarrow (2 \times (2^m \bmod 5)) \bmod 5$.

A few of the state transitions are given below. Complete the rest.

```
(1,0) - 0 -> (2,0)   (2,0) - 0 -> (4,0)   (2,1) - 0 -> (4,1)
(1,0) - 1 -> (2,1)   (2,0) - 1 -> (4,2)   (2,1) - 1 -> (4,3) ...
```

8.18. Repeat Exercises 8.16 and 8.17 for k instead of 5 and for arbitrary number base b (with Σ suitably adjusted) instead of binary. Obtain the state transition relation in both these cases.

8.19. Design a DFA for the following language, for various n (up to the limit of your time availability; do it at least up to $n = 4$):

$$\{x0z \mid x \in \{0,1\}^* \text{ and } z \in \{0,1\}^n\}.$$

Prove that the required DFA has exponentially many states (exponential in n).

8.20. Either design a DFA for the following language

$$L = \{(^n\ )^n \mid n \geq 0\},$$

or prove that no such DFA can exist! *Note:* Since this problem's proof was sketched on page 127, please come up with another proof that considers *taking* the state-repeating path (perhaps more than once).

8.21. Write clear pseudocode for a stack-based algorithm to recognize strings in L above.

9

NFA and Regular Expressions

In the last chapter, we have seen many examples of the versatility of regular languages. We have also seen several examples of regular languages for which the DFAs were both simple and intuitive. In Chapters 11 and 20, we will further elaborate on the power of DFAs by showing that they can be used to encode as well as reason about statements from certain decidable branches of mathematical logic. In Section 8.1.3, we also discussed

some of the limitations of DFAs. Their main limitation is, of course, that they cannot serve as recognizers for non-regular languages such as $L = \{(^n)^n \mid n \geq 0\}$, and these languages are very important in computer science. In fact, they are part of the syntax of most computer programming languages. In later chapters, we will develop rigorous proof techniques for proving that certain languages are not regular. We will then study machines that are strictly more powerful than DFAs.

In this chapter, we continue the study of regular languages, and in that context, the main limitation of DFAs is that they can be *unnatural*,[1] *exponentially large* or both. To overcome these limitations, in this chapter we will introduce a new machine type called *nondeterministic finite automata*, or NFA. NFAs subsume DFAs; in fact, every DFA is an NFA (*more precisely said*, every DFA can, with a small adjustment of mathematical definitions, be regarded as an NFA). We also wish to point out that nondeterminism is one of the central ideas in theoretical computer science, going well beyond its role as a 'description size compressor.' In this book, we will repeatedly be revisiting the concept of nondeterminism in the context of various machine types.

---

[1] DFAs can end up being unnatural with respect to highlighting the essential structure of a regular language, as will be illustrated in Illustration 9.2.2.

NFA are specified much like DFA are: through pictures of directed graphs (for human consumption), or through a description of the δ function, say through a tabular function description (for machine consumption). However, often a textual syntax is preferred over these styles. This is precisely what *regular expressions* are. In other words, regular expressions allow us to write down mathematical expressions that denote regular languages, and each such regular expression has a very straightforward interpretation as an NFA. All regular languages over an alphabet Σ can be specified using just the primitive regular expressions $\emptyset, \varepsilon$, and $a \in \Sigma$, and the regular expression building operators $\cup, \circ$ and $*$.

To sum up, the key results we are aiming to establish by the end of Chapter 12 are as follows:

- Regular languages are those recognized by a deterministic finite automaton.
- For every DFA, there is an equivalent NFA; for every NFA there is an equivalent RE; and for every RE there is an equivalent DFA.
- For any regular language, the *minimal* DFA—the DFA with the smallest number of states that serves as a recognizer for the language—is *unique*.
- Regular languages are closed under many operations, including:
  - union, intersection, complementation, star
  - homomorphism, inverse homomorphism
  - reversal, prefix-closure.

In this chapter, we will introduce NFAs and REs through several examples. We explain how NFA process strings by moving from *sets of states* to other *sets of states* either by reading a symbol from an alphabet or without reading any symbols (through ε moves). We explain how to determine the language of an NFA using the notion of *Eclosure*.

We strive to provide an experimental flavor to this chapter by employing the **grail** tool suite for illustrations. In Section 9.3.6, we illustrate through a case study how to use the **grail** tool suite to design and formally debug regular expressions. Virtually all the diagrams in this book are generated using **grail** and a graph drawing package called **dot**. We also try to relate the material to the real world by demonstrating how a simple *lexical analyzer* can be built using the **flex** tool.

9.1 What is Nondeterminism?

Nondeterminism has many uses in computer science. It can help designers describe the systems they are about to build even at stages of the design where they have not determined all the details. In some cases, while these details may be known to designers, they may still choose *not* to include them to avoid inundating their audience with excess information.[2]

To understand nondeterminism in a real-world context, consider the act of summoning an elevator car in a busy building that has multiple elevators. After one presses the call button, it is not entirely predictable what will happen. One may, if lucky, get a car headed in the same direction as they intend to travel. If unlucky, one would get a car going in the opposite direction, or a car that is full, etc. If one had perfect knowledge about the entire building and its occupants, they could predict the outcome with certainty. However, most people[3] *do not* want to keep track of all such information, instead preferring to live according to a nondeterministic protocol that goes as follows: "if I am fated not to get an empty car, I should try again." In short, by employing nondeterminism, one can write system descriptions at a high level, without worrying about pinning down details too early. It has been said that pinning down details too early ("premature optimization") is at the root of all that is evil in software design.

Nondeterministic descriptions have another property: they tend to *over-approximate* the system being described. Over-approximation helps ignore special cases in the behavior of a system (it is akin to packaging a delicate, but odd-shaped, electronic gadget by inserting it between a pair of molded Styrofoam carriers, thus smoothening the overall appearance). By adding behaviors, over-approximation often helps "round" things up, hence simplifying the whole system.

To better understand the ramifications of over-approximation, consider a building where all north side elevators are designed *not* to be interrupted during their upward journey during early mornings.[4] In other words, each elevator control algorithm has an *if-then-else* in it that tests whether it is running inside a north side elevator, checks what time of the day it is, and prevents interruption if the tests confirm 'north' and 'morning.' Now, if one were to hire a formal verifi-

---

[2] In the modern society that suffers from 'information pollution,' nondeterminism can be the breath of oxygen that saves us from asphyxiation.

[3] Except perhaps "control freaks."

[4] Perhaps to ensure the speedy progression of the janitorial staff and their accouterments to top floors.

cation specialist to mathematically verify that all the elevator control algorithms are working correctly, here is how they could employ non-determinism to simplify their activities. They could simply replace the *if-then-else* with a nondeterministic jump to both cases. Such a modified control algorithm has more behaviors than the original, in which *every* elevator could nondeterministically decide whether to ignore the user's interrupt or to heed to it. Now, if the verification specialist is handed the property to verify; "If I ring for a car, it will eventually come to my floor," that property would pass on the nondeterministically over-approximated system. The specialist would end up having verified this property more easily, by ignoring the *if-then-else*. On the other hand, if they are handed another property, namely "if the car is below my floor, is headed to a floor above my floor and I press the 'up' button, it will stop for me," the verification will fail for the over-approximated model because every elevator can exercise the 'ignore interrupt' option. The verification specialist will realize that this is a *false positive*—a false alarm—and then add the missing detail, which is the *if-then-else* statement. During system design and verification, one can then add just enough information to prove each properties of interest. The alternative approach of revealing all internal information,[5] both taxes the mind and adds to verification time. Computer science's essential mission is complexity management, and nondeterminism plays an essential role in this regard.

In this book, we now return to the use of nondeterminism for describing regular languages. We now begin discussing the topic, "what else does nondeterminism affect?" We provide the answers under separate headings.

9.1.1 How nondeterminism affects automaton operations

The presence of nondeterminism affects the ease with which automaton operations (such as union, concatenation, and star) can be carried out. In the context of finite automata, while the operations of union, concatenation, and star become *easier* with the use of nondeterminism, the operations of complementation, intersection, and equivalence become *harder* to perform. We will see the details when we introduce these algorithms.

---

[5] Popular with many legal departments, who do so in fine print.

9.1.2 How nondeterminism affects the power of machines

Nondeterminism also fundamentally affects the power of various machine types, as follows (see also Section 4.3):

- Nondeterministic finite automata and deterministic finite automata are equivalent in power.
- Nondeterministic push-down automata are strictly more powerful than deterministic push-down automata.
- The equivalence between deterministic PDAs is algorithmically decidable. This is actually a recent result. It was obtained 30 years after it was first conjectured to be true.
- The equivalence between nondeterministic and deterministic PDAs (and hence between two nondeterministic PDAs) is undecidable.
- It is still an open problem whether nondeterministic and deterministic linear bounded automata are equivalent in power.
- The power of Turing machines does not change (with respect to decidability properties) through the use of nondeterminism. However, with respect to complexity, certain problems (for instance, *NP-complete problems* which are discussed later) take exponential time on deterministic machines, but can be solved in polynomial time on nondeterministic Turing machines. Also, Turing machine equivalence is *undecidable*.

With this general introduction, we look at regular expressions and NFA in greater detail. We begin with regular expressions because we believe that they will already be familiar to users of computers - if not known by that name.

9.2 Regular Expressions

The idea of using regular expressions to denote whole classes of strings is quite widespread. In many card games, the Joker card is called the "wild-card," in that it can be used in lieu of any other card. In most operating systems, the wild-card is * that matches all file names—as in rm *.* or del *.*. Such expressions are practical examples of regular expressions in day-to-day use. Formally, we define regular expressions as follows. We use the inductive definition style in which a basis case and a list of inductive rules are provided, and define regular expressions to be the least set satisfying these rules:

- ∅ is an RE denoting the empty language ∅.

Most computer-assisted tools have trouble accepting $\emptyset$ directly, and so their syntax for $\emptyset$ varies. It is written as {} in grail. Check with each tool to see how it encodes $\emptyset$. A command man regexp or man egrep issued on most Unix systems usually reveals how Unix encodes REs.

- ε is an RE denoting the language $\{\varepsilon\}$.

  Again, most computer-assisted tools encode ε variously—as "" by grail, for example. In our drawings, we sometimes use e or epsilon in lieu of ε. Unix has several flavors of ε—those occurring within words, at the beginning, at the end, etc.[6]

- $a \in \Sigma$ is an RE denoting the language $\{a\}$.

  Most tools directly encode this regular expression in ASCII - for example, a.

- For REs r, r_1 and r_2, $r_1 r_2$, $r_1 + r_2$, r^*, and (r) are regular expressions. These regular expressions denote the following languages, with the indicated encodings in grail (we write $\mathcal{L}(x)$ to mean '*language of*'):

RE $r_1 r_2$	denotes $\mathcal{L}(r_1) \circ \mathcal{L}(r_2)$ encoded in grail as r1 r2
RE $r_1 + r_2$	denotes $\mathcal{L}(r_1) \cup \mathcal{L}(r_2)$ encoded in grail as r1+r2
RE r^*	denotes $(\mathcal{L}(r))^*$	encoded in grail as r*
RE (r)	denotes $\mathcal{L}(r)$	encoded in grail as (r)

The above regular operators form a *basis set* in the sense that we do not need any other operators such as *complementation, reversal, intersection RE*, etc., to build regular languages. *All regular languages can be specified using only the above regular expression syntax.* In practice, however, there are many languages that are very difficult to specify without the use of these additional operators, especially complementation (also known as negation).[7] The reason for this difficulty is that the regular expression syntax, which only has $*$, $\circ$, and $+$, allows us to build *up* the language from simpler languages, while often it would be much more natural to build *down* a language, say, by specifying two different (but larger) languages and intersecting them. If you attempt Exercises 9.6 and 9.8, you will confront the extreme difficulty of specifying two example languages directly in the RE syntax of the basis set - i.e., without the use of negation.

---

[6] According to the Unix Manual pages for grep(1), the symbol $\backslash b$ matches the empty string at the edge of a word, and $\backslash B$ matches the empty string provided it's not at the edge of a word.

[7] Recall that for two languages L_1 and L_2, the complement of their union, $\overline{(L_1 \cup L_2)}$, is equivalent to the intersection of their complements, $\overline{L_1} \cap \overline{L_2}$, by DeMorgan's law. Hence, if we have *complementation* and *union*, we can obtain the effect of intersection.

Illustration 9.2.1 In previous chapters we have seen how to write out languages using set-theoretic syntax. In particular, we can express language

$$Labc_k = \{xabc \mid x \in \{0,1\}^k \wedge a,b,c \in \{0,1\} \wedge (a = 0 \vee b = 1)\}$$

for various k as follows:

- $k = 0 : \{\varepsilon\} \circ \{00,01,11\} \circ \{0,1\}$,
- $k = 1 : \{0,1\} \circ \{00,01,11\} \circ \{0,1\}$,
- $k = 2 : \{0,1\}^2 \circ \{00,01,11\} \circ \{0,1\}$,
- ...

We can encode these languages using regular expressions as follows:

- $k = 0 : \ \varepsilon \circ (00 + 01 + 11) \circ (0 + 1)$,
- $k = 1 : \ (0 + 1) \circ (00 + 01 + 11) \circ (0 + 1)$,
- $k = 2 : \ (0 + 1)(0 + 1) \circ (00 + 01 + 11) \circ (0 + 1)$,
- ...

In the `grail` syntax, these regular expressions can be encoded as:

- $k = 0 :$ `""(00+01+11)(0+1)`,
- $k = 1 :$ `(0+1)(00+01+11)(0+1)`,
- $k = 2 :$ `(0+1)(0+1)(00+01+11)(0+1)`,
- ...

Illustration 9.2.2 Consider the language

$$L_k = \{x0z \mid x \in \{0,1\}^* \text{ and } z \in \{0,1\}^k\}.$$

For various k, we can express this language using regular expressions (in `grail` syntax) as follows:

- $k = 0$: `(0+1)*0`,
- $k = 1$: `(0+1)*0(0+1)`,
- $k = 2$: `(0+1)*0(0+1)(0+1)`,
- $k = 3$: `(0+1)*0(0+1)(0+1)(0+1)`,
- ...

We see that the size of the regular expression grows linearly with k—a property also shared by nondeterministic automata for this language, as we shall illustrate in Figure 9.2. In contrast, as shown in Figure 9.1, minimal DFAs for this language will grow exponentially with k. In particular, the minimal DFA for each k has 2^{k+1} states. The intuitive reason for this situation is that DFAs have no ability to "postpone decisions," whereas NFAs have this ability "by keeping multiple options open."

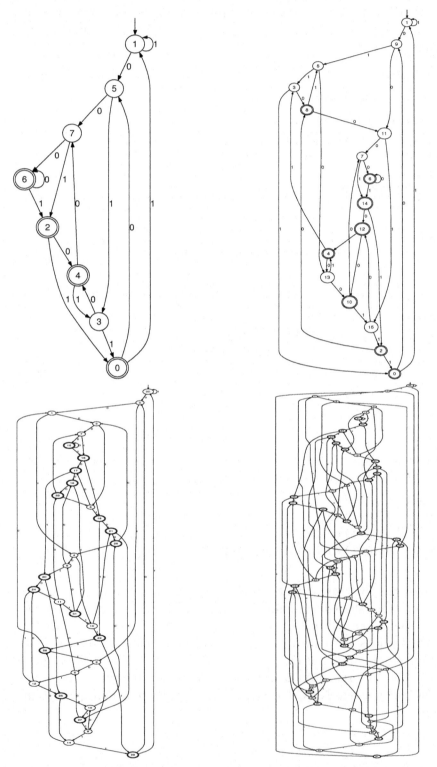

Fig. 9.1. Minimal DFAs for L_k for $k = 2, 3, 4, 5$, with 2^{k+1} states

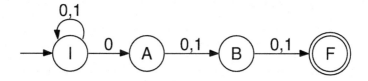

Fig. 9.2. An NFA for $(0+1)^* \, 0 \, (0+1)^k$, for $k = 2$

9.3 Nondeterministic Finite Automata

We now formally introduce nondeterministic finite automata. Let Σ_ε stand for $(\Sigma \cup \{\varepsilon\})$. A *nondeterministic finite-state automaton* N is a structure $(Q, \Sigma, \delta, q_0, F)$, where:

- Q, a *finite non-empty* set of states;
- Σ, a *finite non-empty* alphabet;
- $\delta : Q \times \Sigma_\varepsilon \to 2^Q$, a *total* transition function (note, however, that the range includes $\emptyset$; therefore, if for some state q, $\delta(q, x) = \emptyset$, the move from state q on input x is essentially to a black hole[8]);
- $q_0 \in Q$, an initial state; and
- $F \subseteq Q$, a *finite, possibly empty* set of final states.

The use of nondeterminism allows much more natural-looking as well as compact descriptions of many automata. Contrast Figure 9.2 with the equivalent *minimal* DFAs in Figure 9.1. Notice that these minimal DFAs are minimized as well as unique; in other words, it is impossible to avoid the exponential growth based on k. The NFA, on the other hand, simply accommodates all these figures by progressively elongating its tail. Figure 9.2 for $k = 2$ has a tail labeled with $0, 1$ of length $k = 2$, and this sequence can be made three, four, and five steps long to correspond to the machines in Figure 9.1.

We now discuss various aspects of nondeterministic automata step-by-step, and present a comprehensive, practical example in Section 9.3.5.

9.3.1 Nondeterministic Behavior Without ε

We provide two examples illustrating NFA. The first, provided in Figure 9.2, demonstrates one type of nondeterminism where, from a single state, an NFA can advance to multiple next states upon reading an input symbol—in this case, from state I upon reading 0 to states I and A. The various components of this NFA are as follows:

---

[8] Unlike in recent theories on real black holes by Stephen Hawking where he proves that information *can* escape, nothing escapes an automaton blackhole!

- $Q = \{\texttt{I,A,B,F}\}$
- $\Sigma = \{\texttt{0,1}\}$
- $q_0 = \texttt{I}$
- $F = \texttt{F}$
- $\delta : \{I, A, B, F\} \times \{0, 1, \varepsilon\} \rightarrow 2^{\{I,A,B,F\}}$ is given by the following table:

```
        Input    0     1     e
        -----------------------
States |
   I   |   {I,A}  { I }  { }
       |
   A   |   { B }  { B }  { }
       |
   B   |   { F }  { F }  { }
       |
   F   |    { }    { }   { }
```

Notice that this NFA is assumed to move to the empty set of states, $\emptyset$ (written {}) in the absence of an explicitly given move in this table. To trace out the behavior of this NFA on an input string such as 10101, we place a token in state I and subject it to moves according to its δ function. The use of *instantaneous descriptions* (ID) as employed in Section 8.1.1 suggests itself as a way to keep track of 'where the NFA is in its diagram' and 'what the unconsumed input string is,' except we know that an NFA can be in a set of states after a move. What type of IDs must we employ to keep track of all this? Of the many choices possible, we employ the following type:

$$T_{ID\_NFA} = 2^Q \times \Sigma^*.$$

This ID keeps a *set of states* and the unconsumed string *specific to that set of states*. We now use the $\vdash$ notation to indicate 'one step' of forward progress of the NFA, where a step may be made through a member of Σ or through ε.

$\vdash$ relation for an NFA

We define $\vdash \subseteq T_{ID\_NFA} \times T_{ID\_NFA}$ as follows:

$$(Q_1, \sigma_1) \vdash (Q_2, \sigma_2) \doteq \exists a \in \Sigma_\varepsilon : \sigma_1 = a\sigma_2 \wedge Q_2 = \cup_{q_1 \in Q_1} \delta(q_1, a).$$

In other words, $\vdash$ causes a move from an ID (Q_1, σ_1) to another ID (Q_2, σ_2) by moving every state $q_1 \in Q_1$ via the δ function upon $a \in \Sigma_\varepsilon$, and taking a union of the resulting sets of states. Notice that we allow the choice of ε for a, meaning that σ_1 can be the same as σ_2, and the

NFA can simply move through ε. An example that illustrates the use
of $\vdash$ is as follows:

- Initial ID of the NFA is $(\{I\}, 10101)$.
- $(\{I\}, 10101) \vdash (\{I\}, 0101)$; after reading 1, the NFA is still in the
  set of states $\{I\}$, as given by the δ function of the NFA.
- $(\{I\}, 0101) \vdash (\{I, A\}, 101)$; after reading 0, the NFA is simultaneously in two states, $\{I, A\}$, as given by the δ function.
- $(\{I, A\}, 101) \vdash (\{I, B\}, 01)$; after reading 1, the token in state I
  goes back to state I itself, while the token in state A advances to
  state B. For example, the NFA is now simultaneously in two states,
  $\{I, B\}$. We obtain $\{I, B\}$ by taking the union of $\delta(I, 1)$ and $\delta(A, 1)$.
  Continuing in this manner, we go through the IDs shown below.
- $(\{I, B\}, 01) \vdash (\{I, A, F\}, 1)$; after reading 0, state I goes to the
  set of states $\{I, A\}$ whereas state B goes to state F. *Now the DFA
  has seen a substring that it accepts, namely* 1010. If 1010 were the
  entire string, one token would have reached F, and consequently
  the NFA would have accepted. However, since the entire string is
  10101, we continue as below.
- $(\{I, A, F\}, 1) \vdash (\{I, B\}, \varepsilon)$; after reading the final 1, F "falls out of
  the diagram" (goes to the empty set of states, as shown by $\delta(F, 1)$),
  while the other states, I and A, move to I and B. Since neither of
  these are final states, the NFA does not accept 10101.

9.3.2 Nondeterministic behavior with ε

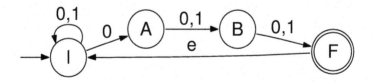

Fig. 9.3. An NFA for $((0+1)^* \, 0 \, (0+1)^k)^+$, for $k = 2$

We now illustrate an NFA's ability to advance through ε moves with
the aid of Figure 9.3, which is a slight modification of Figure 9.2. In
Figure 9.3, the NFA can move from state $\mathbf{F}$ to state $\mathbf{I}$ via ε. In short,
this machine accepts *one or more repetitions* of a string in the language
of Figure 9.2. Let us see how. In terms of instantaneous descriptions,
the sequence of ID changes, when running with input 10101010, are as
follows: As before,

- $(\{I\}, 10101010) \vdash (\{I\}, 0101010)$
- $(\{I\}, 0101010) \vdash (\{I, A\}, 101010)$
- $(\{I, A\}, 101010) \vdash (\{I, B\}, 01010)$
- $(\{I, B\}, 01010) \vdash (\{I, A, F\}, 1010)$.
- One way to proceed is to treat the ID $(\{I, A, F\}, 1010)$ as $(\{I, A, F\}, \varepsilon 1010)$ and allow the ε transition to fire. This sends I and A to the empty set of states, while sending F to $\{I\}$:
- $(\{I, A, F\}, \varepsilon 1010) \vdash (\{I\}, 1010)$.
- We now can perform $(\{I\}, 1010) \vdash (\{I\}, 010)$.
- Let us now accelerate our presentation by using $\vdash^*$.
- $(\{I\}, 010) \vdash^* (\{I, A, F\}, \varepsilon)$. The string is accepted since one of the tokens has reached F.

Another way in which the NFA could proceed from $(\{I, A, F\}, 1010)$ is by moving states I, A, and F on 1 (without inserting the ε):

- $(\{I, A, F\}, 1010) \vdash (\{I, B\}, 010)$, in which move I proceeds to $\{I\}$, A proceeds to $\{B\}$, and F proceeds to $\emptyset$.
- $(\{I, B\}, 010) \vdash (\{I, A, F\}, 10)$.
- $(\{I, A, F\}, 10) \vdash (\{I, B\}, 0)$.
- $(\{I, B\}, 0) \vdash (\{I, A, F\}, \varepsilon)$, also resulting in the string being accepted.

To sum up, using $\vdash$, we have to be prepared to read a string such as 0100 as $x0x1x0x0x$, where each x is any number of εs in sequence. Section 9.3.3 introduces a better alternative for determining the behavior of NFAs using the concept of *Eclosure* or ε-closure.

> *Eclosure*, in effect, considers all[9] ε-laden interpretations of strings in "one fell swoop."

The language of an NFA

The language of an NFA consists of all those sequences of symbols that can be encountered while tracing a path from the start state to some final state. We eliminate all occurrences of ε from such sequences unless the entire sequence consists of εs, in which case, we turn the sequence into a single[10] ε. Formally, given an NFA $N = (Q, \Sigma, \delta, q_0, F)$ with the $\vdash$ relation defined as above, its language

---

[9] While discussing nondeterministic push-down automata (NPDA) in later chapters, we will see that a construct similar to *Eclosure* cannot be defined for NPDAs, as the stack state also has to be kept around.

[10] Please note that ε is *not* part of the alphabet of the NFA.

$$\mathcal{L}(n) = \{w \mid (\{q_0\}, w) \vdash^* (Q, \varepsilon) \land Q \cap F \neq \emptyset\}.$$

In other words, *if* the ID $(\{q_0\}, w)$ can evolve through zero or more $\vdash$ steps to an ID (Q, ε) where Q contains a final state, *then* w is in the language of the NFA.

9.3.3 *Eclosure* (also known as ε-closure)

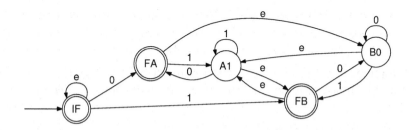

Fig. 9.4. An example NFA

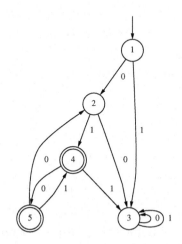

Fig. 9.5. DFA obtained using `grail`

Eclosure(q) obtains, starting from a state q, the set of states reached by an NFA traversing zero or more ε labeled transitions. The best way to see which states are included in $Eclosure(q)$, for any q, is to imagine the following:

Apply a *high-voltage* to state q; imagine that every ε edge is a diode that conducts in the direction of the arrow; now see which are the states that would be fatal to touch due to the high-voltage;[11] all those are in $Eclosure(q)$.

Example: Let us obtain the *Eclosure* of various states in Figure 9.4.

- The Eclosure of state IF is IF itself. The high-voltage spreads from IF to itself.
- What is *Eclosure* of FA? Applying high-voltage to this state, it spreads to state FA, to $B0$, to $A1$, and finally, to FB. For example, $Eclosure(FA) = \{FA, B0, A1, FB\}$.

Unfortunately, an intuitive definition in terms of voltages isn't rigorous enough! Hence, we set up an alternate definition of $Eclosure(q)$ as follows:

- First, we need a way to compute all states which can be reached from a given state by traversing ε edges.
  Let $\rightarrow \subseteq Q \times Q$ be an arbitrary relation over Q. Then, given $\rightarrow \subseteq Q \times Q$, we will define a postfix usage of this operator, namely $q \rightarrow$, to be the *image* of q under $\rightarrow$:

$$q \rightarrow = \{x \mid \langle q, x \rangle \in \rightarrow\}.$$

- Now define the relation $\xrightarrow{\varepsilon}$ which is a subset of $Q \times Q$:
  $$\xrightarrow{\varepsilon} = \{\langle q_1, q_2 \rangle \mid q_1, q_2 \in Q \wedge q_2 \in \delta(q_1, \varepsilon)\}.$$
- Next, define the reflexive and transitive closure of $\xrightarrow{\varepsilon}$ in the usual way. Call it $\xrightarrow{\varepsilon}{}^*$.

Given all this, we define

$$Eclosure(q) = q\xrightarrow{\varepsilon}{}^*.$$

Here, we use the $\xrightarrow{\varepsilon}{}^*$ as a postfix operator. The reflexive part above is very important to ensure that q gets included within $Eclosure(q)$.

The overall effect of employing $\xrightarrow{\varepsilon}{}^*$ as a postfix operator is to force all ε-only paths to be considered.

Example: Redoing our example with respect to Figure 9.4,

$$Eclosure(FA) = FA\xrightarrow{\varepsilon}{}^*.$$

---

[11] An *ideal* non-leaky diode that does not break down.

- This "grabs" state FA itself (the reflexive part of $\overset{\varepsilon}{\rightarrow}{}^*$ does this).
- Next, $B0$ enters this set, as it is one step away. In fact, $B0$ is in $FA\overset{\varepsilon}{\rightarrow}{}^1$.
- Next, $A1$ enters this set, as it is two steps away. In fact, $A1$ is in $FA\overset{\varepsilon}{\rightarrow}{}^2$.
- Finally, FB enters this set, as it is three steps away. In fact, FB is in $FA\overset{\varepsilon}{\rightarrow}{}^3$.
- No more states enter $Eclosure(FA)$.

As said earlier, $Eclosure$ helps define the behavior of an NFA, as well as its language more directly. It also helps us define the NFA to DFA conversion algorithm (Chapter 12) very directly.

9.3.4 Language of an NFA

Having defined $Eclosure$, we can now formally define the language of an NFA. For a string $x \in \Sigma^*$, define $\hat{\delta}(q_0, x)$ of an NFA to be the set of states reached starting from $Eclosure(q_0)$ and traversing all the symbols in x, taking an $Eclosure$ after every step. We are, in effect, taking the image of q_0 under string x, except that we are allowing an arbitrary number of εs to be arbitrarily inserted into x. Also, we will overload $Eclosure$ to work over sets of states in the obvious manner, as follows:

$$Eclosure(S) = \{x \mid \exists s \in S : x \in Eclosure(s)\}.$$

This definition can also be written as

$$Eclosure(S) = \cup_{s \in S} Eclosure(s).$$

Likewise, we overload δ to work over sets of states:

$$\delta(S, a) = \{x \mid \exists s \in S : x \in \delta(s, a)\}.$$

Now we define $\hat{\delta}(q, x)$, the 'string transfer function,' for state q and string x inductively as follows:

$\hat{\delta}(q, \varepsilon) = Eclosure(q)$.
For $a \in \Sigma$ and $x \in \Sigma^*$,
$\hat{\delta}(q, ax) = \{y \mid \exists s \in Eclosure(\delta(Eclosure(q), a)) : y \in \hat{\delta}(s, x)\}$.

In other words, for every symbol a in ax, we *Eclose* state q, "run" a from each one of these states, and *Eclose* the resulting states. From each state s that results, we recursively run x. Notice that we apply *Eclosure* before as well as afterward. While this is strictly redundant, it leads to definitions that are simpler and more general, and hence easier to reason about.[12] Specifically, in the definition of $\hat{\delta}(q, ax)$, we need not assume that q is an already *Eclosed* state, even though in the current context of its usage, we will be inductively guaranteeing that to be the case because: (i) we begin with $\hat{\delta}(q, \varepsilon) = Eclosure(q)$, and (ii) in $\hat{\delta}(q, ax)$, we restore "Eclosedness."

Finally, the language of an NFA N is

$$\mathcal{L}(N) = \{w \mid \hat{\delta}(q_0, w) \cap F \neq \emptyset\}.$$

In other words, after running the NFA from state q_0 with input w, we see whether any 'token' has reached a final state.

A good way to intuitively understand the above definitions pertaining to NFAs is through the following 'token game:'

- Place a token in state q_0. Spread one copy of the token to each state in $Eclosure(q_0)$.
- For each symbol a from Σ that is entered, advance each token to its set of a successors.
- *Eclose* the tokens and continue.
- If and when one of the tokens reaches some final state, the string seen so far is accepted.

Illustration 9.3.1 The NFA in Figure 9.4 has string 0001 in its language. First, $Eclosure(IF) = \{IF\}$. After the first 0, the NFA is in state FA, whose *Eclosure* is $\{FA, B0, A1, FB\}$. After the second 0, $A1$ goes to FA, and the *Eclosure* results in $\{FA, B0, A1, FB\}$. The same happens after the third 0. After the 1, FA and $A1$ go to $A1$, $B0$ goes to FB, and the token in FB goes to $\emptyset$, resulting in $\{A1, FB\}$, which is also its own *Eclosure*. This matches the definition

$$\hat{\delta}(q, ax) = \{y \mid \exists s \in Eclosure(\delta(Eclosure(q), a)) \ : \ y \in \hat{\delta}(s, x)\}$$

as argued below:

- $Eclosure(\delta(Eclosure(IF), 0))$ is $\{FA, B0, A1, FB\}$.
- We recursively process the remaining input 001 from these states to reach $\{A1, FB\}$.

---

[12] Another way to set up the definitions would have been to start the NFA in the *Eclosure* of its start state, and at each stage perform a δ step followed by one *Eclosure* step.

9.3.5 A detailed example: telephone numbers

```
--------------        --------------
Legal                 Illegal
--------------        --------------
[201]221-1221         [201]2211221
201-221-1221          201-2211221
2012211221            201221-1221
221-1221              2211-221
2211221
--------------        --------------
```

Fig. 9.6. Legal and illegal phone numbers in our example

We now point out that in some cases we prefer NFAs (or regular expressions) *not* because they are smaller, but simply because they are much easier to specify. Also, the results would be much more convincing than if we were to draw a DFA directly or simply write a natural language description of the language in question. Our example also concretely illustrates the ability of ε transitions to make descriptions clearer.

Illustration 9.3.2 Suppose we want to develop a lexical analyzer[13] for telephone numbers. We want to be maximally flexible, and allow multiple syntaxes for phone numbers for users' convenience. First, let us assume a world where we have not invented the terminology of non-deterministic automata or regular expressions. We will then be forced to describe the set of legal telephone numbers in natural language prose (in our case, English). Here is how an initial attempt might look:

A telephone number consists of three parts: an area code (3 digits), a middle part (3 digits), and the extension (4 digits), written in that order. We consider only digits in the range of 0 to 2, to keep things simple. All three parts may be written in juxtaposition or may be separated by exactly one dash, '-'. If the area code is separated from the middle part by a dash, the middle part must also be separated from the extension by a dash. The area code may also be surrounded by brackets,[14] '[' and '].' In this case, the middle part *must* be separated

---

[13] Also known as a "scanner" - a tool that digests input at the character level and converts them into meaningful entities called *tokens*.

[14] We use brackets [and] in lieu of the more traditional parentheses '(' and ')' because the **grail** tool we use assigns special significance to parentheses - and we wanted to avoid work-arounds to keep things simple.

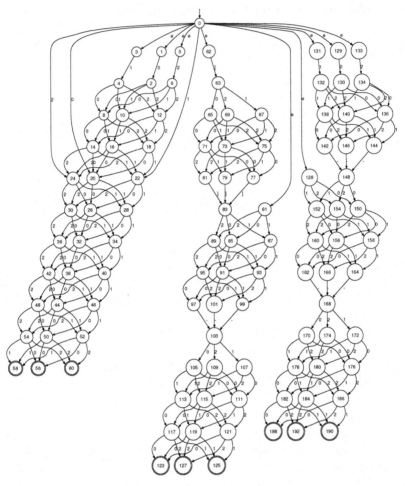

```
(""+(0+1+2)(0+1+2)(0+1+2))(0+1+2)(0+1+2)(0+1+2)(0+1+2)(0+1+2)
  (0+1+2)(0+1+2)+
(""+[(0+1+2)(0+1+2)(0+1+2)])(0+1+2)(0+1+2)(0+1+2)-(0+1+2)(0+1+2)
  (0+1+2)(0+1+2)+
(""+(0+1+2)(0+1+2)(0+1+2)-)(0+1+2)(0+1+2)(0+1+2)-(0+1+2)(0+1+2)
  (0+1+2)(0+1+2)
```

Fig. 9.7. An NFA formally specifying allowed telephone numbers, and the RE describing it

from the extension by a dash. The area code is optional; if not present, neither the brackets surrounding the area code nor the dash separating the area code from the middle part must be employed.

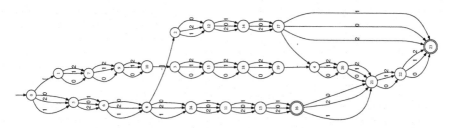

Fig. 9.8. A minimal DFA for the NFA of Figure 9.7

Example telephone numbers that are legal and illegal according to these conventions are listed in Figure 9.6 (please verify that these indeed follow the rules stated in English).

Unfortunately, as is well-known, natural language descriptions are often prone to misinterpretation. Let us therefore provide a *formal* specification of legal and illegal telephone numbers. We do this by writing a *regular expression* also shown in Figure 9.7. This RE can be turned into an NFA as shown in Figure 9.7. This NFA exhibits three classes of strings in the form of "three lobes" of state clusters. The correspondence between the RE and the NFA must be quite apparent. The minimal DFA for this NFA is given in Figure 9.8. We can then use a modern lexical analyzer generator (**flex** in our case) that generates a recognizer for this language, as sketched in Figure 9.9. The commands to generate the lexical analyzer and a few interactions with the generated analyzer (including an error scenario) are also included. The concept behind scanner programs was illustrated in Figure 8.2; modern scanners are, however, far more efficient than the simple-minded program used in Figure 8.2.

Writing regular expressions is a *highly error-prone activity*. The next section illustrates how to formally verify regular expressions through *putative* queries ("challenge queries"). The main idea we stress is that often the *complement* of the language we are interested in is far easier to characterize—at least to a large approximation. We can then intersect the complement of the language with the language of interest to see if the intersection is empty.

9.3.6 Tool-assisted study of NFAs, DFAs, and REs

In this section, we illustrate the use of the Grail tools to help understand the material so far. Basically, **grail** and the support scripts provide us the following commands that can be invoked from Unix:

```
/*---------- Definitions space ----------*/
D            [0-2]      /* Digit */
TD           {D}{D}{D}  /* Three Digits */
FD           {D}{TD}    /* Four Digits */
BAC          "["{TD}"]" /* Bracketed Area Code */
DAC          {TD}"-"    /* Dashed Area Code */
BACOpt       {BAC}|""   /* Optional Bracketed Area Code */
DACOpt       {DAC}|""   /* Optional Dashed Area Code */
ACOpt        {TD}|""    /* Optional Area Code */

Tele         {ACOpt}{TD}{FD}|{BACOpt}{TD}"-"{FD}|{DACOpt}{TD}"-"{FD}
/* (""+(0+1+2)(0+1+2)(0+1+2)) (0+1+2)(0+1+2)(0+1+2)(0+1+2)(0+1+2)(0+1+2)+
   (""+[(0+1+2)(0+1+2)(0+1+2)])(0+1+2)(0+1+2)(0+1+2)-(0+1+2)(0+1+2)(0+1+2)(0+1+2)+
   (""+(0+1+2)(0+1+2)(0+1+2)-)(0+1+2)(0+1+2)(0+1+2)-(0+1+2)(0+1+2)(0+1+2)(0+1+2)   */
/*---------- Rules space ----------*/
%%
{Tele} { printf("A number:%s\n",yytext); }

[\t\n]+ /* eat up whitespace */

. printf("Unrecognized character:%s\n",yytext);
/*---------- User code space ----------*/
%%
int yywrap();
main()
{ yylex();}

int yywrap()
{ return 1;}
/*---------- End ----------*/
 > flex telephone.l
 > cc lex.yy.c
 > a.out
 > 0120120120
 A number:0120120120
 > 000-0000
 A number:000-0000
 > [000]111-2222
 A number:[000]111-2222
 > 000-000-0000
 A number:000-000-0000
 > 0001111
 A number:0001111
 > [122] 111-1111
 Unrecognized character:[
 Unrecognized character:1
 Unrecognized character:2
 Unrecognized character:2
 Unrecognized character:]
 Unrecognized character:
 A number:111-1111
```

Fig. 9.9. Skeleton of an encoding of the telephone number syntax in flex

- fa2grail.perl, a Perl script
- grail2ps.perl, a Perl script
- dot, a graphics package
- ghostview or gv, a postscript viewer
- One of several grail-specific commands. We include some of these commands below (for details, kindly see the grail user manual):

- **retofm** for converting regular expressions into grail-specific internal "fm" format,
- **fmdeterm** for determinizing fm-formatted NFAs,
- **fmmin** for minimizing DFAs,
- **fmcment** for complementing DFAs, and
- **fmcross** for obtaining the intersection of two DFAs.

A quick tool check

In order to check that all the tools are available in one's search path, type

```
> echo '(a+b)*' | retofm | fmdeterm | fmmin \
            | perl grail2ps.perl - | gv -
```

Here > stands for the operating system prompt (**grail** currently runs under Linux). This command has the following effect. The regular expression (ab)*+ would be turned into an NFA, this NFA determinized and minimized, and finally picked up by the script **grail2ps.perl** which, with the help of the **dot** tool, produces postscript and draws the DFA for you! Note how we quote the command using ' and ' to prevent Unix from assigning a special interpretation to *. Also note the use of the Unix pipe |, and the character – that denotes "standard input."

Tool-assisted RE debugging

While regular expressions are often easy to write, they can also prove to be error-prone to write. We highly recommend that one always applies tools similar to **grail** for debugging proposed REs. In this section, we present a systematic method of debugging REs using **grail**.

Proposed RE

Someone proposes the following RE to express the language L, where "L = the set of strings of 0s and 1s with at most one pair of consecutive 1s."

$$(0^+ + \varepsilon) \, (10^+)^* \, (11 + 1) \, (0^+1)^* \, (0^+ + \varepsilon)$$

We assume that the reader did not find anything wrong about this RE, but nevertheless wanted to make sure. Let's *formally* debug by applying *putative queries* to study the expected properties of such an RE. We

know that the language of this machine must not have any 11(0+1)*11
or any 111 in it. One can often identify and specify whole classes of
such "bad strings" much more reliably than one can do with the RE of
interest. Capitalizing these facts, we proceed to debug as follows:

1. First, obtain a minimal DFA corresponding to the given RE:

```
> echo '(00*+"")(100*)*(11+1)(00*1)*(00*+"")' | retofm
  | fmdeterm | fmmin >! a1.txt
> cat a1.txt | perl grail2ps.perl - >! a1.ps
```

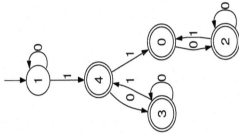

2. Second, get "the machine of all bad strings" built. (Please make
   sure you understand that this machine contains *all* the bad strings;
   it contains two patterns, namely (i) at least two pairs of consecutive
   11s, and (ii) the 111 pattern):

```
> echo '(0+1)*11(0+1)*11(0+1)*+(0+1)*111(0+1)*' | retofm
  | fmdeterm | fmmin >! a2.txt
> cat a2.txt | perl grail2ps.perl - >! a2.ps
```

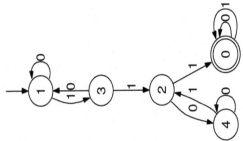

3. Now we take a look at the *complement* of the machine of bad strings:

```
> echo '(0+1)*11(0+1)*11(0+1)*+(0+1)*111(0+1)*' | retofm
  | fmdeterm | fmcment | fmmin >! a2c.txt
> cat a2c.txt | perl grail2ps.perl - >! a2c.ps
```

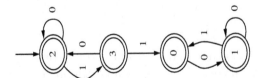

We find that the complement of the "bad machine" is not the same as "our machine." This is clearly indicative of something being wrong with our original RE (assuming that the complement machine was correct). Since the description of the complement machine was so direct and simple, we have every room to suspect the former. In the following sequel, we will proceed to *formally* determine where exactly the difference lies. This is the gist of the concept known as *formal verification*, where we have every room to trust the "property" (namely a description that is extremely easy to trust being correct) and formally compare the system against it.

4. Let's at least make sure that the original RE doesn't contain any bad strings.

```
> fmcross a1.txt a2.txt | perl grail2ps.perl - >! cross.ps
> gv cross.ps
```

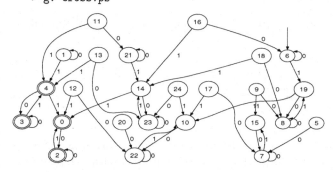

Here we see the unminimized DFA. Apparently none of its states are reachable. We now proceed to minimize and view this machine:

```
> fmcross a1.txt a2.txt | fmmin | perl grail2ps.perl - >!
                                                   crossmin.ps
> gv crossmin.ps
```

The last command (`gv crossmin.ps`) displays virtually *nothing* on the screen—indicative of an empty automaton. This experiment indicates that the original RE, indeed, does not include any bad strings. Therefore, its error must lie in the fact that it *is leaving out some of the good strings*. The next quest is to find out which these are.

5. So what is in the complement of the bad machine that's not in the original RE (i.e. what are we leaving out)?

```
> cat a1.txt | fmcment | fmmin >! a1c.txt
> fmcross a2c.txt a1c.txt | fmmin >! a1ca2ccrossmin.txt
> cat a1ca2ccrossmin.txt | perl grail2ps.perl - >! a1ca2ccrossmin.ps
```

Viola! We found the omitted set of strings!

Clearly we left out the all zeros case! Let's fix that omission in our original RE, and view that RE after this fix (notice that the fix lies in using (11+1+"") now, instead of (11+1), which is what our original RE contained):

```
> echo '(00*+"")(100*)*(11+1+"")(00*1)*(00*+"")' | retofm
  | fmdeterm | fmmin >! a1fixed.txt
> cat a1fixed.txt | perl grail2ps.perl - >! a1fixed.ps
```

Now, this *exactly* matches the complement of the bad machine.

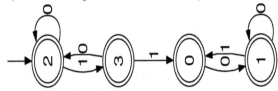

Since these perspectives agree, we consider our original RE as having been formally verified *against* the bad machine RE that was simpler to fathom.

Inputting DFAs into the Grail pipeline

We have created a script to intuitively encode DFAs in an ASCII syntax and input into **grail**. We illustrate this through an example. Let file **expt1** contain the following:

```
I - 0 -> A
I - 1 -> Dead
A - 0 -> Dead
A - 1 -> Fbfi
Fbfi - 0 -> Ffia
Fbfi - 1 -> Dead
Ffia - 1 -> Fbfi
Ffia - 0 -> A
Dead - 0,1 -> Dead
```

Now, if we type

```
> fa2grail.perl expt1 | fmdeterm | perl grail2ps.perl - | gv -
```

we can obtain the DFA in Figure 9.5. We will introduce further features
of the **grail** pipeline in subsequent chapters.

Chapter Summary

This chapter began with a discussion of why nondeterminism is such a
powerful idea in computer science. In addition to succinct descriptions
that are made possible (a dramatic illustration of which is provided
in Figure 9.1), the use of nondeterminism *may* make certain machine
types strictly more powerful, as in the case of push-down automata.
We also introduce the notion of regular expressions, hand in hand with
our discussion of nondeterminism. We introduce the notion of ε moves
in NFA: basically, ε allows machines to move without reading their
input. The notion of $\varepsilon - closure$ was also introduced. This operator
allows NFAs to be converted into language equivalent DFAs. Several
NFAs were illustrated. In particular, Section 9.3.6 discussed how NFAs
for given regular expressions may be obtained, as well as verified, in a
tool-assisted manner.

Exercises

In all problems involving verification using **grail***, clearly explain your
verification method. If you checked for properties such as the pres-
ence/absence of sub-languages (as was shown in Section 9.3.6), please
elaborate on those details. You are also encouraged to use JFLAP to
perform these verification steps.*

9.1. It seems as if when adding the ε transition from F to I, we didn't
change the language of the NFA in Figure 9.3. Argue that this is true.

9.2. For the NFA in Figure 9.4, step through its IDs starting with input
101101. Also, answer whether this string is accepted or not.

9.3. This exercise may be attempted if you read Chapter 7, "Dealing
with Recursion." An alternate definition of *Eclosure* is via recursion:

$$\text{For } q \in Q, \; Eclosure(q) = \{q\} \cup \{y \mid y \in \delta(q, \varepsilon)$$
$$\vee \; \exists z \in \delta(q, \varepsilon) \wedge y \in Eclosure(z)\}$$

Notice that the use of recursion here is "well defined." In particular,
we can compute the *least fixed-point* of the *Eclosure* function. We start
with *Eclosure*(q) returning the empty set $\emptyset$ for every state. Plug this
definition of *Eclosure* on the right-hand side and "pump" the recursion
one step, obtaining the *Eclosure* function for the left-hand side for

every state. This adds q itself to *Eclosure*, and also adds those states
that are directly reachable from q via ε edges. Next plug in this version
of *Eclosure* on the right-hand side and pump up again until no more
states are added to any *Eclosure*. This process comes to a halt since
we have a *finite* number of states. Argue that this least fixed-point
based definition of *Eclosure* is equivalent to the transitive-closure based
definition.

9.4. What is the language of the NFA of Figure 9.4? List all strings of
length six or less in the language of this NFA. For each string, provide
all possible reasons for that string to be in the language.

9.5. Give an NFA accepting the set of strings over $\Sigma = \{0, 1\}$ such that
every block of five consecutive symbols contains at least two 0s.

9.6. Attempt to write a regular expression *directly* for the language over
$\Sigma = \{0, 1\}$ in which every block of five consecutive symbols contains
at least two 0s. Spend some time directly encoding the said condition
using regular expressions, and verify your results using `grail`. If you
fail, you may try again later by drawing a DFA and converting that
DFA to an RE.

9.7. Design an NFA to recognize the language of strings over $\{0, 1\}$,
beginning with a 0, ending with a 1, and having an occurrence of 0101
somewhere in every string.

9.8. Attempt to *directly* write an RE for the language over $\Sigma = \{0, 1\}$
in which no string has a 0101 occurring in it. Spend some time directly
encoding the 'no 0101' condition using regular expressions, and verify
your results using `grail`. If you fail, you may try again later by drawing
a DFA and converting that DFA to an RE.

9.9. Draw an NFA to recognize the *star* of the language of the NFA in
Figure 9.4.

9.10. Design an NFA to recognize the *reverse* of the language of the
NFA in Figure 9.4.

9.11. Design an NFA to capture the syntax of floating-point numbers
over alphabet $\{0, 1, +, -, E, .\}$ (for the sake of simplicity). A floating-
point number begins with an optional sign character (one of $+$ or $-$),
followed by a *mantissa*, the letter E, an optional sign, and an *exponent*.
While the exponent is an integer, the mantissa is a decimal fraction of
the form *whole.frac* where *whole* or *frac* (but not both) may be empty.

10

Operations on Regular Machinery

In this chapter, we introduce algorithms that operate on 'regular machinery,' meaning various representations of regular sets.

- We present algorithms to convert between NFA, DFA, and regular expressions, thus establishing the equivalence of their expressive power.
- We present algorithms for performing the following operations on machines, and discuss which machine type (NFA or DFA) each operation would be natural to perform on:
  - union, intersection, complementation, star;
  - homomorphism, inverse homomorphism;
  - reversal, and Prefix-closure.

  The existence of these algorithms establishes the closure of regular languages under these operations.
- We present an algorithm to minimize DFAs, arguing that the results are unique (up to isomorphism). We then briefly discuss a similar operation on NFAs.
- Finally, we discuss the notion of *ultimate periodicity*. Ultimate periodicity not only helps understand regular languages better, but also sets the stage to prove that certain languages are not regular. For instance, it helps us argue very directly that sets such as $\{a^{i^2} \mid i \geq 0\}$ are not regular.

10.1 NFA to DFA Conversion

Given an NFA $N = (Q, \Sigma, \delta, q_0, F)$, a DFA that is language-equivalent to N is given by

$$D = (2^Q, \Sigma, \delta_D, Eclosure(q_0), \{f \mid f \in 2^Q \land f \cap F \neq \emptyset\}).$$

```
                NFA                                      DFA
                ===                                      ===

      Input   0     1     e            Input    0         1
      ---------------------            ---------------------
States |                       States |
   I   | {I,A}  { I }  { }        {I} |  {I,A}      { I }
       |                              |
   A   | { B }  { B }  { }      {I,A} |  {I,A,B}   {I,B}
       |                              |
   B   | { F }  { F }  { }      {I,B} |  {I,A,F}   {I,F}
       |                              |
   F   |  { }   { }  { I }    {I,A,B} |  {I,A,B,F} {I,B,F}
                                      |
                          {I,A,B,F} |  {I,A,B,F} {I,B,F}
                                      |
                            {I,B,F} |  {I,A,F}   {I,F}
                                      |
                            {I,A,F} |  {I,A,B}   {I,B}
                                      |
                              {I,F} |  {I,A}      {I}
                                      |
```

Fig. 10.1. NFA to DFA conversion illustrated on the NFA of Figure 9.3

This DFA D is designed to mimic the acceptance behavior of the NFA N as follows:

- N can have its tokens in (potentially) any combination of its states. Therefore, the states of D are chosen to be 2^Q, the powerset of the states of N.
- When we place a token in the initial state of N, it will spread (ε-close) to all states reachable through ε moves. To model this, D's starting state is chosen to be $Eclosure(q_0)$.
- N accepts a string when one of its tokens reaches a state in F. To model this, the final states of D are chosen to be those subsets of 2^Q that include a state of F.
- The last (and crucial) detail is δ_D:

$$\delta_D(S, a) = \{y \mid \exists s \in S : y \in Eclosure(\delta(Eclosure(s), a))\}.$$

In other words, δ_D takes a state S of D, first $Ecloses$ all the states it contains, runs δ of the NFA from each resulting state, and finally $Ecloses$ and unions the results.

Please refer to Exercise 10.1 that addresses the "double Eclosure" we are performing.

Another way to define δ_D is through the $\hat{\delta}$ (string transition) function of the NFA which, in effect, does the same thing as the above definition:

$$\delta_D(S, a) = \{y \mid \exists s \in S : y \in \hat{\delta}(s, a)\}.$$

It is straightforward to see that this construction is correct; by construction, whenever N accepts a string w, D also accepts w, and vice versa. Since NFAs can be converted to DFAs, the language of an NFA is also *regular*. Also, D can potentially have $2^{|Q|}$ states.

Illustration 10.1.1 Let us convert the NFA of Figure 9.3 to its equivalent DFA. The results are shown in Figure 10.1. We succinctly demonstrate the conversion by listing their state transition tables side by side as below. *We introduce only those states of 2^Q that are reachable.* A systematic way to proceed is now illustrated with respect to an example. Suppose the DFA is in state {I,A,B}. Suppose the input be 0. Then we walk through the column 0 in the NFA table, union what we see against I with what we see against A, and what we see against B. These sets are, respectively, {I,A}, {B}, and {F}. The union of these sets is {I,A,B,F}. Now, we perform the *Eclosure* step, which applies only to state F which has an ε move to state I. But since I is already in {I,A,B,F}, the final result is {I,A,B,F}. We proceed to do the same for every other state of the DFA.

10.2 Operations on Machines

We now take operations one at a time and illustrate then on regular machinery. Under each operation, we comment on the relative difficulty of performing the operation directly on a certain machine type. More importantly, we point out those algorithms that are *incorrect* to carry over to the 'wrong' machine type. The operations will be illustrated on DFA D_1 and D_2, NFA N_1 and N_2, and REs R_1 and R_2, as the case may be. For the sake of notational uniformity, we employ alphabet names Σ_1 and Σ_2, where in fact $\Sigma_1 = \Sigma_2 = \Sigma$. Specifically, we use

$D_1 = (Q_1, \Sigma_1, \delta_1, q_0^1, F_1)$, and $D_2 = (Q_2, \Sigma_2, \delta_2, q_0^2, F_2)$, for binary operations, and D_1 alone for unary operations.

Similarly, we will employ NFA

$N_1 = (Q_1, \Sigma_1, \delta_1, q_0^1, F_1)$ and $N_2 = (Q_2, \Sigma_2, \delta_2, q_0^2, F_2)$

and regular expressions R_1 and R_2. The results will be presented in terms of

$D = (Q, \Sigma, \delta, q_0, F)$ for DFA, $N = (Q, \Sigma, \delta, q_0, F)$ for NFA, and R for regular expressions.

In some cases, the operation changes the type of the machine, and if this happens we will explicitly point it out. For example, concatenation of two DFA can be directly accomplished if we treat the DFA as NFA and concatenate them; the result will then be an NFA. We show the results for REs along with NFAs. Through minor notational abuse, we also will call the resulting machines $\overline{D_1}$ for the complement of DFA D_1, $D_1 \cup D_2$ for the union of two DFAs, etc.

10.2.1 Union

DFA Union:

The union of two DFAs is accomplished through the *product construction* algorithm. The same algorithm also works for intersection, except for a minor detail (Section 10.2.2).

Given DFAs D_1 and D_2, a DFA D that models the marching of D_1 and D_2 in lock-step, is constructed. Basically, the initial states of D_1 and D_2 are paired, and made the initial state of D. Next, for each symbol $a \in \Sigma$, the resulting next states of the individual DFAs are paired, and the pair is regarded as the next state of D. Any paired state where one of the DFAs is in one of its final states is considered a final state of D. In particular,

$D = D_1 \cup D_2 = (Q_1 \times Q_2, \Sigma, \delta, \langle q_0^1, q_0^2 \rangle, (Q_1 \times F_2 \cup F_1 \times Q_2))$, where
$\delta(\langle x, y \rangle, a) = \langle \delta_1(x, a), \delta_2(y, a) \rangle$.

As an example, the result of applying the union algorithm on the DFA shown in Figures 10.2(a) and 10.2(b) is given in Figure 10.2(c). Notice that the result is a DFA, and that its state names are derived from the names of the states of the individual DFA by gluing the state names with an underscore '_' - a convenient way to convert pairs of states to state names.

NFA and RE Union:

Now we illustrate how to perform union on two NFAs. We introduce a new start state I_0 not present in Q_1 or Q_2, from which ε transitions are

(a)

(b)

(c)

(d)

(e)

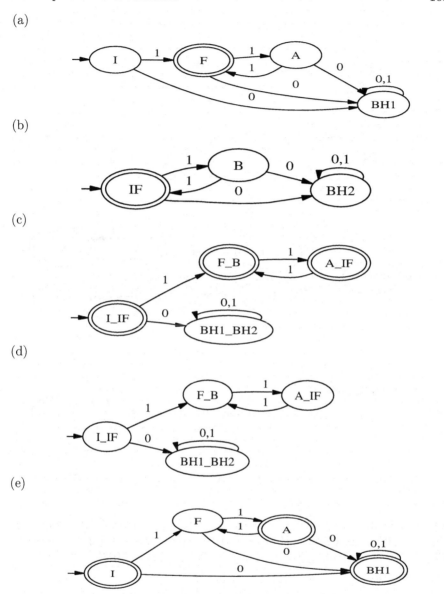

Fig. 10.2. $\text{DFA}_{(a)} \cup \text{DFA}_{(b)} = \text{DFA}_{(c)}$; $\text{DFA}_{(a)} \cap \text{DFA}_{(b)} = \text{DFA}_{(d)}$; $\overline{\text{DFA}_{(a)}} = \text{DFA}_{(e)}$

(a)

(b)

(c)

(d)

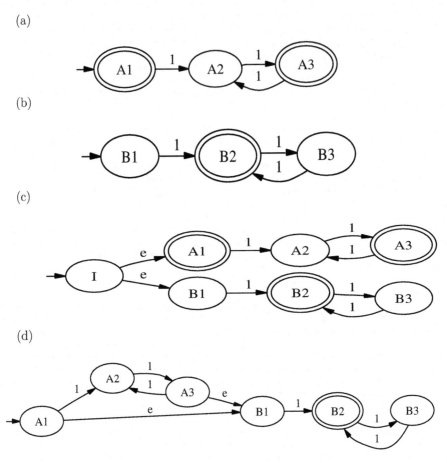

Fig. 10.3. $\text{NFA}_{(c)} = \text{NFA}_{(a)} \cup \text{NFA}_{(b)}$; $\text{NFA}_{(d)} = \text{NFA}_{(a)} \circ \text{NFA}_{(b)}$

introduced into both the start states q_0^1 and q_0^2. The result of applying this algorithm on Figures 10.3(a) and 10.3(b) is given in Figure 10.3(c). This NFA is equivalent to the DFA given in Figure 10.2(c).

In symbols:

$$N = N_1 \cup N_2 = (Q_1 \cup Q_2 \cup \{I_0\}, \Sigma, \delta, I_0, F_1 \cup F_2), \text{ where}$$
$$\delta = \delta_1 \cup \delta_2 \cup \{\langle I_0, \varepsilon, \{q_0^1, q_0^2\}\rangle\}.$$

As can be seen, with NFA and RE, union is a linear-time operation, while with a DFA, it is an $O(N^2)$ operation.

For regular expressions, union is achieved using the RE builder '+.'

10.2.2 Intersection

DFA Intersection:

With DFA, intersection works exactly like union, except that $F = F_1 \times F_2$, as shown in Figure 10.2(d), which is the result of intersecting the DFAs of Figures 10.2(a) and 10.2(b). Notice that this machine has no final states, because the component DFAs are never simultaneously in one of their respective final states.

NFA Intersection:

Performing intersection directly on NFA or RE is not straightforward (unless an extended RE notation containing intersection is employed—but then, conversion to the basic RE syntax is not straightforward). Therefore, before performing intersection, it is customary to convert the NFA or RE (as the case may be) to a DFA. This can, of course, incur an exponential cost, as the number of states after such conversion may be exponential (see Figure 9.1 of Chapter 9, for example).

10.2.3 Complementation

DFA Complementation:

This operation is most easily performed on DFA, resulting in

$$D = \overline{D_1} = (Q_1, \Sigma_1, \delta_1, q_0^1, Q_1 \setminus F_1),$$

as illustrated in Figure 10.2(e).

NFA and RE Complementation:

With NFA and RE, the operation is most commonly done after conversion to a DFA. In fact, with NFA, doing complementation directly by exchanging final and nonfinal states is incorrect (see Exercise 10.10).

10.2.4 Concatenation

DFA Concatenation:

Concatenation on DFAs involves treating them as NFAs and applying the NFA concatenation algorithm. So, in effect, a *direct* DFA concatenation algorithm is not usually given.

NFA and RE Concatenation:

With NFAs, concatenation involves introducing an ε transition from all the final states of the first NFA to the start state of the second NFA, and regards the final states of the second NFA as the final states of the resulting NFA; see Figure 10.3(d). In symbols:

$$N = N_1 \circ N_2 = (Q_1 \cup Q_2, \Sigma, \delta, q_0^1, F_2), \text{ where}$$
$$\delta = \delta_1 \cup \delta_2 \cup \{\langle x, \varepsilon, \{q_0^2\}\rangle \mid x \in F_1\}.$$

With RE, $R = R_1 \circ R_2$ or simply $R = R_1 R_2$ (recall that we can omit $\circ$ and simply use juxtaposition).

10.2.5 Star

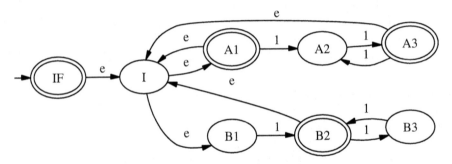

Fig. 10.4. The result of Starring Figure 10.3(c)

DFA Kleene star:

With DFAs, Kleene star $(()^*)$ is best performed by treating the DFA as an NFA. So, in effect, a *direct* DFA star algorithm is not usually given.

NFA and RE Star:

With NFA, we introduce a new start state I_0, regard this state also as a final state (thus introducing ε into the language of N^*), and introduce a transition on ε from I_0 to q_0. Finally, to introduce iteration, we introduce a transition from every state in F to q_0. In symbols:

$$N = N_1^* = (Q_1 \cup \{I_0\}, \Sigma_1, \delta, I_0, F_1 \cup \{I_0\}), \text{ where}$$
$$\delta = \delta_1 \cup \{\langle I_0, \varepsilon, \{q_0^1\}\rangle\} \cup \{\langle x, \varepsilon, \{q_0^1\}\rangle \mid x \in F_1\}.$$

Figure 10.4 illustrates this construction.

With REs, naturally we have $R = R_1^*$.

10.2.6 Reversal

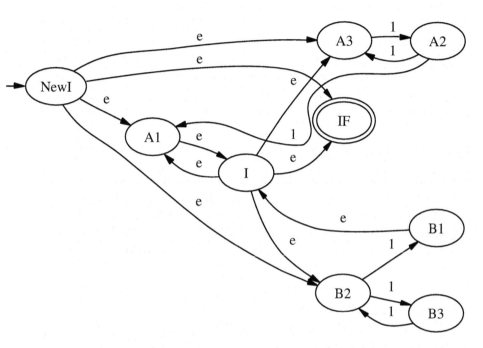

Fig. 10.5. The result of reversing the NFA of Figure 10.4

DFA Reversal:

With DFA, reversal is best performed by treating the DFA as an NFA.

NFA and RE Reversal:

With REs, reversal is best performed after converting to an NFA. With NFA, to do reversal, we introduce a new state I_0, make that the initial state, and introduce a transition on ε from I_0 to the set of states F_1. We also reverse every edge in the NFA diagram, and regard $\{q_0^1\}$ as the set of final states. In symbols:

$$N = rev(N_1) = (Q_1 \cup \{I_0\}, \Sigma_1, \delta, I_0, \{q_0^1\}), \text{ where}$$
$$\delta = \{\langle I_0, \varepsilon, F_1 \rangle\} \cup$$
$$\{\langle r, a, \{q \mid q \in Q_1 \wedge r \in \delta(q, a)\} \rangle \mid r \in Q_1 \wedge a \in \Sigma_1 \cup \{\varepsilon\}\}.$$

Figure 10.5 illustrates this construction. Notice that in this example, we have a jump from state NewI to state IF because IF was both a final state and start state in the original machine.

The reason why δ looks so complicated is because we try to reverse the transitions that, for an NFA, are specified via its delta function as moves from a state to a *set of next states*. We must also preserve the 'state to set of states' nature of the mapping for the δ of the reversed machine. In the δ for the reversed machine, we introduce a move from every state $r \in Q_1$ for every $a \in \Sigma_1 \cup \{\varepsilon\}$ to all those states q, such that those q states had at least one 'forward' transition through a to r.

10.2.7 Homomorphism

In Chapter 7, we introduced the notion of a homomorphism as a modular mapping with signature $h : \Sigma^* \to \Gamma^*$. In general, a homomorphism can be a function that, in place of each symbol in Σ_1, substitutes a regular language over Γ_1. Suppose for $x \in \Sigma_1$, we must substitute language L_x. In the context of an NFA, such a homomorphism can be easily implemented as follows:

- Build a separate NFA N_x corresponding to each instance of x in the given NFA N_1.
- Wherever a move on x appears in N_1 from state s_1 to state s_2, replace it with a jump from s_1 to the starting state of N_x via ε, and a jump from every final state of N_x via ε to s_2. Make every state of N_x nonfinal.

10.2.8 Inverse Homomorphism

We illustrate this construction through an example. Consider h to be

$$h(\varepsilon) = \varepsilon$$
$$h(a) = 0$$
$$h(b) = 1$$
$$h(c) = h(d) = 2.$$

Then, $h^{-1}(012) = \{abc, abd\}$ because 2 could have come either from c or from d. Therefore, given an NFA over alphabet $\{0, 1, 2\}$, we can label each move on a symbol (such as 2) with a move that is labeled with all the symbols that map to 2 (in our case, c and d).

10.2.9 Prefix-closure

This operation is straightforward to perform either on an NFA or on a DFA. We simply turn every state along the way from the start state towards one of the final states into a final state.

10.3 More Conversions

In this section, we define additional interconversions between REs, NFAs, and DFAs, thus establishing the equivalence of their expressive power.

10.3.1 RE to NFA

We first specify how to convert the *basic* regular expressions $\emptyset$, ε, and $a \in \Sigma$ into an NFA. For all other REs, we can recursively convert their constituent basic REs into NFA and then apply the corresponding NFA-building operator. The conversion of basic REs goes as follows:

- $\emptyset$ is an RE denoting $\emptyset$ The corresponding NFA is

$$N = (\{q^{\emptyset}\}, \Sigma, \emptyset, q^{\emptyset}, \emptyset),$$

  where $q^{\emptyset}$ is the only state of the NFA. The transition function is $\emptyset$, i.e., has no moves.
- ε is an RE denoting $\{\varepsilon\}$. The corresponding NFA is

$$N = (\{q^{\varepsilon}\}, \Sigma, \emptyset, q^{\varepsilon}, \{q^{\varepsilon}\}),$$

  where q^{ε} is the only state of the NFA that also happens to be a final state.
- $a \in \Sigma$ is an RE denoting $\{a\}$. The corresponding NFA is

$$N = (\{q_0^a, q_F^a\}, \Sigma, \delta^a, q_0^a, \{q_F^a\}),$$

  where δ is $\{\langle q_0^a, a, \{q_F^a\}\rangle\}$.
- The NFA for $r_1\, r_2$, $r_1 + r_2$, and r_1^* are obtained by first obtaining the NFA for r_1 and r_2, and applying, respectively, the algorithms for NFA concatenation, union, and star.

Illustration 10.3.1 For the RE of Figure 10.14, an NFA can be obtained in a very straightforward manner. The result is very easy to imagine (although left out due to space considerations). It will consist of an NFA for the embedding head sequence, branches labeled by ε to the various cases of errors, finally converging on an NFA fragment for the embedding tail sequence. The cases of errors will also have NFA fragments that directly track the RE syntax. $\qquad\square$

10.3.2 NFA to RE

Given NFA $N = (Q, \Sigma, \delta, q_0, F)$, we can convert it to a regular expression by successively eliminating its states. This is often called the *generalized NFA* or GNFA approach, where we build GNFA whose transitions are labeled using regular expressions, instead of members of Σ_ε.

- Preprocess N by adding one new initial state B and one new final state E. In the following steps, we will eliminate every state of N, leaving only B and E behind. The transition connecting B and E will be labeled with the desired RE upon termination of our algorithm.
- Make all states F nonfinal, and introduce a transition from every state in F to E via ε.
- Introduce an ε transition from B to q_0.
- Repeatedly eliminate a state from Q. In all the steps below, whenever any pair of states p and q has two transitions going from p to q, labeled with regular expressions R_1 and R_2, replace them by a single transition labeled $R_1 + R_2$.
- Suppose
  - state p has a transition into s labeled with RE R_{ps},
  - state s has a transition to itself labeled R_{ss}, and
  - state s has a transition out of it to state q labeled R_{sq}.

  Then, we can eliminate the ps and sq transitions, and introduce a direct transition from p to q labeled $R_{ps}(R_{ss})^* R_{sq}$. We keep repeating these steps until state s is disconnected from the rest of the graph, at which point, it can be eliminated.

We now illustrate the NFA to RE conversion algorithm on the example NFA given in Figure 10.6. The result of the preprocessing step is in Figure 10.7. This machine is represented as a GNFA.

Notice that state B0 reaches state A1 via (1+e). Therefore, we replace the two edges going from B0 to A1 by a single edge labeled with (1+e). No other immediate simplifications are possible.

Now, let us eliminate state A1. Introduce the following paths:

FA to FB labeled by (11*) (this is to be understood as (1(1*)) or as 1(1*), as * binds tighter than concatenation).
FB to FA labeled by (1*).
B0 to FA labeled by (1+e)(1*).
B0 to FB labeled by (1+e)(1*).

The result appears in Figure 10.8.
Now, let us eliminate state B0. Introduce the following paths:

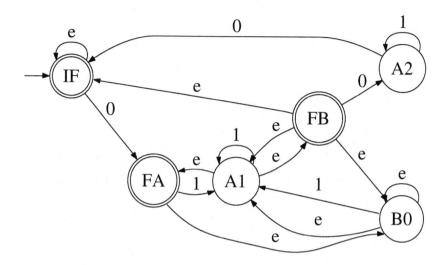

Fig. 10.6. An example NFA to be converted to a regular expression

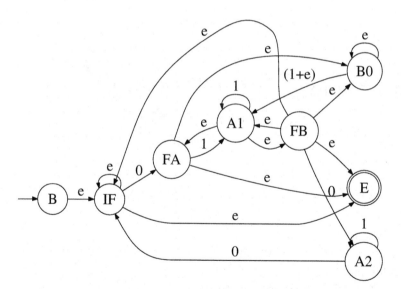

Fig. 10.7. Result of the preprocessing step

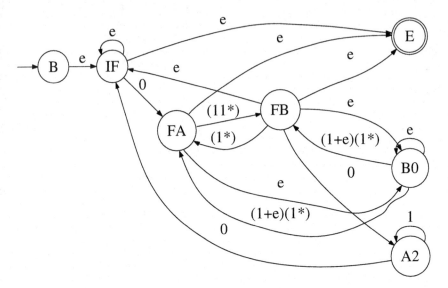

Fig. 10.8. Result of eliminating state A1

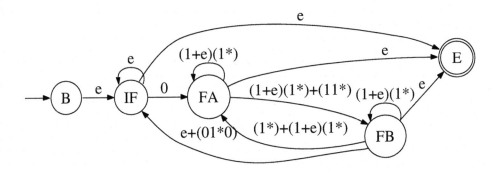

Fig. 10.9. Result of Eliminating B0 and A2

FB to IF labeled by e.
FB to FA labeled by ((1*)+(1+e)(1*)).
FA to FB labeled by ((1+e)(1*)+(11*)).
Self-loop FA to FA labeled by (1+e)(1*).
Self-loop FB to FB labeled by (1+e)(1*).

We don't depict this result yet. Let us also eliminate state A2 and then depict the combined results of eliminating B0 and A2. In eliminating A2, we merge the resulting FB to IF path with the existing one, resulting in the single **FB** to **IF** path labeled by the following label (with no other changes in the GNFA):

    e+(01*0)

The combined result of eliminating B0 and A2 appear in Figure 10.9.

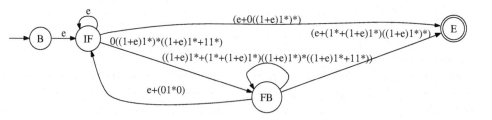

Fig. 10.10. Result of Eliminating FA

Now, let us choose to eliminate state FA. Introduce the following paths:

IF to E labeled by (e+0((1+e)1*)*).
Self-loop FB to FB labeled by ((1+e)1*+(1*+(1+e)1*)((1+e)1*)* ((1+e)1*+11*)).
IF to FB labeled by 0((1+e)1*)*((1+e)1*+11*).
FB to E labeled by (e+(1*+(1+e)1*)((1+e)1*)*).

The results appear in Figure 10.10.
We leave the final result (obtainable by eliminating FB and IF) as Exercise 10.17.

From this example, it should be evident that

- NFAs can *often* express regular languages far more intuitively than corresponding regular expressions can. The intuitiveness is due to the use of intermediate states that help split the behavior into various categories.
- Sometimes, minimal DFAs (Section 10.3.3) are not very intuitive (e.g., Figure 10.13), while regular expressions (see Section 10.4.2) and their corresponding NFAs (see Section 10.3.1) are quite intuitive.

10.3.3 Minimizing DFA

The most important result with regard to DFA minimization is the *Myhill-Nerode Theorem*.

Theorem 10.1. *The result of minimizing DFAs is unique, up to isomorphism.*[1]

The theorem says that given two DFAs over the same alphabet that are language-equivalent, they will result in identical DFAs when minimized, up to the renaming of states. One very dramatic illustration of the Myhill-Nerode Theorem will be in Chapter 13, where it will be shown that *Binary Decision Diagrams* (BDDs)—an efficient data structure for Boolean functions—are minimized DFAs for certain finite languages of binary strings. These finite strings, in fact, encode the truth assignment for Boolean formulas according to certain conventions that will be explained in Chapter 13. Because of this uniqueness, equality testing between two Boolean functions can be reduced to pointer-equality in a representation of BDDs using hash tables. Another illustration is provided in Section 10.4. We now discuss the minimization algorithm itself.

The basic idea behind DFA state minimization is to consider all pairs of states systematically by constructing a table. For each pair of states, we consider all strings of length zero and up, and see if they can distinguish any pair of states. Initially, we distinguish all pairs of states $\langle p, q \rangle$ such that p is a final state and q is nonfinal. We enter an x in the table to record that these states are ε-distinguishable. In essence, at the beginning of the algorithm we are treating all final states as belonging to one equivalence class, and all non-final states as belonging to another.

Thereafter, in the ith iteration of table filling, we see if any of the state pairs $\langle p, q \rangle$ that are not yet distinguished have a move on some $a \in \Sigma$ such that they go to states $\langle p', q' \rangle$ that are distinguishable (at the end of the $i - 1$st iteration); if so, distinguish $\langle p, q \rangle$. The algorithm stops when two iterations k and $k + 1$ result in the same table.

Illustration: Consider the DFA in Figure 10.11 (from [71]) where the final states are 2, 3, 6, and $\Sigma = \{a, b\}$.

1. The initial blank table that permits all pairs of states to be compared is in Figure 10.11.

---

[1] The concept of isomorphism comes from graph theory. Two directed graphs G_1 and G_2 are isomorphic if there is a bijection b between their nodes that preserves the graph connectivity structure. In other words, if n_1 and n_2 are nodes of G_1 and G_2, respectively, and if the bijection relates n_1 and n_2, then the list of successors of n_1 in G_1 are also bijective with the list of successors of n_2 in G_2.

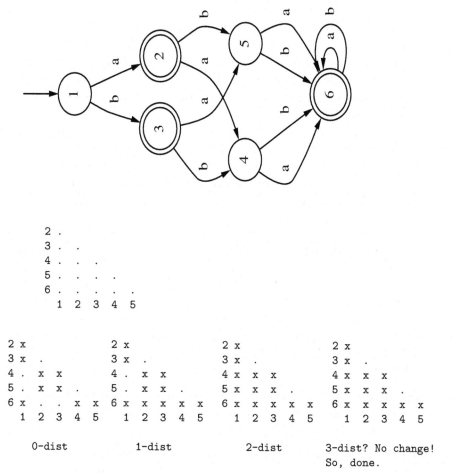

```
2 .
3 .  .
4 .  .  .
5 .  .  .  .
6 .  .  .  .  .
  1  2  3  4  5
```

```
2 x                 2 x                 2 x                 2 x
3 x  .              3 x  .              3 x  .              3 x  .
4 .  x  x           4 .  x  x           4 x  x  x           4 x  x  x
5 .  x  x  .        5 .  x  x  .        5 x  x  x  .        5 x  x  x  .
6 x  .  .  x  x     6 x  x  x  x  x     6 x  x  x  x  x     6 x  x  x  x  x
  1  2  3  4  5       1  2  3  4  5       1  2  3  4  5       1  2  3  4  5

     0-dist              1-dist              2-dist          3-dist? No change!
                                                            So, done.
```

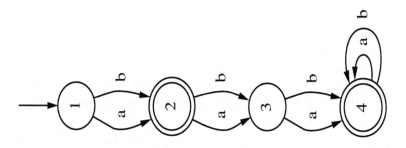

Fig. 10.11. (i) Example for DFA minimization, (ii) initial table for DFA minimization, (iii) steps in DFA minimization, and (iv) the final minimized DFA

2. All 0-length string distinguishable states are all pairs of states that consist of *exactly* one accept state (see below). All subsequent steps identifying *i*-distinguishable states for all *i* are also in Figure 10.11.
3. Let us understand how one x was added. In Figure 10.11, we put an x (to distinguish between) states 2 and 6. Why is this so? This is because 2 has a move upon input a to state 4, while state 6 moves upon a to state 6 itself. From the 0-dist table, we know that states 4 and 6 are distinguishable: one being a final and the other being a nonfinal state.
4. At the 3-dist step, there was no change from the previous table (in other words, a *fixed-point has been reached*, as described in Chapter 6). At this point, state pairs 3,2 and 5,4 still have a '.' connecting them (they could not be distinguished). Therefore, we merge these states, resulting in the minimized DFA of Figure 10.11.

10.4 Error-correcting DFAs

Consider another experiment that shows the value of tool-assisted debugging of machines. In this experiment, a DFA has to be designed to recognize all strings that are a Hamming distance of 2 away from the set of strings denoted by the regular expression (0101)+. For instance, 010101 and 010110 are a Hamming distance of 2 apart, as are 010111 and 000110.

Definition 10.2. *(Hamming Distance)* Given two strings V_1 and V_2, of equal length and over $\{0,1\}^*$, they are a Hamming distance of d apart if they differ in exactly d positions.

The DFA we are to design can be regarded as an *error-correcting* DFA which corrects two-bit errors. We shall derive this DFA using two distinct approaches, each time by following two different lines of logic:

1. The first approach will be to develop a cyclic DFA that performs transitions between states I, A, B, C, F, and back to A upon seeing 01010. However, upon seeing any erroneous symbol—for instance, seeing a 1 in state I, it goes to a cycle at a lower "stratum" labeled with states A1, B1, etc. This is presented in Figure 10.12.
2. Another approach will be to write a regular expression that captures all possible zero-bit errors, all possible one-bit errors, and all possible two-bit errors.

We shall find the minimal DFAs corresponding to these constructions. If correctly performed, we must obtain *isomorphic* minimal DFAs, again serving to verify our construction methods. We discuss these constructions in the following sections.

10.4.1 DFA constructed using error strata

```
I - 0 -> A
I - 1 -> A1

A  - 1 -> B     A1 - 1 -> B1    A2 - 0 -> BH
A  - 0 -> B1    A1 - 0 -> B2    A2 - 1 -> B2
B  - 0 -> C     B1 - 0 -> C1    B2 - 0 -> C2
B  - 1 -> C1    B1 - 1 -> C2    B2 - 1 -> BH
C  - 1 -> F     C1 - 1 -> F1    C2 - 1 -> F2
C  - 0 -> F1    C1 - 0 -> F2    C2 - 0 -> BH
F  - 0 -> A     F1 - 0 -> A1    F2 - 0 -> A2
F  - 1 -> A1    F1 - 1 -> A2    F2 - 1 -> BH    BH - 0,1 -> BH
```

Fig. 10.12. A DFA that has two error strata implementing all strings that are a Hamming distance of 2 away from the language (0101)+

The DFA corresponding to the use of error-correcting strata is captured in Figure 10.12. By running the command **perl fa2grail.perl h2 > h2fa**,[2] we convert this ASCII input into a **grail** representation. Following that, we apply the command

```
cat h2fa | fmdeterm | fmmin | perl grail2ps.perl - > h2fa.ps
```

The result is shown in Figure 10.13 on the left-hand side.

10.4.2 DFA constructed through regular expressions

A regular expression that captures all possible zero, one, and two-bit errors is in Figure 10.14. We have added spaces and newlines, as well as comments beginning with "--" to enhance the readability of the above RE. We then perform the command sequence:

```
cat h2re | retofm | fmdeterm | fmmin | perl grail2ps.perl - > h2fa1.ps
```

The result is shown in Figure 10.13 on the right-hand side. Contrasting it with the other DFA in this figure, we can see that barring node numberings as well as the layout (under the control of the 'dot' drawing package), these DFAs are isomorphic.

---

[2] We present this file in an intuitively layered manner; before running the script, one must present all the entries to occupy one column.

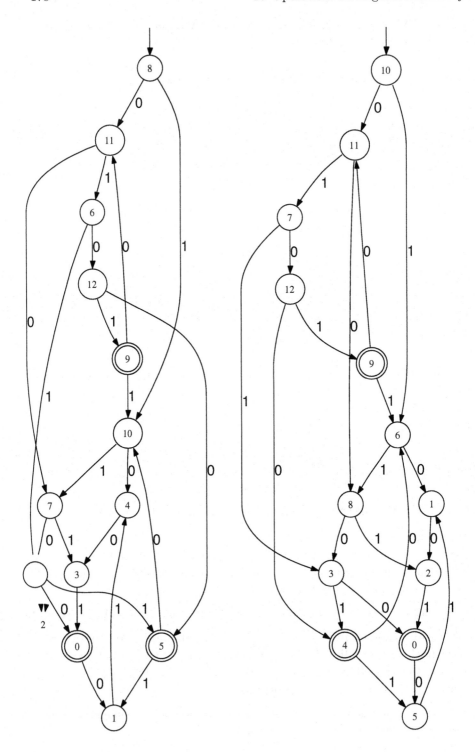

```
((0101)*)                                -- Embedding head sequence

        ( 0101                           -- 0-bit error option

            + 1101+0001+0111+0100        -- 1-bit  error option

        +
            (1101+0001+0111+0100)        -- One possibility for a
                ((0101)*)                -- 2-bit error as two one-
            (1101+0001+0111+0100)        -- bit errors with a
                                         -- correct mid-sequence
        +

        1001+1111+1100+0011+0000+0110    -- Another possibility for
                                         -- a 2-bit error as two
        )                                -- erroneous  bits within
                                         -- a block of four bits

((0101)*)                                -- Embedding tail sequence
```

Fig. 10.14. A regular expression for all 0-, 1-, and 2-bit errors

10.5 Ultimate Periodicity and DFAs

Ultimate periodicity is a property that captures a central property of regular sets (recall the definition of ultimate periodicity given in Definition 5.1, page 76).

Theorem 10.3. *If L is a regular language over an alphabet Σ, then the set $Len = \{length(w) \mid w \in L\}$ is ultimately periodic.*

Note: The converse is *not* true. For example, the language $\{0^n 1^n \mid n \geq 0\}$ has strings whose lengths are ultimately periodic, and yet this language is not regular.

A good way to see that Theorem 10.3 is true is as follows. Given any DFA D over some alphabet Σ, consider the NFA N obtained from D by first replacing every transition labeled by a symbol in $\Sigma \setminus \{a\}$ by a. In other words, we are applying a homomorphism that replaces every symbol of the DFA by a. Hence, the resulting machine is bound to be an NFA, and the *lengths* of strings in its language will be the same as those of the strings in the language of the original DFA (in other words, what we have described is a *length-preserving* homomorphism). Now, if we convert this NFA to a DFA, we will get a machine that starts out in a start state and proceeds for some number (≥ 0) of steps before it "coils" into itself. In other words, it attains a *lasso shape*. It *cannot* have any other shape than the 'coil' (why?). This coiled DFA shows that the length of strings in the language of any DFA is *ultimately periodic* with the period defined by the size of the coil.

We now take an extended example to illustrate these ideas. Consider the DFA built over $\Sigma = \{a, b, c, d, f\}$ using the following command pipeline:

```
echo '(ad)*+((abc)((acf)*+(da)*)d)' | retofm | fmdeterm
   | fmmin | grail2ps.perl - | gv -
```

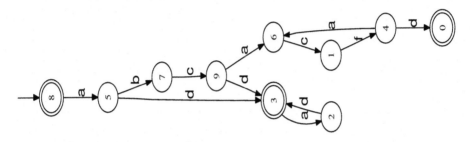

The DFA generated by converting every symbol to a and determinizing the result is as follows.

```
echo '(aa)*+((aaa)((aaa)*+(aa)*)a)' | retofm | fmdeterm
   | fmmin | grail2ps.perl - | gv -
```

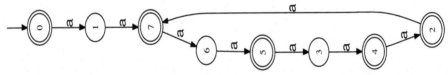

The length of strings in the language of this DFA is ultimately periodic, with the values of the constants $n = 2$ and $p = 6$ as per the definition of UP appearing in Section 5.2.1. Based on all these observations, we can state another theorem:

Theorem 10.4. *A language over a singleton alphabet is regular if and only if the length of strings in this language is ultimately periodic.*

Chapter Summary

This chapter covered quite a bit of important ground in terms of conversions between machine types. It also illustrated two very fascinating topics assisted by the **grail** tools. The first is that minimal DFAs for the same language are isomorphic. The second is that DFAs with infinite languages over a singleton alphabet always have a "lasso" shape to them, and accepting states are sprinkled along the lasso. This has the effect of making the *string lengths* of strings in this language *ultimately periodic*.

Exercises

10.1. Suppose an alternate definition of δ_D is offered:

$$\delta_D(S, a) = \{y \mid \exists s \in S : y \in Eclosure(\delta(s, a))\}.$$

Does it change the behavior of the resulting DFA? Justify your answer. Why do we perform *Eclosure* before and after δ in case of the NFA, on page 161?

10.2. Convert the NFA in Exercise 9.4 into an equivalent DFA, showing all the steps.

10.3. Convert the NFA of Figure 9.2 to a DFA for $k = 2$ and $k = 5$. Repeat for the modified NFA in Figure 9.3.

10.4. The token game of an NFA can be succinctly stated as follows: an NFA accepts a string x if there *exists* a path labeled by z from the start state to some final state, and x is z projected onto the alphabet Σ. Consider a variant of an NFA called *all-paths NFA*. In an all-paths NFA, a string x is accepted if and only if *all* such z paths that are in the machine actually lead to some final state. Formally define the all-paths NFA as a five-tuple mathematical structure, and prove that its language is regular, by converting it to an equivalent DFA.

10.5. What are the languages of the machines in Figure 10.2 and Figure 10.3?

10.6. Argue that L_{add} is regular, where

$$L_{add} = \{a_0b_0c_0a_1b_1c_1 \ldots a_{k-1}b_{k-1}c_{k-1} \mid k > 0 \wedge \text{AddOK}\}$$

where AddOK $= (a_{k-1} \ldots a_0) + (b_{k-1} \ldots b_0) = (c_{k-1} \ldots c_0)$. In other words, the addition of the unsigned binary words $(a_{k-1} \ldots a_0)$ and $(b_{k-1} \ldots b_0)$ yields $(c_{k-1} \ldots c_0)$, where a_{k-1} is the MSB and a_0 the LSB (and likewise for b and c).

10.7.
Let $\Sigma = \{0, 1\}$ and let D be a DFA over Σ. Obtain an NFA N for the language of strings that are the first third of all the strings accepted by D (the "first-third" language). Formally,

$$L_{x--} = \{x \mid x \in \Sigma^* \wedge \exists y, z \in \Sigma^* : |x| = |y| = |z| \wedge xyz \in L(D)\}$$

10.8. Repeat Exercise 10.7 for the middle-third language.

10.9. Call the DFA in Figure 8.5 D. Obtain the complement DFA $\overline{D}$ by the complementation algorithm. Then obtain a DFA corresponding to the union of D and $\overline{D}$. Repeat for the intersection of D and $\overline{D}$. Check that you are indeed obtaining the right answers.

10.10. Give an example of an NFA on which performing complementation, as with DFAs, (exchanging final and nonfinal states) is correct, and another example where it is incorrect. *This shows that exchanging final and non-final states does* **not** *complement NFAs!*

10.11. Section 10.2.5 describes a construction for star. Describe an alternate construction for star that results in an NFA with exactly one final state.

10.12. Notice that the NFA of Illustration 10.3.1 had a very direct correspondence with the corresponding RE. In fact, if we apply the GNFA method to convert it to an RE, we will obtain an RE that is very close to that in Figure 10.14. However, the NFA of Section 10.3.2 when converted to an RE resulted in an extremely complex RE. Now, if we were to convert this RE back to an NFA using the procedure described in Illustration 10.3.1, we would obtain something quite different from the NFA of Figure 10.6. Intuitively describe the kinds of NFAs and REs for which close correspondence will be maintained during conversions, and those NFAs and REs where such correspondence will not be obtained.

10.13. Instantiate the NFA in Figure 9.3 for $n = 2$ and $n = 3$, calling these machines N_2 and N_3, respectively. Perform union and concatenation. With respect to both these results, list eight strings in lexicographic order.

10.14. For the N_3 machine in Exercise 10.13, perform the star operation. How does the result differ from the machine in Figure 9.3 for $k = 3$?

10.15. Reverse the NFAs in Figure 9.2 and Figure 9.3 for $n = 3$. Convert each resulting NFA to a DFA and compare their languages.

10.16. Modify the regular expression in Figure 10.14 to account for the constraint that no two consecutive bits may be in error. Perform this modification in two ways:

1. By directly editing the RE of Figure 10.14
2. By constraining the RE of Figure 10.14 suitably

Compare the results using `grail` by obtaining isomorphic minimal DFAs.

10.17. Complete the derivation in Page 173.

10.18. Express the NFA to RE conversion algorithm through recursive pseudocode. Assume that you are given a preprocessed NFA. Check whether this NFA is "done" by seeing that it has a direct path from B to E, and if so, output the RE that labels this path. Else, express the choice of a state s at random using $\exists$ or *choose*. Then, eliminate s, updating the REs of all the states that are directly reachable from s or directly reach s. Recurse on the resulting automaton.

10.19. Convert the DFA of Figure 9.5 into a regular expression using the conversion procedure that you just now pseudocoded in Exercise 10.18. Now convert the RE you obtain to an NFA. Determinize this NFA, and compare the resulting DFA to the one you started from.

10.20. Another way to convert NFAs to RE uses Arden's Lemma [68], which is:

> A language equation of the form $X = AX \cup B$, where $\varepsilon \notin A$ has a unique solution $X = A^*B$.

1. Write a system of recursive equations corresponding to the example DFA in Figure 9.5. Some of the equations are the following, where L_1, L_2, etc, denote the languages of states 1, 2, etc. (meaning, if the start state were to be set to these states, these would be the languages of the DFA):
   $$L_1 = 0\, L_2$$
   $$L_2 = 1\, L_4$$
   $$L_2 = 1\, L_4$$
   $$L_4 = 0\, L_5 \cup \{\varepsilon\}$$
2. Convert this mutually recursive system of equations into a self-recursive equation in terms of one variable, and solve it using Arden's lemma. You may refer to a method such as Gaussean elimination which is used to solve simultaneous equations over Reals.
3. Solve the self-recursion to a closed form solution using Arden's lemma, and back substitute the result to obtain a closed form solution to all languages.

10.21. Use the least fixed-point approach introduced in Chapter 6 to derive Arden's Lemma.

10.22. Apply the DFA minimization algorithm to the DFA of Figure 9.5.

10.23. Perform the `grail` command sequence

```
> echo '(00*+"")(100*)*(11+1+"")(00*1)*(00*+"")' | retofm
  | fmdeterm >! a1fixed-unmin.txt
```

Then, hand-minimize the result (meaning, construct the DFA from the regular expression by inspection and then hand-minimize it), comparing it with the minimized version `a1fixed.txt` discussed on page 156.

10.24. Express the DFA minimization algorithm neatly using pseudocode. Analyze its time complexity.

10.25. Describe a language over a singleton alphabet such that the length of strings in this language is *not* ultimately periodic.

10.26. Apply the homomorphism $1 \to 0$, $0 \to 0$, and $\varepsilon \to \varepsilon$ to the DFA of Figure 9.5. Convert the resulting NFA into a DFA. Show that the length of strings in this DFA is ultimately periodic by finding n and p parameters.

The Automaton/Logic Connection, Symbolic Techniques

Most believe that computer science is a very young subject. In a sense, that is true - there was the theory of relativity, vacuum tubes, radio, and Tupperware well before there were computers. However, from another perspective, computer science is at least 150 years old! Charles Babbage[1] started building his Analytic Engine in 1834 which remained unfinished till his death in 1871. His less programmable Difference Engine No. 2 was designed between 1847 and 1849, and built to his specifications in 1991 by a team at London's Science Museum. As for the 'theory' or 'science' behind computer science, George Boole published his book on Boolean Algebra[2] in 1853.

Throughout the entire 150 years (or so) history of computer science, one can see an attempt on part of researchers to understand *reasoning* as well as *computation* in a unified setting. This direction of thinking is best captured by Hilbert in one of his famous speeches made in the early 1900s in which he challenged the mathematical community with 23 open problems. Many of these problems are still open, and some were solved only decades after Hilbert's speech. One of the conjectures of Hilbert was that the entire body of mathematics could perhaps be "formalized." What this meant is basically that mathematicians had no more creative work to carry out; if they wanted to discover a new result in mathematics, all they had to do was to program a computer to systematically crank out all possible proofs, and check to see whether the theorem whose proof they are interested in appears in one of these proofs!

---

[1] Apparently, Babbage is also credited with the invention of the 'cow-catcher' that you see in front of locomotive engines!

[2] Laws of thought. (You might add: to prevent loss of thought through loose thought).

In 1931, Kurt Gödel dropped his 'bomb-shell.[3] He formally stated and proved the result, "Such a device as Hilbert proposed is impossible!" By this time, Turing, Church, and others demonstrated the true limits of computing through concrete computational devices such as Turing machines and the Lambda calculus. The rest "is history!"

11.1 The Automaton/Logic Connection

Scientists now have a firm understanding of how computation and logic are inexorably linked together. The work in the mid 1960s, notably that of J.R. Büchi, furthered these connections by relating branches of mathematics known as *Presburger arithmetic* and branches of logic known as WS1S[4] with deterministic finite automata. Work in the late 1970s, notably by Pnueli, resulted in the adoption of *temporal logic* as a formal logic to reason about concurrency. Temporal logic was popularized by Manna and Pnueli through several textbooks and papers. Work in the 1980s, notably by Emerson, Clarke, Kurshan, Sistla, Sifakis, Vardi, and Wolper established deep connections between temporal logic and automata on infinite words (in particular Büchi automata). Work in the late 1980s, notably by Bryant, brought back yet another thread of connection between logic and automata by the proposal of using binary decision diagrams, essentially minimized deterministic finite automata for the finite language of satisfying instances of a Boolean formula, as a data structure for Boolean functions. The *symbolic model checking algorithm* proposed by McMillan in the late 1980s hastened the adoption of BDDs in verification, thus providing means to tackle the correctness problem in computer science. Also, spanning several decades, several scientists, including McCarthy, Wos, Constable, Boyer, Moore, Gordon, and Rushby, led efforts on the development of mechanical theorem-proving tools that provide another means to tackle the correctness problem in computer science.

11.1.1 DFA can 'scan' and also 'do logic'

In terms of practical applications, the most touted application domain for the theory of finite automata is in string processing – pattern matching, recognizing tokens in input streams, scanner construction, etc. However, the theory of finite automata is much more fundamental

---

[3] Some mathematicians view the result as their salvation.

[4] WS1S stands for *the weak monadic second-order logic of one successor.*

to computing. Most in-depth studies about computing in areas such as concurrency theory, trace theory, process algebras, Petri nets, and temporal logics rest on the student having a solid foundation on *classical automata*, such as we have studied so far. This chapter introduces some of the less touted, but nevertheless equally important, ramifications of the theory of finite automata in computing. It shows how the theory of DFA helps arrive at an important method for representing *Boolean functions* known as *binary decision diagrams*. The efficient representation as well as manipulation of Boolean functions is central to automated reasoning in several areas of computing, including computer-aided design and formal verification. In Chapter 21, we demonstrate how exploiting the "full power" of DFAs, one can represent logics with more power than propositional logic. In Chapter 22, we demonstrate how automata on infinite words can help reason about finite as well as infinite computations generated by finite-state devices. In this context, we briefly sketch the connections between automata on infinite words as well as temporal logics. We now turn to binary decision diagrams, the subject of this chapter.

Note: We use $\vee$ and $+$ interchangeably, depending on what looks more readable in a given context; they both mean the same (the *or* function).

11.2 Binary Decision Diagrams (BDDs)

Binary Decision Diagrams (BDDs) are bit-serial DFA for satisfying instances of Boolean formulas.[5] To better understand this characterization, consider the *finite* language

$$L_1 = \{abcd \mid d \vee (a \wedge b \wedge c)\}.$$

Since all finite languages (a finite number of finite strings in this case) are regular, a regular expression describing this language can be obtained by spelling out all the satisfying instances of $d \vee (a \wedge b \wedge c)$. This finite regular language is denoted by the following regular expression:

```
(1110+1111+0001+0011+0101+0111+1001+1011+1101)
```

By putting this regular expression in a file called a.b.c+d, we can use the following **grail** command sequence to obtain a minimal DFA for it, as shown in Figure 11.1(a):

---

[5] BDDs may also be viewed as optimized decision trees. We view BDDs as DFA following the emphasis of this book. Also note that strictly speaking, we must say *Reduced Ordered Binary Decision Diagrams* or ROBDDs. We use "BDD" as a shorthand for ROBDD.

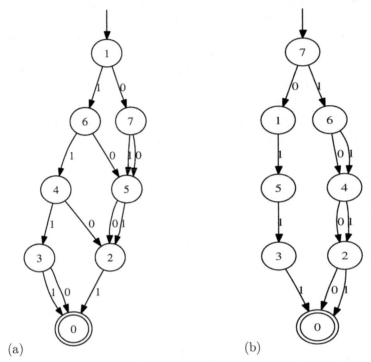

(a)   (b)

Fig. 11.1. Minimal DFAs for $d \vee (a \wedge b \wedge c)$ for (a) variable ordering $abcd$, and (b) $dabc$. The edges show transitions for inputs arriving according to this order.

```
cat a.b.c+d | retofm | fmdeterm | fmmin | perl grail2ps.perl -
    > a.b.c+d.ps
```

Now consider the language that merely changes the bit-serial order in which the variables are examined from $abcd$ to $dabc$:

$$L_2 = \{dabc \mid d \vee (a \wedge b \wedge c)\}.$$

Using the regular expression

$$(0111+1000+1001+1010+1011+1100+1101+1110+1111)$$

as before, we obtain the minimal DFA shown in Figure 11.1(b). The two minimal DFAs seem to be of the same size. Should we expect this in general? The minimal DFAs in Figure 11.1 and Figure 11.2, are suboptimal as far as their role in decoding the binary decision goes, as they contain redundant decodings. For instance, in Figure 11.1(a),

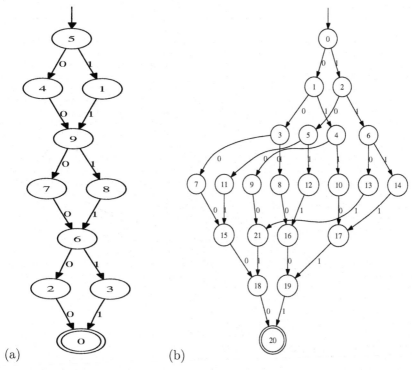

(a)   (b)

Fig. 11.2. Minimal DFAs where the variable ordering matters

after $abc = 111$ has been seen, there is no need to decode d; however, this diagram redundantly considers 0 and 1 both going to the accepting state 0. In Figure 11.1(b), we can make node 6 point directly to node 0. Eliminating such redundant decodings, Figures 11.1(a) and (b) will, essentially, become BDDs; the only difference from a BDD at that point would be that BDDs explicitly include a 0 node to which all falsifying assignments lead to.

Let us now experiment with the following two languages where we shall discuss these issues even more, and actually present the drawing of a BDD.

$$L_{interleaved} = \{abcdef \mid a = b \wedge c = d \wedge e = f\}$$

has a regular expression of satisfying assignments

(000000+001100+000011+110000+001111+110011+111100+111111)

and

$$L_{noninterleaved} = \{acebdf \mid a = b \wedge c = d \wedge e = f\}$$

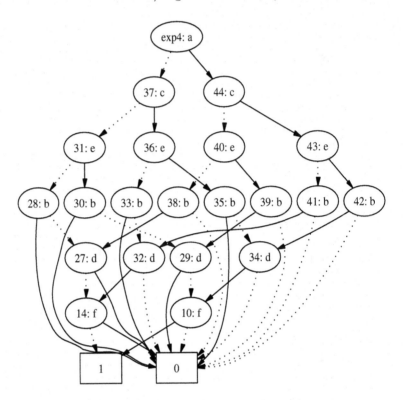

Fig. 11.3. BDD for $a = b \wedge c = d \wedge e = f$ for variable order *acebdf*

has

$$(000000+010010+001001+100100+011011+101101+110110+111111).$$

When converted to minimized DFAs, these regular expressions yield Figures 11.2(a) and (b), where the size difference due to the *variable orderings* is very apparent. The BDD for Figure 11.2(b) created using the BED tool appears in Figure 11.3. The commands used to create this BDD were:

```
bed> var a c e b d f               % declares six variables
bed> let exp4 = (a=b) and (c=d) and (e=f) % defines the desired expn.
bed> upall exp4                    % builds the BDD -
bed> view exp4                     % displays the BDD
```

By comparing Figure 11.3 against Figure 11.2(b), one can see how, in general, BDDs eliminate redundant decodings.[6]

---

[6] The numbers inside the BDD nodes—such as the "14:" and "10:" in the nodes for variable f—may be ignored. They represent internal numberings chosen by the BED tool.

BDDs are efficient data structures for representing Boolean functions and computing the reachable states of state transition systems. In these applications, they are very often 'robust,' *i.e.*, their sizes remain modest as the computation advances. As many of these state transition systems have well over 2^{150} states (just to pick a large number!), this task cannot be accomplished in practice by explicitly enumerating the states. However, BDDs can often very easily represent such large state-spaces by capitalizing on an implicit representation of states as described in Section 11.3. However, BDDs can deliver this 'magic' only if a "good" variable ordering is chosen.

One also has to be aware of the following realities when it comes to using BDDs:

- The problem of determining an optimal variable ordering is NP-complete (see Chapter 20 for a definition of NP-completeness). [42]; this means that the best known algorithms for this task run in exponential worst-case time.
- In many problems, as the computation proceeds and new BDDs are built, variable orderings must be recomputed through *dynamic variable re ordering* algorithms, which are never ideal and add to the overhead.
- For certain functions (e.g., the middle bits of the result of multiplying two N-bit numbers), the BDD is provably exponentially sized, no matter which variable ordering is chosen.

Even so, BDDs find extensive application in representing as well as manipulating state transition systems realized in hardware and software. We now proceed to discuss how BDDs can be used to represent state transition relations and also how to perform reachability analysis.

11.3 Basic Operations on BDDs

BDDs are capable of efficiently representing transition relations of finite-state machines. In some cases, transition relations of finite-state machines that have of the order of 2^{100} states have been represented using BDDs. For example, a BDD that represents the transition relation for a 100-bit digital ripple-counter can be built using about 200 BDD nodes.[7] Such compression is, of course, achieved by *implicitly* representing the state space; an explicit representation (*e.g.*, using pointer based

---

[7] Basically, each bit of such a counter toggles when all the lower order bits are a 1, and thus all the BDD basically represents is an *and* function involving all the bits.

data structures) of a state-space of this magnitude is practically impossible. Given a transition relation, one can perform *forward* or *backward* *reachability* analysis. 'Forward reachability analysis' is the term used to describe the act of computing reachable states by computing the forward image ("image") of the current set of states (starting from the initial states). Backward reachability analysis computes the *pre-image* of the current set of states. One typically starts from the current set of states *violating* the desired property, and attempts to find a path back to the initial state. If such a path exists, it indicates the possibility of a computation that violates the desired property.

Each step in reachability analysis takes the current set of states represented by a BDD and computes the next set of states, also represented by a BDD. It essentially performs a *breadth-first* traversal, generating each breadth-first frontier in one step from the currently reached set of states. The commonly used formulation of traversal is in terms of computing the least fixed-point as explained in Section 11.3.2. When the least fixed-point is reached, one can query it to determine the overall properties of the system. One can also check whether desired system invariants hold in an incremental fashion (without waiting for the fixed-point to be attained) by testing the invariant after each new breadth-first frontier has been generated. Here, an *invariant* refers to a property that is true at every reachable state.

We will now take up these three topics in turn, first illustrating how we are going to represent state transition systems.

11.3.1 Representing state transition systems

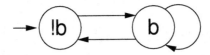

Fig. 11.4. Simple state transition system (example SimpleTR)

We capture transition systems by specifying a binary state transition relation between the *current* and *next* states, and also specifying a predicate capturing the initial states. If inputs and outputs are to be modeled, they are made part of the state vector. Depending on the problem being modeled, we may not care to highlight which parts of the state vector are inputs and which are outputs. In some cases, the entire state of the system will be captured by the states of inputs and outputs. Figure 11.4 presents an extremely simple state transition system, called SimpleTR. Initially, the state is 0. Whenever the state is 0, it can become 1. When it is 1, it can either stay 1 or become 0. These requirements can be captured using a single Boolean variable b representing the current state, another Boolean variable b' representing the next state,[8] and an initial state predicate and a state transition relation involving these variables, as follows:

- The initial state predicate for SimpleTR is $\lambda b.\neg b$, since the initial state is 0. Often, instead of using the lambda syntax, initial state predicates are introduced by explicitly introducing a named initial state predicate I and defining it by an equation such as $I(b) = \neg b$. For brevity,[9] we shall often say "input state represented by $\neg b$."
- The state transition relation for SimpleTR is $\lambda(b,b').\neg bb' + bb' + b\neg b'$, where each product term represents one of the transitions. The values of b and b' for which this relation is satisfied represent the present and next states in our example. In other words,
  - a move where b is false now and true in the next state is represented by $\neg bb'$.
  - a move where b is true in the present and next states is represented by bb'.
  - finally, a move where b is true in the present state and false in the next state is represented by $b\neg b'$.

  This expression can be simplified to $\lambda(b,b').(b + b')$. The above relation can also written in terms of a transition relation T defined as $T(b,b') = b + b'$. We shall hereafter say "transition relation $b + b'$." Notice that this transition relation is false for $b = 0$ and $b' = 0$, meaning there is no move from state 0 to itself (all other moves are present).

11.3.2 Forward reachability

The set of reachable states in SimpleTR starting from the initial state $\neg b$ can be determined as follows:

- Compute the set of states in the initial set of states.
- Compute the set of states reachable from the initial states in n steps, for $n = 1, 2, \ldots$.

---

[8] The 'primed variable' notation was first used by Alan Turing in one of the very first program proofs published by him in [89].
[9] Syntactic sugar can cause cancer of the semi-colon – Perlis

In other words, we can introduce a predicate P such that a state x is in P if and only if it is reachable from the initial state I through a finite number of steps, as dictated by the transition relation T. The above recursive recipe is encoded as

$$P(s) = (I(s) \lor \exists x.(P(x) \land T(x, s))).$$

This formula says that s is in P if it is in I, or there exists a state x such that x is in P, and the transition relation takes x to s.
Rewriting the above definition, we have

$$P = \lambda s.(I(s) \lor \exists x.(P(x) \land T(x, s)))).$$

Rewriting again, we have

$$P = (\lambda G.(\lambda s.(I(s) \lor \exists x.(G(x) \land T(x, s)))))\, P.$$

In other words, P is a fixed-point of

$$\lambda G.(\lambda s.(I(s) \lor \exists x.(G(x) \land T(x, s)))).$$

Let us call this Lambda expression H:

$$H = \lambda G.(\lambda s.(I(s) \lor \exists x.(G(x) \land T(x, s)))).$$

In general, H can have multiple fixed-points. Of these, the *least fixed-point* represents exactly the reachable set of states, as next explained in Section 11.3.3.

11.3.3 Fixed-point iteration to compute the least fixed-point

As shown in Section 6.1, the least fixed-point can be obtained by "bottom refinement" using the functional obtained from the recursive definition. In the same manner, we will determine P, the least fixed-point of H, by computing its approximants that, in the limit, become P. Let us denote the approximants $P_0, P_1, P_2, \dots$. We have $P_0 = \lambda x.false$, the "everywhere false" predicate. The next approximation to P is obtained by feeding P_0 to the "bottom refiner" (as illustrated in Section 6.1):

$$P_1 = \lambda G.(\lambda s.(I(s) \lor \exists x.(G(x) \land T(x, s))))P_0$$

which becomes $\lambda s.I(s)$. This approximant says that P is a predicate true of s whenever $I(s)$. While this is not true (P must represent the reachable state set and not the initial state alone), it is certainly a better answer than what P_0 denotes, which is that there are *no* states in the reachable state set! We now illustrate all the steps of this computation, taking SimpleTR for illustration. We use the abbreviation of not showing the lambda abstracted variables in each step.

- $I = \lambda b. \neg b$.
- $T = \lambda(b, b'). \ (b + b')$.
- $P_0 = \lambda s. false$, which encodes the fact that "we've reached nowhere yet!"
- $P_1 = \lambda G.(\lambda s.(I(s) \vee \exists x.(G(x) \wedge T(x, s))))P_0$.
  This simplifies to $P_1 = I$, which is, in effect, an assertion that we've "just reached" the initial state, starting from P_0.
- Let's see the derivation of P_1 in detail. Expanding T and P_0, we have
  $P_1 = \lambda G.(\lambda s.(I(s) \vee \exists x.(G(x) \wedge \ (x + s) \ ))) \ (\lambda x. false)$.
- The above simplifies to $\neg b$.
- By this token, we are expecting P_2 to be all states that are zero or one step away from the start state. Let's see whether we obtain this result.
- $P_2 = \lambda G.(\lambda s.(I(s) \vee \exists x.(G(x) \wedge T(x, s))))P_1$.
  $= \lambda s.(\neg s \vee \exists x.(\neg x \wedge (x + s)))$.
  $= \lambda s.1$.
- This shows that the set of states reached by all the breadth-first frontiers (combined) that are zero and one step away from the start state, includes every state. Another iteration would not change things; the[10] least fixed-point has been reached.

BED Commands for SimpleTR:

The BED commands given in Figure 11.5 compute the reachable set of states using forward reachability in our example. We can see that P2, the least fixed-point, is indeed *true* — namely, the characteristic predicate for the set of all states. (*Note: In BED, the primed variables must be declared immediately after the unprimed counterparts*). In addition to the explicit commands to calculate the least fixed-point, BED also provides a single command called **reach**. Using that, one can calculate the least fixed-point in one step. In our present example, RS and P2 end up denoting the BDD for *true*.

```
let RS = reach(I,T)
upall RS
view RS
```

Section 11.3.4 discusses another example where the details of the fixed-point iteration using BED are discussed.

---

[10] We do not discuss many of the theoretical topics associated with computing fixed-points in the domain of state transition systems — such as why least fixed-points are unique, etc. For details, please see [20].

```
var b bp                 % Declare b and b'
let I = !b               % Declare init state
let t1 = !b and bp       % 0 --> 1
upall t1                 % Build BDD for it
view t1                  % View it
let t2 = b and bp        % 1 --> 1
let t3 = b and !bp       % 1 --> 0
let T = t1 or t2 or t3   % All three edges
upall T                  % Build and view the BDD
view T                   %

let    P0 = false
upall P0
view   P0

let    P1 = I or ((exists b. (P0 and T))[bp:=b])
upall P1
view   P1

let    P2 = I or ((exists b. (P0 and T))[bp:=b])
upall P2
view   P2
```

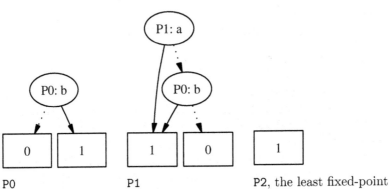

PO   P1   P2, the least fixed-point

Fig. 11.5. BED commands for reachability analysis on SimpleTR, and the fixed-point iteration leading up to the least fixed-point that denotes the set of reachable states starting from I

Why Stabilization at a Fixed-Point is Guaranteed

In every finite-state system modeled using a finite-state Boolean transition system, the least fixed-point is always reached in a finite number of steps. Let us try to argue this fact first using only a simple observation. The observation is that all the Boolean expressions generated during the course of fixed-point computation are over the same set of variables. Since there are exactly 2^{2^N} Boolean functions over N Boolean variables (see Illustration 4.5.2), eventually two of the approximants in the fixed-point computation process will have to be the same Boolean function. However, this argument does not address whether it is possible to have "evasive" or "oscillatory" approximants $P_i, P_{i+1}, \ldots, P_j$ such that $i \neq j$ and $P_j = P_i$. If this were possible, it would be possible to cycle through $P_i, \ldots, P_j$ without ever stabilizing on a fixed-point. Fortunately, this is not possible! Each approximant P_{i+1} is *more defined* than the previous approximant P_i, in the sense defined by the implication lattice defined in Illustration 4.5.3. With this requirement, the number of these ascending approximants is finite, and one of these would be the least fixed-point. See Andersson's paper [7] for additional examples of forward reachability. The book by Clarke et.al. [20] gives further theoretical insights.

11.3.4 An example with multiple fixed-points

Consider the state transition system in Figure 11.6 with initial state s0 (called MultiFP). The set of its reachable states is simply {s0} (and is characterized by the formula $a \wedge b$), as there is no reachable node from s0. Now, a fixed-point iteration beginning with the initial approximant for the reachable states set to $P0 = false$ will converge to the fixed-point $a \wedge b$. What are the other fixed-points one can attain in this system? Here they are:

- With the initial approximant set to {s0,s1}, which is characterized by b, the iteration would reach the fixed-point of $a \vee b$, which characterizes {s0,s1,s2}.
- Finally, we may iterate starting from the initial approximant being 1, corresponding to {s0,s1,s2,s3}. The fixed-point attained in this case is 1, which happens to be the greatest fixed-point of the recursive equation characterizing reachable states.

Hence, in this example, there are three distinct fixed-points for the recursive formula defining reachable states. Of these, the *least* fixed-point is $a \wedge b$, and truly characterizes the set of reachable states; $a \vee b$ is the intermediate fixed-point, and 1 is the greatest fixed-point. It is clear that $(a \wedge b) \Rightarrow (a \vee b)$ and $(a \vee b) \Rightarrow 1$, which justifies these fixed-point orderings. Figure 11.6 also describes the BED commands to produce this intermediate fixed-point.

```
var a ap b bp

let T = (a   and b   and ap   and bp)   or /* S0 -> S0 */
        (!a and b   and !ap and bp)   or /* S1 -> S1 */
        (a   and !b and ap   and !bp) or /* S2 -> S2 */
        (!a and !b and !ap and !bp) or /* S3 -> S3 */
        (!a and b   and ap   and !bp) or /* S1 -> S2 */
        (a   and !b and !ap and bp)   or /* S2 -> S1 */
        (!a and b   and ap   and bp)   or /* S1 -> S0 */
        (a   and !b and ap   and bp)     /* S2 -> S0 */

upall T
view T                    /* Produces BDD for TREL 'T' */

let I = a and b
let P0 = b
let P1 = I or ((exists a. (exists b. (P0 and T)))[ap:=a][bp:=b])

upall P1
view P1
```

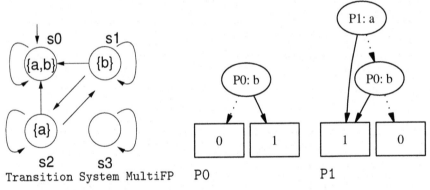

Fig. 11.6. Example where multiple fixed-points exist. This figure shows attainment of a fixed-point $a \vee b$ which is between the least fixed-point of $a \wedge b$ and the greatest fixed-point of 1. The figure shows the initial approximant P0 and the next approximant P1

11.3.5 Playing tic-tac-toe using BDDs

What good are state-space traversal techniques using BDDs? How does one obtain various interesting answers from real-world problems? While we cannot answer these questions in detail, we hope to leave this chapter with a discussion of how one may model a game such as tic-tac-toe and, say, compute the set of all draws in one fell swoop. Following through this example, the reader would obtain a good idea of how to employ mathematical logic to specify a transition system through constraints, and reason about it. We assume the reader knows the game of tic-tac-toe (briefly explained in the passing).

Modeling the players and the board:

We model two players, A and B. The state of the game board is modeled using a pair of variables $a_{i,j}, b_{i,j}$ (we omit the pairing symbols $\langle \rangle$ for brevity) for each square i, j where $i \in 3$ and $j \in 3$. We assume that player A marks square i, j with an o, by setting $a_{i,j}$ and resetting $b_{i,j}$, while player B marks square i, j with an x, by resetting $a_{i,j}$ and setting $b_{i,j}$. We use variable *turn* to model whose turn it is to play (with $turn = 0$ meaning it is A's turn). The state transition relation for each square will be specified using the four variables $a_{i,j}, a_{i,j}p, b_{i,j}$, and $b_{i,j}p$. We model the conditions for a row or column remaining the same, using predicates $samerow_i$ and $samecol_i$. We define nine possible moves for both A and for B. For example, M00 model's A's move into cell $0, 0$; Similarly, we employ N00 to model B's move into cell $0, 0$, and so on for the remaining cells. The transition relation is now defined as a disjunction of the $M_{i,j}$ and $N_{i,j}$ moves. We now capture the constraint *atmostone* that says that, at most one player can play into any square. We then enumerate the gameboard for all possible wins and draws. In the world of BDDs, these computations are achieved through "symbolic breadth first" traversals. We compute the reachable set of states, first querying it to make sure that only the correct states are generated. Then we compute the set of states defining draw configurations. The complete BED definitions are given in Appendix B.

Chapter Summary

This chapter briefly reviewed the history of mathematical logic and pointed out the fact that in the early days of automata theory, mathematical logic and automata were discussed in a unified setting. This approach has immense pedagogical value which this book tries to restore to some extent. A practitioner who works on advanced hardware/software debugging method needs to know *both* of these topics well. For instance, automata theory has, traditionally, been considered an essential prerequisite for an advanced class on compilation. However, recent publications in systems/compilers (e.g., [121]) indicate the central role played by BDDs (see below) and related notions in mathematical logic.

We then discuss how Boolean formulas can be represented in a canonical fashion using the so-called 'reduced ordered binary decision diagrams,' or "BDDs" for short. We then present how finite-state machines can be represented and manipulated using BDDs. We show how reachable states starting from a set of start states can be computed using forward reachability, by using the notion of fixed-points introduced in Chapter 6. We finish the chapter with an illustration of how the game of tic-tac-toe may be modeled using BDDs, and how a tool called BED may be used to compute interesting configurations, such as all the *draw* positions, all possible *win* positions, etc.

BDDs are far richer in scope and application than we have room to elaborate here. The reader is referred to [14, 13] for an exposition of how BDDs are used in hardware and software design, how BDDs may be combined using Boolean operations through the *apply* operator, etc. An alternate proof of canonicity of BDDs appears in [14]. Our presentation of BDDs as automata draws from [22], and to some extent from [111].

Exercises

11.1. Similar to Figure 11.3, draw a BDD for all 16 Boolean functions over variables x and y. (Some of these functions are $\lambda(x, y).true$, $\lambda(x, y).false$, $\lambda(x, y).x$, $\lambda(x, y).y$, etc. Down this list, you have more "familiar" functions such as $\lambda(x, y).nand(x, y)$, and so on. Express these functions without the "lambda" part in BED syntax, and generate the BDDs using BED.)

11.2.
1. Obtain an un-minimized DFA (in the form of a binary tree) for the language
$$L = \{abc \mid a \Rightarrow b \wedge c\}$$
picking the best variable ordering (in case two variable orderings are equal, pick the one that is in lexicographic order). Show the black-hole state also.
2. Minimize this DFA, and then show the additional steps that cast the minimized DFA into a BDD.

11.3. Consider the examples given in Figure 11.2. Construct similar examples for the *addition* operation. More specifically, consider the binary addition of two unsigned two-bit numbers $a_1 a_0$ and $b_1 b_0$, resulting in answer $c_2 c_1 c_0$. Generate a BDD for the carry output bit, c_2. Choose a variable ordering that minimizes the size of the resulting BDD and experimentally confirm using BED.

11.4. Repeat Exercise 11.3 to find out the variable ordering that maximizes the BDD size.

11.5. Represent the behavior of a nand gate, under the inertial delay model, as a state transition system. Encode this transition system using a BDD. Here are some general details on how to approach this problem.

The behavior of an inverter can be modeled using a pair of bits representing its input and output. (For a nand gate, we will need to employ three bits.) In the *transport delay* model, every input change, however short, is faithfully copied to the output, but after a small delay. There is another delay model called the *inertial* delay model in which "short" pulses may not make it to the output.

The behavior of an inverter under these delay models are shown in figures (a) and (b) below.

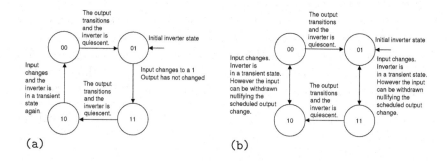

(a)   (b)

11.6. Draw a BDD for the transition relation of a two-bit binary counter with output bits $a_1 a_0$ for initial state 00, counting in the usual $0, 1, 2, 3$ order. Repeat for a two-bit gray-code counter that counts 00, 01, 11, 10, and back to 00.

11.7.
1. With respect to the state transition relation of Figure 11.6(a), identify all the fixed-points of the recursive equation for reachability.
2. Given a state transition system (say, as a graph, as in Figure 11.6(a)), what is a general algorithm to determine the number of fixed-points of its recursive equation for reachability?

11.8. Consider a three-bit shift register based counter with the indicated next-state relation for its three bits:

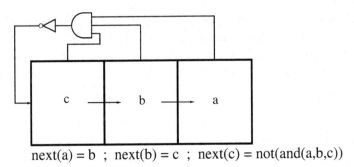

next(a) = b ; next(b) = c ; next(c) = not(and(a,b,c))

1. Represent the next-state relation of this counter using a single ROBDD. Choose a variable ordering that minimizes the size of your ROBDD and justify your choice.
2. Compute the set of all reachable states using forward reachability analysis, using the **reach** command, starting at state 000.
3. Justify the correctness of the answer you obtain. The answer you obtain must be a Boolean formula over a, b, c. Show that this formula is satisfied exactly for those states reachable by the system.

11.9. A three-bit Johnson counter[11] consists of a three-bit shift register where the final $\overline{Q}$ output is connected to the first D input. Starting from a reset state of 000, this counter will go through the sequence 100, 110, 111, 011, 001, and back to 000. For this counter, repeat what Exercise 11.8 asks.

11.10. Using BED, determine the shortest number of steps to win in Tic-Tac-Toe. Appendix B has a full description of the problem encoding.

11.11. Check two conjectures concerning Tic-Tac-Toe, using BED: (i) if a player starts by marking the top-left corner, he/she may lose; (ii) if a player starts by marking the middle square, he/she may win.

11.12. Construct an example with four distinct fixed-points under forward reachability, and verify your construction similar to that explained in Figure 11.6.

11.13. Encode the Man-Wolf-Goat-Cabbage problem using BDDs. In this problem, a man has to carry a wolf, goat, and cabbage across a river. The man has to navigate the boat in each direction. He may carry no more than one animal/object on the boat (besides him) at a time. He must not leave the wolf and goat unattended on either bank,

---

[11] Named after Emeritus Prof. Bob Johnson, University of Utah.

nor must he leave the goat and cabbage unattended on either bank. The number of moves taken is to be minimal. Use appropriate Boolean variables to model the problem.

12

The 'Pumping' Lemma

Theorem 10.3 reiterates why regular languages are so called – their strings are "regular" in length. This fact can be taken advantage of in reasoning about languages. The specific approach taken is based on the fact that long strings meandering through finite-state structures cannot avoid revisiting states. Hence, if such a string goes from the start state to a final state, one can traverse the loop described by the re-visitation an arbitrary number of times, including zero times, and find other strings that also go from the start to the same final state. Expressed rigorously, this idea forms the basis of showing that certain languages are *not* regular, and takes the curious name of the "Pumping" Lemma.

The most common usage comes in the form of *incomplete Pumping Lemmas* or *one-way Pumping Lemmas* that help prove that certain languages are *not* regular. There are also *complete Pumping Lemmas* that can help *prove* that certain languages *are* regular. We now discuss one incomplete Pumping Lemma in depth and show many usages of the same. This will be followed by a brief discussion of one *complete* Pumping[1] Lemma.

12.1 Pumping Lemmas for Regular Languages

One *incomplete Pumping Lemma* for regular languages is as follows. If a language L is regular, then there exists a number n (typically equal in magnitude to the number of states of the minimal DFA D of L) such that for any string $w \in L$ exceeding n in length, w will have a loop somewhere in it. More specifically, when a DFA makes n

---

[1] Rumor has it that this is the most favorite lemma of a certain California governor.

state transitions, it must go through $n + 1$ states;[2] all of these states cannot be distinct (since there are altogether only n states). This causes the DFA to revisit at least one state, thus describing a path such as $s_1, s_2, \ldots, s_i, \ldots, s_i, \ldots, s_{n+1}$. Now, break w into three distinct pieces. Let x be the maximal prefix of w in which no states repeat ($s_1, \ldots, s_i$ in our example). Following x, we will have a segment of w that begins and ends at some specific state; this segment would form a loop, such as $s_i, \ldots, s_i$ in our example. Call this segment y. Now, the rest of w is considered to be the string z, which leads w to one of the accepting states $s \in F$ of D. It is then clear that the portion y can be repeated any number $k \geq 0$ of times in going to s, thus ensuring that strings of the form $xy^k z$ are also in L. Using mathematical logic, and following Lamport's style, discussed on Page 79 in Chapter 5, we write:

$Regular(L) \Rightarrow$
$\qquad \exists n \in N :$
$\qquad \forall w \in L : |w| \geq n$
$\qquad\qquad \Rightarrow$
$\qquad\qquad \exists x, y, z \in \Sigma^* :$
$\qquad\qquad \wedge \; w = xyz$
$\qquad\qquad \wedge \; y \neq \varepsilon$
$\qquad\qquad \wedge \; |xy| \leq n$
$\qquad\qquad \wedge \; \forall k \geq 0 : xy^k z \in L.$

Illustration 12.1.1 *(Quantifier alternation)* The Pumping Lemma resembles the following example English assertion: "A zoo Z is interesting if *forall* giraffes g in Z whose right rear leg is more than n feet in length, there *exists* a reticulation patch on g's skin of exactly $\sqrt{n}$ feet circumference, such that within this reticulation patch, *forall* hair h, the color of h is brown."

To show that Z is *uninteresting*, we have to find one giraffe of height $\geq n$ such that for all patches, either the patch is not $\sqrt{n}$ feet in circumference, or (it is, and) there exists a non-brown hair in it.

To use the incomplete Pumping Lemma in proving that a language $L_{suspect}$ is *non*-regular, we proceed as follows. Assume $Regular(L_{suspect})$. Then, use the incomplete Pumping Lemma, obtaining as a consequence, the following formula C:[3]

---

[2] If the n moves are compared to n webs on the foot of a duck, then the duck must have $n + 1$ digits!

[3] Note that C, being a fully quantified formula, or *sentence*, is either true or false.

$\exists n \in N :$
$\forall w \in L_{suspect} : |w| \geq n$
$$\Rightarrow$$
$$\exists x, y, z \in \Sigma^* :$$
$$\wedge\ w = xyz$$
$$\wedge\ y \neq \varepsilon$$
$$\wedge\ |xy| \leq n$$
$$\wedge\ \forall k \geq 0 :\ xy^k z \in L_{suspect}.$$

Now, we try to show that formula C is false (or that $\neg C$ is true). If we succeed in doing so, we can conclude using proof by contradiction that $\neg Regular(L_{suspect})$. What does showing $\neg C$ involve? Let $D = \neg C$. We can now write D as follows:

$\forall n \in N :$
$\exists w \in L_{suspect} : |w| \geq n$
$$\wedge$$
$$\forall x, y, z \in \Sigma^* :$$
$$\vee\ w \neq xyz$$
$$\vee\ y = \varepsilon$$
$$\vee\ |xy| > n$$
$$\vee\ \exists k \geq 0 :\ xy^k z \notin L_{suspect}.$$

Now, our goal is to show that D is true (if we were to achieve this goal, we would have proved $\neg C$, or that $\neg Regular(L_{suspect})$), which is our original proof goal. To make D true, we must clearly satisfy the "bullet disjunction" embedded in it. That disjunction would be made true by making *any one of the following disjuncts* true *for every* $x,y,z \in \Sigma^*$:

1. pick x, y, z such that $w \neq xyz$,
2. pick $y = \varepsilon$,
3. pick x, y such that $|xy| > n$, or
4. find a $k \geq 0$ such that $xy^k z \notin L_{suspect}$.

Now, for many x, y, z, it will be possible to satisfy one of disjuncts 1 or 3. This is clear because we can quite easily find $xyz \neq w$, find $y = \varepsilon$, or find xy, such that $|xy| > n$. *So we don't even bother with these selections of x, y, z in the rest of this sequel.* What about x, y, z that falsify disjuncts 1 through 3? For that case, we *must* find a $k \geq 0$ such that $xy^k z \notin L_{suspect}$. That surviving case is now spelled out fully, below. This listing incorporates the fact that the first three disjuncts are false.

Pumping recipe: These steps below constitute the *Pumping recipe* we shall follow in attacking problems using the Pumping Lemma.

PR1: Consider x, y, z such that $w = xyz$ and $y \neq \varepsilon$ (thus falsifying disjuncts 1 and 2), and ensure that $|xy| \leq n$ (look for a loop within the first n moves in w), thus falsifying disjunct 3.

PR2: Find a $k \geq 0$ such that $xy^k z \notin L_{suspect}$ (thus satisfying disjunct 4).

One should, however, bear in mind the following frequently committed mistake, and avoid it:

If, instead of showing that formula C of page 206 is false, one ends up showing C, i.e, that C is true, then we cannot draw *any* conclusion about $L_{suspect}$. It could either be regular or non-regular! Refer to the discussion on page 74 around proving $5 = 5$.

We shall now illustrate these steps as well as related methods with several examples.

Illustration 12.1.2 Example: Consider $L = \{0^m 10^m 1 \mid m \geq 0\}$. To show L is not regular:

PR1:
1. Choose $w = 0^n 10^n 1$.
2. Choose $y \neq \epsilon$.
3. Choose x, y, *such that* $|xy| \leq n$ – aha! Observe that y must contain a 0.

PR2:
1. Now, does there exist a $k \geq 0$ such that $xy^k z \notin L$?
2. Sure! For $k = 0$, we lose one 0, giving rise to a string of the form $0^m 1^n$ where $m < n$. This satisfies the "D formula" associated with this example. Hence, L is not regular.

Illustration 12.1.3 Example: Consider $L = \{10^m 10^m \mid m \geq 0\}$. To show L is not regular:

PR1:
1. Choose $w = 10^n 10^n$.
2. Choose $y \neq \epsilon$.
3. Choose x, y, *such that* $|xy| \leq n$.

PR2:
1. We have three choices for y:
    a) $y = 1$,

    b) $y = 10^l$ for $l < n$, or
    c) $y = 0^l$ for $0 < l < n$.
2. Does there exist a $k \geq 0$ such that $xy^k z \notin L$ for all these choices?
3. Sure!
    a) For $y = 1$, choose $k = 0$ (other choices work too; see Exercise 12.5).
    b) $y = 10^l$ for $l < n$, choose $k = 0$ (other choices work too).
    c) $y = 0^l$ for $0 < l < n$ also, choose $k = 0$ (other choices work too).
    d) In all these cases, the assertion $\exists\, k \geq 0$ such that $xy^k z \notin L$ is satisfied.
    e) Therefore, we get *full contradiction*, and hence L is not regular.

12.1.1 A stronger incomplete Pumping Lemma

There is a stronger version of the Pumping Lemma which allows strings "in the middle" to be pumped. We now state this Pumping Lemma semiformally, and illustrate its power on a simple example:

$Regular(L) \Rightarrow$
    $\exists n \in N :$
    $\forall w \in L : |w| \geq n$
        $\Rightarrow$
        $\forall x, y, z \in \Sigma^* :$
        $\wedge\, w = xyz$
        $\wedge\, |y| \geq n$
           $\exists u, v, w \in \Sigma^* :$
           $\wedge\, v \neq \varepsilon$
           $\wedge\, \forall k \geq 0 : xuv^k wz \in L.$

As can be seen, the pumping can occur "in the middle."

Illustration 12.1.4 Consider the language L_{if} defined on page 212. By applying the ordinary Pumping Lemma, we cannot derive a contradiction starting from string $ab^n c^n$ because the possibilities include $x = \varepsilon$, $y = a$, and $z = b^n c^n$, and by pumping y, we do not go outside the language. However, with the stronger Pumping Lemma, we can pick x, y, and z suitably, with $|y| \geq n$. Observe that by letting x be a, we can situate u, v, w in the $b^n c^n$ region, obtaining a violation in all cases.

12.2 An adversarial argument

The Pumping Lemma provides a concrete setting to understand adversarial arguments. Consider proving, *directly using the Pumping Lemma*, that the language

$$L = \{0^i 1^j \mid i \neq j\}$$

is not regular. Here is how the proof goes as an adversarial argument. Suppose an adversary (Y) claims that this is a regular language. You (U) want to prove it is not. Here is how you can argue and win:

1. U: "OK if L is regular, you have a DFA D with you right?"
2. Y: "Yes."
3. U: "How many states in it?"
4. Y: "n".
5. U: "OK, describe to me the sequence of states that D goes through upon seeing the first n symbols of the string $0^n 1^{(n+n!)}$." Here, n is chosen to be the number of states in D. Since $n \neq (n + n!)$, this string surely must be in L. (The choice of $(n + n!)$ as the exponent of 1s is rather purposeful — and *very astute* on the part of U — as we shall see momentarily).
6. Y (Straight-faced): "It visits s_0, s_1, ..., *all of which are different from one another.*"
7. U: (Red-faced): "Lie! If there are n states in your DFA, then seeing n symbols, the DFA must have traversed a loop, and hence you must have listed two states that are the same. Don't you know that this follows from the pigeon-hole principle?"
8. Y: (Blue-faced): "OK, you are right, it is "$s_0, s_1, \ldots, s_i, s_{i+1}, s_{i+2}, \ldots, \ldots, s_i, s_j \ldots$."" Notice that s_i is repeating in such a sequence.
9. U: "Aha! I'm going to call the pieces of the above sequence as follows."
   a) the piece that leads up to the loop, $s_0, s_1, \ldots, s_i$, will be called x,
   b) the piece that traverses the loop, $s_i, s_{i+1}, s_{i+2}, \ldots, s_i$, will be called y, and
   c) the piece that exits the loop and visits the final state, $s_i, s_j \ldots$, will be called z.
   "You may pick any such x, y, z and I'm going to confound you."
10. Y: "How?"
11. U: "Watch me!" (private thoughts of U now follow...)
    a) Since I have no idea what $|y|$ is, I must ensure that by pumping y, no matter what its length, I should be able to create a string of 0s equal in length to $(n + n!)$.

b) So, by pumping, if I can create an overall string $0^{(n+n!)}1^{(n+n!)}$, I would have created the desired contradiction.

c) The initial distribution of 0s along the path xyz is as follows:

   i. x has $|x|$ 0s,

   ii. y has $|y|$ 0s, and

   iii. z has $(n - |y| - |x|)$ 0s.

d) Hence, by pumping y k-times, for integral k, we must be able to attain $n + n! = |x| + k \times |y| + (n - |y| - |x|)$.

e) Simplifying, we should be able to satisfy $n! = (k - 1)|y|$. Since $|y| \leq n$, such a k exists!

12. U now begins his animated conversation: "See the above argument. I can now pump up the y of your string k times where $k = n!/|y| + 1$. Then you get a string $0^{n+n!}1^{n+n!}$ that is not in L. This path also exists in your DFA. So your DFA cannot be designed exactly for L— it also accepts illegal strings. Admit defeat!"

13. Y: Tries for an hour, furiously picking all possible x, y, z and goes back to step 8. For each such choice, U defeats Y[4] in the same fashion. Finally Y admits defeat.

14. U: "Thank you. Next victim please."

12.3 Closure Properties Ameliorate Pumping

The use of closure properties can simplify the application of the Pumping Lemma. However, caution is to be exercised to avoid unsound arguments. We now provide a few illustrations and exercises. First, let us rework Problem 12.1.3 as follows:

1. The reverse of $L = \{10^m10^m \mid m \geq 0\}$ is $L' = \{0^m10^m1 \mid m \geq 0\}$
2. Now, L' was proved to be not regular in Problem 12.1.2
3. Since *reverse* preservers regularity, the original language L isn't regular either.

As a general approach, here is how we use regularity preserving operations to help make our arguments:

1. Suppose $M \cap L(0^* 1^*) = N$
2. Suppose we can show (thru Pumping Lemma) that N is not regular
3. *Then* we can conclude that M is not regular.

---

[4] I promise to make Y win in my next *two* books—and meanwhile, offer to put replacement pages on my web-page for the benefit of anyone wishing that Y trounce U in this very book!

Here is an **abuse** of the incomplete Pumping Lemma (from [111]). Consider the language

$$L_{if} = \{a^i b^j c^k \mid i \geq 0,\ j,k > 0,\ and\ if\ i = 1\ then\ j = k\}.$$

While this language is not regular, we can still show that the C formula of page 206 that results from the incomplete Pumping Lemma will be a tautology ("can pump k without causing any violations"). This is because

PR1: for *every* choice of w of the form $a^i b^j c^k$, and a way to split it into x, y, z that abide by the PR1 conditions,

PR2: we must find a $k \geq 0$ such that $xy^k z \notin L_{suspect}$. Basically, for any such x, y, z, there must always be a choice of y such that pumping causes us to stay in the language $L_{suspect}$, thus deriving no contradictions. Exercise 12.8 asks you to spell out this argument, and offers another attack on the same problem.

12.4 Complete Pumping Lemmas

There are many *complete* Pumping Lemmas of the form "*Regular*(L) if and only if *conditions*," i.e., a language is regular if and only if certain conditions hold. We present two popular versions, one due to Jaffe [65] and the other due to Stanat and Weiss [113]. Possible uses of these *complete* Pumping Lemmas include showing that certain languages *are* regular (we do not pursue such proofs of regularity in this book).

12.4.1 Jaffe's complete Pumping Lemma

For a language L over a finite alphabet Σ, Jaffe's Pumping Lemma is the following:

$Regular(L) \Leftrightarrow$
        $\exists k \in N :$
        $\forall y \in \Sigma^* : |y| = k$
            $\Rightarrow$
            $\exists u, v, w \in \Sigma^* :$
            $\wedge\ y = uvw$
            $\wedge\ v \neq \varepsilon$
            $\wedge\ \forall z \in \Sigma^* :$
                $\forall i \in N : (yz \in L \Leftrightarrow uv^i wz \in L).$

Notice that for a "long string" $y = uvw$ with a pump-able middle portion v, it is expressed that we can follow the original string y with an arbitrary z and stay within L, if and only if we can pump the middle and still follow it with that same z and stay within L. In [65], a proof of this Pumping Lemma is provided.

12.4.2 Stanat and Weiss' complete Pumping Lemma

For a language L over a finite alphabet Σ, Stanat and Weiss' Pumping Lemma is the following:

$$Regular(L) \Leftrightarrow$$
$$\exists p \in N :$$
$$\forall x \in \Sigma^* : |x| \geq p$$
$$\Rightarrow$$
$$\exists u, v, w \in \Sigma^* :$$
$$\wedge\ x = uvw$$
$$\wedge\ v \neq \varepsilon$$
$$\wedge\ \forall r, t \in \Sigma^* :$$
$$\forall i \in N : (rut \in L \Leftrightarrow ruv^i t \in L).$$

Notice that this Pumping Lemma does not require the pump-able string to be part of the prefix; an arbitrary string r can lead off, and an arbitrary tail t can follow. In [113], a proof of this Pumping Lemma using Jaffe's Pumping Lemma is provided.

Chapter Summary

We discuss the so-called Pumping Lemmas that characterize regular sets. We also discuss operations that *preserve* regularity; given one or more sets, these operations are guaranteed to deliver only regular sets. This chapter shows how one may exploit these facts to disprove that certain languages are *not* regular. For the sake of completeness, we also very briefly discuss the so-called *complete* Pumping Lemmas that actually help establish that certain languages *are regular*. While we do not utilize these complete Pumping Lemmas to carry out any proofs, the fact that such lemmas exist is important to know.

Exercises

12.1. Argue that L_{badd} is non-regular, where L_{badd} is almost similar to L_{add} of Exercise 10.6 except for what is shown below:

$$L_{badd} = \{a_0 a_1 \dots a_{k-1} b_0 b_1 \dots b_{k-1} c_0 c_1 \dots c_{k-1} \mid \dots same \dots\}$$

12.2. Consider the example language of Section 12.2 again:

$$L = \{0^i 1^j \mid i \neq j\}.$$

Using closure properties, show that this set is not regular.

12.3. Prove that if L is not regular, then LL is also not regular.

12.4. $L_{0n1n} = \{0^n 1^n \mid n \geq 0\}$ is easily shown to be non-regular. Now, show

$$L_{eq} = \{x \mid x \in \{0,1\}^* \text{ and } \#_0(x) = \#_1 x\}$$

is not regular. (Hint: intersection with 0* 1*).

12.5. Show how to solve the problems presented in Illustration 12.1.2 and 12.1.3 by choosing a $k \neq 0$. Write out the complete proof using such k values.

12.6. What's wrong with this argument?

1. Suppose $M \cap L(0^* 1^*) = N$.
2. Suppose we can show (through Pumping Lemma) that M is not regular.
3. Conclude that N is not regular.

12.7. Among the assertions below, identify those that are true, and justify them. For those that are false, provide a counterexample.

1. The union of two non-regular sets is always non-regular.
2. The intersection of two non-regular sets is always non-regular.
3. A regular set can have a non-regular subset.
4. A regular set can have a non-regular superset.
5. Every regular set has a regular subset.
6. The star of a non-regular set can never be regular.
7. The prefix-closure of a non-regular set can never be regular.
8. The reverse of a non-regular set can never be regular.
9. The union of a regular and a non-regular set can be regular.
10. The union of a regular and a non-regular set can be non-regular.
11. The concatenation of a regular and a non-regular set can sometimes be regular.
12. There is a finite non-regular set.
13. It is possible to apply a homomorphism to turn a non-regular set into a regular set.

12.8. Consider the language L_{if} of page 12.3. Show that using the initial incomplete Pumping Lemma, we can pump, *i.e.*, prove the Pumping Lemma condition C to be true, thus being unable to conclude anything. Now, apply a closure property and use the incomplete Pumping Lemma to show that L_{if} is non-regular.

12.9. Using the stronger incomplete Pumping Lemma of Section 12.1.1, show that L_{if} is non-regular.

12.10. Prove the following languages to be non-regular:

1. $L_{sq} = \{0^{i^2} \mid i \geq 0\}$.
2. $L_{()} = \{x \mid x \in \{(,)\}^* \text{ and } x \text{ is well} - \text{parenthesized}\}$.
3. The set of palindromes over $\{0,1\}^*$ is not regular.

The definition of well-parenthesized is as follows (see also page 224):

1. The number of (and) in x is the same.
2. In any prefix of x, the number of (is greater than or equal to the number of).

13

Context-free Languages

A *context-free language* (CFL) is a language accepted by a *push-down automaton* (PDA). Alternatively, a context-free language is one that has a *context-free grammar* (CFG) describing it. This chapter is mainly about context-free grammars, although a brief introduction to push-down automata is also provided. The next chapter will treat push-down automata in greater detail, and also describe algorithms to convert PDAs to CFGs and vice versa. The theory behind CFGs and PDAs has been directly responsible for the design of *parsers* for computer languages, where parsers are tools to analyze the syntactic structure of computer programs and assign them meanings (e.g., by generating equivalent machine language instructions).

A CFG is a structure (N, Σ, P, S) where N is a set of symbols known as *non-terminals*, Σ is a set of symbols known as *terminals*, $S \in N$ is called the *start symbol*, and P is a finite set of *production rules*. Each production rule is of the form

$$L \rightarrow R_1 \ R_2 \ \dots R_n,$$

where $L \in N$ is a non-terminal and each R_i belongs to $N \cup \Sigma_\varepsilon$. We will now present several context-free grammars through examples, and then proceed to examine their properties.

Consider the CFG $G_1 = (\{S\}, \{0\}, P, S)$ where P is the set of rules shown below:

Grammar G_1:
```
S -> 0.
```

A CFG is machinery to produce strings according to production rules. We start with the start symbol, find a rule whose left-hand side matches the start symbol, and derive the right-hand side. We then repeat the process if the right-hand side contains a non-terminal. Using grammar G_1, we can produce only one string starting from S, namely 0, and so the derivation stops. Now consider a grammar G_2 obtained from G_1 by adding one extra production rule:

Grammar G_2:
```
S -> 0
S -> 1 S.
```

Using G_2, an infinite number of strings can be derived as follows:

1. Start with S, calling it a *sentential form.*
2. Take the current sentential form and for one of the non-terminals N present in it, find a production rule of the form $N \rightarrow R_1 \ldots R_m$, and replace N with $R_1 \ldots R_m$. In our example, S -> 0 matches S, resulting in sentential form 0. Since there are no non-terminals left, this sentential form is called a *sentence.* Each such sentence is a member of the *language* of the CFG - in symbols, $0 \in \mathcal{L}(G_2)$. The step of going from one sentential form to the next is called a *derivation step.* A sequence of such steps is a *derivation sequence.*
3. Another derivation sequence using G_2 is

$$S \Rightarrow 1S \Rightarrow 11S \Rightarrow 110.$$

To sum up, given a CFG G with start symbol S, S is a sentential form. If $S_1 S_2 \ldots S_i \ldots S_m$ is a sentential form and there is a rule in the production set of the form $S_i \rightarrow R_1\ R_2\ \ldots R_n$, then $S_1 S_2 \ldots\ R_1\ R_2\ \ldots R_n\ \ldots S_m$ is a sentential form. We write

$$S \Rightarrow \ldots \Rightarrow S_1 S_2 \ldots S_i \ldots S_m \Rightarrow S_1 S_2 \ldots\ R_1\ R_2\ \ldots R_n\ \ldots S_m \Rightarrow \ldots.$$

As usual, we use $\Rightarrow^*$ to denote a multi-step derivation.

Given a CFG G and one of the sentences in its language, w, a *parse tree* for w with respect to G is a tree with frontier w, and each interior node corresponding to one derivation step. The parse tree for string 110 with respect to CFG G_2 appears in Figure 13.1(a).

13.1 The Language of a CFG

The language of a CFG G, $\mathcal{L}(G)$, is *the set of all sentences* that can be derived starting from S. In symbols, for a CFG G,

$$\mathcal{L}(G) = \{w \mid S \Rightarrow^* w \land w \in \Sigma^*\}.$$

According to this definition, for a CFG G_3, with the only production

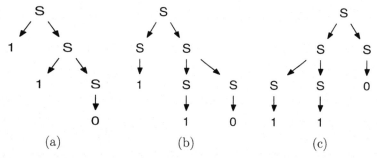

Fig. 13.1. (a) The parse tree for string 110 with respect to CFG G_2; (b) and (c) are parse trees for 110 with respect to G_4.

Grammar G_3:
$S \to S$.

we have $\mathcal{L}(G_3) = \emptyset$. The same is true of a CFG all of whose productions contain a non-terminal on the RHS, since, then, we can never get rid of all non-terminals from any sentential form.

A derivation sequence, in which the leftmost non-terminal is selected for replacement in each derivation step, is known as a *leftmost derivation*. A rightmost derivation can be similarly defined. Specific derivation sequences such as the leftmost and rightmost derivation sequences are important in compiler construction. We will employ leftmost and rightmost derivation sequences for pinning down the exact derivation sequence of interest in a specific discussion. This, in turn, decides the shape of the *parse tree*. To make this clear, consider a CFG G_4 with three productions

Grammar G_4:
$S \to SS \mid 1 \mid 0$.

The above notation is a compact way of writing three distinct *elementary productions* $S \to SS$, $S \to 1$, and $S \to 0$. A string 110 can now be derived in two ways:

- Through the leftmost derivation $S \Rightarrow SS \Rightarrow 1S \Rightarrow 1SS \Rightarrow 11S \Rightarrow 110$ (Figure 13.1(b)), or
- Through the rightmost derivation $S \Rightarrow SS \Rightarrow S0 \Rightarrow SS0 \Rightarrow S10 \Rightarrow 110$ (Figure 13.1(c)).

Notice the connection between these derivation sequences and the parse trees.

Now consider grammar G_5 with production rules

Grammar G_5:
$S \rightarrow aSbS \mid bSaS \mid \varepsilon.$

The terminals are $\{a, b\}$. What CFL does this CFG describe? It is easy
to see that in each replacement step, an S is replaced with either ε or
a string containing an a and a b; and hence, all strings that can be
generated from G_5 have the same number of a's and b's. Can *all* strings
that contain equal a's and b's be generated using G_5? We visit this
(much deeper) question in the next section. If you try to experimentally
check this conjecture out, you will find that no matter what string of
a's and b's you try, you can find a derivation for it using G_5 so long as
the string has an equal number of a's and b's.

> **Note:** We employ ε, e, and `epsilon` interchangeably, often for
> the ease of type-setting.   □

13.2 Consistency, Completeness, and Redundancy

Consider the following CFG G_6 which has one extra production rule
compared to G_5:

Grammar G_6:
$S \rightarrow aSbS \mid bSaS \mid SS \mid \varepsilon.$

As with grammar G_5, all strings generated by G_6 also have an equal
number of a's and b's. If we identify this property as *consistency*, then
we find that grammars G_5 and G_6 satisfy consistency. What about
completeness? In other words, will *all* such strings be derived? Does it
appear that the production $S \rightarrow SS$ is essential to achieve completeness? It turns out that it is not - we can prove that G_5 is complete,
thus showing that the production $S \rightarrow SS$ of G_6 is *redundant*.

How do we, in general, prove grammars to be complete? The general
problem is undecidable,[1] as we shall show in Chapter 17. However,
for *particular* grammars and *particular* completeness criteria, we can
establish completeness, as we demonstrate below.

Proof of completeness:

The proof of completeness typically proceeds by induction. We have to
decide between arithmetic or complete induction; in this case, it turns
out that complete induction works better. Using complete induction,
we write the *inductive hypothesis* as follows:

---

[1] The undecidability theorem that we shall later show is that for an *arbitrary*
grammar G, it is not possible to establish whether $L(G)$ is equal to Σ^*.

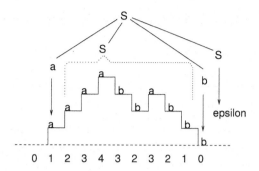

Fig. 13.2. A string that does not cause zero-crossings. The numbers below the string indicate the running difference between the number of a's and the number of b's at any point along the string

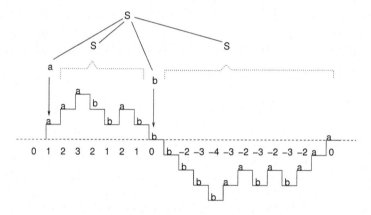

Fig. 13.3. A string that causes zero-crossings

Suppose G_5 generates all strings less than n in length having an equal number of **a**'s and **b**'s.

Consider now a string of length $n + 2$ – the next longer string that has an equal number of **a**'s and **b**'s. We can now draw a graph showing the running difference between the number of **a**'s and the number of **b**'s, as in Figure 13.2 and Figure 13.3. This plot of the running difference between #a and #b is either fully above the x-axis, fully below, or has zero-crossings. In other words, it can have many "hills" and "valleys." Let us perform a case analysis:

1. The graph has no zero-crossings. There are further cases:
   a) it begins with an **a** and ends with a **b**, as in Figure 13.2.

b) it begins with a b and ends with an a (this case is symmetric and hence will not be explicitly argued).
2. It has zero-crossings, as in Figure 13.3. Again, we consider only one case, namely the one where the first zero-crossing from the left occurs after the curve has grown in the positive direction (i.e., after more a's occur initially than b's).

Let us consider case 1a. By induction hypothesis, the shorter string in the middle can be generated via S. Now, the entire string can be generated as shown in Figure 13.2 using production S -> aSbS, with a matching the first a, the first S matching 'the shorter string in the middle,' the b matching the last b in the string, and the second S going to ε. Case 1b may be similarly argued. If there is a zero-crossing, then we attack the induction as illustrated in Figure 13.3, where we split the string into the portion before its last zero-crossing and the portion after its last zero-crossing. These two portions can, by induction, be generated from G_5, with the first portion generated as aSb and the second portion generated as an S, as in Figure 13.3.   □.

Illustration 13.2.1 *Consider*

$$L_{a^m b^n c^k} = \{a^m b^n c^k \mid m, n, k \geq 0 \text{ and } ((m = n) \text{ or } (n = k))\}$$

Develop a context-free grammar for this language. Prove the grammar for consistency and completeness.
Solution: The grammar is given below. We achieve "equal number of a's and b's" by growing "inside out," as captured by the rule M -> a M b. We achieve zero or more c's by the rule C -> c C or e. Most CFGs get designed through the use of such "idioms."

    S -> M C | A N

    M -> a M b | e

    N -> b N c | e

    C -> c C | e

    A -> a A | e

Consistency: No string generated by S must violate the rules of being in language $L_{a^m b^n c^k}$. Therefore, if M generates matched a's and b's, and C generates only c's, consistency is guaranteed. The other case of A and N is very similar.

Notice that from the production of M, we can see that it generates matched a's and b's in the e case. Assume by induction hypothesis that in the M occurring on the right-hand side of the rule, M -> a M b, respects consistency. Then the M of the left-hand side of this rule has an extra a in front and an extra b in the back. Hence, it too respects consistency.

Completeness: We need to show that any string of the form $a^n b^n c^k$ or $a^k b^n c^n$ can be generated by this grammar. We will consider all strings of the kind $a^n b^n c^k$ and develop a proof for them. The proof for the case of $a^k b^n c^n$ is quite similar and hence is not presented.

We resort to arithmetic induction for this problem. Assume, by induction hypothesis that the particular $2n + k$-long string $a^n b^n c^k$ was derived as follows:

- S ⟹ M C.
- M ⟹* $a^n b^n$ through a derivation sequence that we call S1, and
- C ⟹* c^k through derivation sequence S2.
- S ⟹ M C ⟹* $a^n b^n$ C ⟹* $a^n b^n$ c^k. Notice that in this derivation sequence, the first ⟹* derivation sequence is what we call S1 and the second ⟹* derivation sequence is what we call S2.

Now, consider the next legal longer string. It can be either $a^{n+1} b^{n+1} c^k$ or $a^n b^n c^{k+1}$. Consider the goal of deriving $a^{n+1} b^{n+1} c^k$. This can be achieved as follows:

- S ⟹ M C ⟹ a M b C.
- Now, invoking the S1 sequence, we get ⟹* $a\ a^n b^n\ b$ C.
- Now, invoking the S2 sequence, we get ⟹* $a\ a^n b^n\ b\ c^k$; and hence, we can derive $a^{n+1} b^{n+1} c^k$.

Now, $a^n b^n c^{k+1}$ can be derived as follows:

- S ⟹ M C ⟹ M c C.
- Now, invoking the S1 derivation sequence, we get ⟹* $a^n b^n$ c C.
- Finally, invoking the S2 derivation sequence, we get ⟹* $a^n b^n$ c^{k+1}.

Hence, we can derive any string that is longer than $a^n b^n c^k$, and so by induction we can derive all legal strings.

13.2.1 More consistency proofs

In case of grammars G_5 and G_6, writing a consistency proof was rather easy: we simply observed that all productions introduce equal a's and b's in each derivation step. Sometimes, such "obvious" proofs are not possible. Consider the following grammar G_9:

```
S -> ( W S | e
W -> ( W W | ).
```

It turns out that grammar G_9 generates the language of the set of *all* well-parenthesized strings, even though three of the four productions appear to introduce *unbalanced* parentheses. Let us first formally define what it means to be well-parenthesized (see also Exercise 12.10), and then show that G_9 satisfies this criterion.

Well-parenthesized strings

A string x is well-parenthesized if:

1. The number of (and) in x is the same.
2. In any prefix of x, the number of (is greater than or equal to the number of).

With this definition in place, we now show that G_9 generates only consistent strings. We provide a proof outline:

1. Conjecture about S: *same number of (and)*. Let us establish this via induction.
   a) Epsilon (e) satisfies this.
   b) How about (W S ?
   c) OK, we need to "conquer" W.
      i. Conjecture about W: *it generates strings that have one more) than (.*
      ii. This is true for both arms of W.
      iii. Hence, the conjecture about W is true.
   d) Hence, the conjecture about S is true.
   e) Need to verify one more step: *In any prefix, is the number of (more than the number of)?*
   f) Need conjecture about W: *In any prefix of a string generated by W, number of) at most one more than the number of (.* Induct on the W production and prove it. Then S indeed satisfies consistency.   □.

13.2.2 Fixed-points again!

The language generated by a CFG was explained through the notion of a derivation sequence. Can this language also be obtained through fixed-points? The answer is 'yes' as we now show.

Consider recasting the grammar G_5 as a language equation (call the language $L(S_5)$):

$$L(S_5) = \{a\}\, L(S_5)\, \{b\}\, L(S_5) \ \cup \ \{b\}\, L(S_5)\, \{a\}\, L(S_5) \ \cup \ \{\varepsilon\}.$$

Here, juxtaposition denotes language concatenation. What solutions to this language equation exist? In other words, find languages to plug in place of $L(S_5)$ such that the right-hand side language becomes equal to the left-hand side language. We can solve such language equations using the fixed-point theory introduced in Chapter 7. In particular, one can obtain the least fixed-point by iterating from "bottom" which, in our context, is the empty language $\emptyset$. *The least fixed-point is also the language computed by taking the derivation sequence perspective.*

To illustrate this connection between fixed-points and derivation sequences more vividly, consider a language equation obtained from CFG G_6:

$$L(S_6) = \{a\}L(S_6)\{b\}L(S_6) \cup \{b\}L(S_6)\{a\}L(S_6) \cup \{b\}L(S_6)L(S_6) \cup \{\varepsilon\}$$

It is easy to verify that this equation admits two solutions, one of which is the desired language of equal a's and b's, and the other language is a completely uninteresting one (see Exercise 13.5). The language equation for $L(S_5)$ does not have multiple solutions. It is interesting to note that the reason why $L(S_6)$ admits multiple solutions is because of the redundant rule in it. To sum up:

- Language equations can have multiple fixed-points.
- The least fixed-point of a language equation obtained from a CFG also corresponds to the language generated by the derivation method.

These observations help sharpen our intuitions about least fixed-points. The notion of least fixed-points is central to programming because programs also execute through "derivation steps" similar to those employed by CFGs. Least fixed-points start with "nothing" and help attain the least (most irredundant) solution. As discussed in [20] and also in Chapter 22, in Computational Tree Logic (CTL), both the least fixed-point and the greatest fixed-point play significant and meaningful roles.

13.3 Ambiguous *Grammars*

As we saw in Figures 13.1(b) and 13.1(c), even for one string such as 1110, it is possible to have two distinct parse trees. In a compiler, the existence of two distinct parse trees can lead to the compiler producing two *different* codes - or can ascribe two different meanings - to the same sentence. This is highly undesirable in most settings. For example, if we

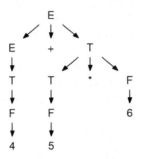

Fig. 13.4. Parse tree for $4 + 5 * 6$ with respect to G_8.

have an arithmetic expression $4 + 5 * 6$, we want it parsed as $4 + (5 * 6)$ and not as $(4 + 5) * 6$. If we write the expression grammar as G_7,

Grammar G_7:
$E \rightarrow E + E \mid E * E \mid number,$

then *both* these parses (and their corresponding parse trees) would be possible, in effect providing an *ambiguous* interpretation to expressions such as $4 + 5 * 6$. It is necessary to *disambiguate* the grammar, for example by rewriting the simple expression grammar above to grammar G_8. One such disambiguated grammar is the following:

Grammar G_8:
$E \rightarrow E + T \mid T$
$T \rightarrow T * F \mid F$
$F \rightarrow number \mid (E).$

In the rewritten grammar, for any expression containing $+$ and $*$, the parse trees will situate $*$ deeper in the tree (closer to the frontier) than $+$, thus, in effect, forcing the evaluation of $*$ first, as illustrated in Figure 13.4.

13.3.1 If-then-else ambiguity

An important practical example of ambiguity arises in the context of grammars pertaining to `if` statements, as illustrated below:

```
STMT ->   if EXPR then STMT
        | if EXPR then STMT else STMT
        | OTHER

OTHER -> p
```

```
EXPR  -> q
```

The reason for ambiguity is that the `else` clause can match either of the *then* clauses. Compiler writers avoid the above if-then-else ambiguity by modifying the above grammar in such a way that the `else` matches with the closest unmatched `then`. One example of such a rewritten grammar is the following:

```
STMT -> MATCHED | UNMATCHED

MATCHED -> if EXPR then MATCHED else MATCHED | OTHER

UNMATCHED ->  if EXPR then STMT
            | if EXPR then MATCHED else UNMATCHED

OTHER -> p

EXPR  -> q
```

This forces the `else` to go with the closest previous unmatched `then`.

13.3.2 Ambiguity, inherent ambiguity

In general, it is impossible to algorithmically decide whether a given CFG is ambiguous (see Chapter 17 and Exercise 17.2 that comes with hints). In practice, this means that there cannot exist an algorithm that can determine whether a given CFG is ambiguous. To make things worse, there are *inherently ambiguous languages* – languages for which every CFG is ambiguous. If the language that one is dealing with is inherently ambiguous, it is *not* possible to eliminate ambiguity in all cases, such as we did by rewriting grammar G_7 to grammar G_8.

> Notice that the terminology is not *inherently ambiguous grammar* but *inherently ambiguous language* - what we are saying is *every grammar* is ambiguous for *certain CFLs*.

An example of an inherently ambiguous language is

$$\{0^i 1^j 2^k \mid i, j, k \geq 0 \ \wedge \ i = j \ \vee \ j = k\}.$$

The intuition is that every grammar for this language must have productions geared towards matching 0s against 1s and 1s against 2s. In this case, given a string of the form $0^k 1^k 2^k$, *either* of these options can be exercised. A formal proof may be found in advanced papers in this area, such as [82].

13.4 A Taxonomy of Formal Languages and Machines

Machines	Languages	Nature of grammar
DFA/NFA	Regular	Left-linear or Right-linear productions
DPDA	Deterministic CFL	Each LHS has one non-terminal The productions are deterministic
NPDA (or "PDA")	CFL	Each LHS has only one non-terminal
LBA	Context Sensitive Languages	LHS may have length > 1, but $\mid$ LHS$\mid \leq \mid$RHS$\mid$, ignoring ε productions
DTM/NDTM	Recursively Enumerable	General grammars ($\mid$LHS$\mid \geq \mid$RHS$\mid$ allowed)

Fig. 13.5. The Chomsky hierarchy and allied notions

We now summarize the formal machines, as well as languages, studied in this book in the table given in Figure 13.5. This is known as the *Chomsky hierarchy* of formal languages. For each machine, we describe the nature of its languages, and indicate the nature of the grammar used to describe the languages. It is interesting that simply by varying the nature of production rules, we can obtain *all* members of the Chomsky hierarchy. This single table, in a sense, summarizes some of the main achievements of over 50 years of research in computability, machines, automata, and grammars. Here is a summary of the salient points made in this table, listed against each of the language classes:[2]

Regular languages:

DFAs and NFAs serve as machines that recognize regular languages. Context-free grammars written with only left-linear or only right-linear productions can generate or recognize regular languages. The linearity

---

[2] We prefer to highlight the language classes as they constitute the more abstract concept, while machines and grammars are two different syntactic devices that denote languages.

of the production rules means that there is only *one* non-terminal allowed on the RHS of each production, which appears leftmost or rightmost. Hence, these non-terminals can be regarded as states of an NFA, as discussed in Section 13.6.

Deterministic context-free languages (DCFL):

Push-down automata are machines that recognize DCFLs, as illustrated in Illustration 13.4.2. In effect, they can parse sentences in the language without backtracking (deterministically). As for grammars, the fact that each context-free production specifies the expansion of *one* and only one non-terminal on the left-hand side means that this expansion is good *wherever the non-terminal appears—i.e.*, regardless of the context (hence "context-free"). The grammars are *deterministic*, as illustrated in Illustration 13.4.2.

Context-free languages (CFL):

These are more general than DCFLs, as the constraint of determinism is removed in the underlying machines and grammars.

Context-sensitive languages (CSL):

CSLs can be recognized by *linear bounded automata* which are described in Section 15.2.3. Basically, they are restricted Turing machines which can write only on that portion of the input tape on which the input was originally laid out. In particular, given any LBA M and a string w, it can be conclusively answered as to whether M accepts w or not. This is impossible with a Turing machine.

As for grammars, CSLs are recognized by productions in which the length of a left-hand side is allowed to be more than 1. Such a *context-sensitive* production specifies a *pattern* on the LHS, and a sentential form on the RHS. In a sense, we can have a rule of the form a A d -> a a c d and another of the form a A e -> a c a d. Notice that A's expansion when surrounded by a and d can be different from when surrounded by a and e, thus building in context sensitivity to the interpretation of A. The length of the RHS is required to be no less than that of the LHS (except in the ε case) to ensure decidability in some cases.

Recursively enumerable or Turing recognizable (RE or TR) languages:

These form the most general language class in the Chomsky hierarchy. Notice that Turing machines as well as unrestricted productions, form the machines and grammars for this language class.

CFGs and CFLs are fundamental to computer science because they help describe the structure underlying programming languages. The basic "signature" of a CFL is "nested brackets:" for example, nesting occurs in expressions and in very many statically scoped structures in computer programs. In contrast, the basic signature of regular languages is "iteration (looping) according to some ultimate periodicity."

Illustration 13.4.1 Let us argue that the programming language C is not regular. Let there be a DFA for C with n states. Now consider the C program

$$C_{NOP} = \{main()\{^n\}^n \mid n \geq 0\}.$$

Clearly, the DFA for C will loop in the part described by $main()\{^n$, and by pumping this region wherever the loop might fall, we will obtain a malformed C program. Some of the pumps could, for instance, result in the C program $maiaiain()\{\ldots$, while some others result in strings of the form $main\{\{\{\}\}$, etc.

Using a CFG, we can describe C_{NOP} using production rules, as follows:

```
L_C_nop -> main Paren Braces
Paren -> ()
Braces -> epsilon | { Braces }.
```

13.4.1 Non-closure of CFLs under complementation

It may come as a surprise that most programming languages are *not* context-free! For instance, in C, we can declare function *prototypes* that can introduce an arbitrary number of arguments. Later, when the function is defined, the same arguments must appear in the same order. The structure in such "define/use" structures can be captured by the language

$$L_{ww} = \{ww \mid w \in \{0,1\}^*\}.$$

As we shall sketch in Section 13.8 (Illustration 13.8.1), this language is *not* context-free. It is a context-sensitive language which can be accepted by a *linear bounded automaton* (LBA). Basically, an LBA has a tape, and can sweep across the tape as many times as it wants, writing "marking bits" to compare across arbitrary reaches of the region of the tape where the original input was presented. This mechanism can easily spot a w and a later w appearing on the tape. The use of *symbol tables* in compilers essentially gives it the power of LBAs, making compilers able to handle C prototype definitions.

While L_{ww} is not context-free, its complement, $\overline{L_{ww}}$, is indeed a CFL. This means that CFLs are *not* closed under complementation!

$\overline{L_{ww}}$ is generated by the following grammar $G_{\overline{ww}}$:

Grammar $G_{\overline{ww}}$:
```
S -> AB | BA | A | B
A -> CAC | 0
B -> CBC | 1
C -> 0 | 1.
```

Illustration 13.4.2 For each language below, write

- R if the language is regular,
- $DCFL$ if the language is deterministic context-free (can be recognized by a DPDA),
- CFL if it can be recognized by a PDA **but not** a DPDA,
- IA if the language is CFL but is inherently ambiguous, and
- N if not a CFL.

Also provide a one-sentence justification for your answer. *Note:* In some cases, the language is described using the set construction, while in other cases, the language is described via a grammar ("L(G)").

1. $\{x \mid x \text{ is a prefix of } w \text{ for } w \in \{0,1\}^*\}$.
   Solution: This is R, because the language is nothing but $\{0,1\}^*$.
2. L(G) where G is the CFG $\quad S \to 0\,S\,0 \mid 1\,S\,1 \mid \varepsilon$.
   Solution: CFL, because nondeterminism *is required* in order to guess the midpoint.
3. $\{a^n b^m c^n d^m \mid m, n \geq 0\}$.
   Solution: The classification is N, because comparison using a single stack is not possible. If we push a^n followed by b^m, it is no longer possible to compare c^n against a^n, as the top of the stack contains b^m. Removing b^m "temporarily" and restoring it later isn't possible, as it is impossible to store away b^m in the *finite-state* control.
4. $\{a^n b^n \mid n \geq 0\}$.
   Solution: $DCFL$, since we can deterministically switch to matching b's.
5. $\{a^i b^j c^k \mid i, j, k \geq 0 \text{ and } i = j \text{ or } j = k\}$.
   Solution: IA, because for $i = j = k$, we can have two distinct parses, one comparing a's and b's, and the other comparing b's and c's (the capability for these two comparisons must exist in any grammar, because of the "or" condition).

Illustration 13.4.3 Indicate which of the following statements pertaining to closure properties is true and which is false. For every true assertion below, provide a one-sentence supportive answer. For every false assertion below, provide a counterexample.

1. The union of a CFL and a regular language is a CFL.
   Solution: True, since regular languages are also CFLs. Write the top-level production of the new CFL as S -> A | B where A generates the given CFL and B generates the given regular language.
2. The intersection of any two CFLs is always a CFL.
   Solution: False. Consider $\{a^m b^m c^n \mid m, n \geq 0\} \cap \{a^m b^n c^n \mid m, n \geq 0\}$. This is $\{a^n b^n c^n \mid n \geq 0\}$, which is not a CFL.
3. The complement of a CFL is always a CFL.
   Solution: False. Consider $L_{ww} = \{ww \mid w \in \{0,1\}^*\}$ which was discussed on page 230. Try to come up with another example yourself.

Illustration 13.4.4 Describe the CFL generated by the following grammar using a regular expression. Show how you handled each of the non-terminals.
$$S \rightarrow TT$$
$$T \rightarrow UT \mid U$$
$$U \rightarrow 0U \mid 1U \mid \varepsilon.$$

It is easy to see that U generates a language represented by regular expression $(0 + 1)^*$, while T generates U^+. Note that for any regular expression R, it is the case that $(R^*)^+$ is $R^* \cup R^* R^* \cup R^* R^* R^* \ldots$ which is R^*. Therefore, T generates $(0 + 1)^*$. Now, S generates TT, or $(0+1)^*(0+1)^*$, which is the same as $(0+1)^*$. Therefore, $L(S) = \{0, 1\}^*$.

13.4.2 Simplifying CFGs

We illustrate a technique to simplify grammars through an example.

Illustration 13.4.5 Simplify the following grammar, explaining why each production or non-terminal was eliminated:

$$S \rightarrow A B \mid D$$
$$A \rightarrow 0 A \mid 1 B \mid C$$
$$B \rightarrow 2 \mid 3 \mid A$$
$$D \rightarrow A C \mid B D$$

$E \rightarrow 0$.

Solution: Grammars are simplified as follows. First, we determine which non-terminals are *generating* - have a derivation sequence to a terminal string (if a non-terminal is non-generating, the language denoted by it is $\emptyset$, and we can safely eliminate all such non-terminals, as well as, recursively, all other non-terminals that use them). We can observe that in our example, B is generating. Therefore, A is generating. C is an undefined non-terminal, and so we can eliminate it. Now, we observe that S is generating, since AB is generating; so we had better retain S (!). D is reachable ('reachable' means that it appears in at least one derivation sequence starting at S) but non-generating, so we can eliminate D. Finally, E is *not* reachable from S through any derivation path, and hence we can eliminate it, all productions using it (none in our example) and all productions expanding E (exactly one in our example). Therefore, we obtain the following simplified CFG:

$$S \rightarrow A\,B$$
$$A \rightarrow 0\,A \ \mid\ 1\,B$$
$$B \rightarrow 2 \ \mid\ 3 \ \mid\ A.$$

Here, then, are sound steps one may employ to simplify a given CFG (it is assumed that the productions are represented in terms of *elementary productions* in which the disjunctions separated by | on the RHS of a production rule are expressed in terms of separate productions for the same LHS):

- A non-generating non-terminal is useless, and it can be eliminated.
- A non-terminal for which there is no rule defined (does not appear on the left-hand side of any rule) is non-generating in a trivial sense.
- The property of being non-generating is 'infectious' in the following sense: if non-terminal N_1 is non-generating, and if N_1 appears in every derivation sequence of another non-terminal N_2, then N_2 is also non-generating.
- A non-terminal that does not appear in any derivation sequence starting from S is unreachable.
- Any CFG production rule that contains either a non-generating or an unreachable non-terminal can be eliminated.

Illustration 13.4.6 Simplify the following grammar, clearly showing how each simplification was achieved (name criteria such as 'generating' and 'reaching'):

```
S -> A B | C D
A -> 0 A | 1 B
B -> 2   | 3
D -> A C | B D E
E -> 4 E | D | 5.
```

B is generating. Hence, A is generating. S is generating. B, A, and S are reachable. Hence, S, A, and B are essential to preserve, and therefore C and D are reachable; however, C is not generating. Hence, production CD is useless. Hence, we are left with:

```
S -> A B
A -> 0 A | 1 B
B -> 2 | 3.
```

$\square$

We now examine *push-down automata* which are machines that recognize CFLs, and bring out some connections between PDAs and CFLs.

13.5 Push-down Automata

A push-down automaton (PDA) is a structure $(Q, \Sigma, \Gamma, \delta, q_0, z_0, F)$ where Q is a finite set of states, Σ is the input alphabet, Γ is the stack alphabet (that usually includes the input alphabet Σ), q_0 is the initial state, $F \subseteq Q$ is the set of accepting states, z_0 the initial stack symbol, and

$$\delta : Q \times (\Sigma \cup \{\varepsilon\}) \times \Gamma \to 2^{Q \times \Gamma^*}.$$

In each move, a PDA can *optionally* read an input symbol. However, in each move, it *must* read the top of the stack (later, we will see that this assumption comes in handy when we have to convert a PDA to a CFG). Since we will always talk about an NPDA by default, the δ function returns a set of nondeterministic options. Each option is a next-state to go to, and a stack string to push on the stack, with the first symbol of the string appearing on top of the stack after the push is over. For example, if $\langle q_1, ba \rangle \in \delta(q_0, x, a)$, the move can occur when x can be read from the input and the stack top is a. In this case, the PDA moves over x (it cannot read x again). Also, an a is removed from the stack. However, as a result of the move, an a is promptly pushed back on the stack, and is *followed by* a push of b, with the machine going to state q_1. The transition function δ of a PDA may be either *deterministic* or nondeterministic.

13.5.1 DPDA versus NPDA

A push-down automaton can be deterministic or nondeterministic. DPDA and NPDA are *not equivalent* in power; the latter are strictly more powerful than the former. Also, notice that unlike with a DFA, a deterministic PDA can move on ε. Therefore, the exact specification of what *deterministic* means becomes complicated. We summarize the definition from [60]. A PDA is deterministic if and only if the following conditions are met:

1. $\delta(q, a, X)$ has at most one member for any q in Q, a in Σ_ε, and X in Γ.
2. If $\delta(q, a, X)$ is non-empty, for some a in Σ, then $\delta(q, \varepsilon, X)$ must be empty.

In this book, I will refrain from giving a technically precise definition of DPDAs. It really becomes far more involved than we wish to emphasize in this chapter, at least. For instance, with a DPDA, it becomes necessary to know when the string ends, thus requiring a *right end-marker* ⊢. For details, please see [71, page 176].

13.5.2 Deterministic context-free languages (DCFL)

Current State	Input	Stack top	String pushed	New State	Comments
q0	0	z0	0 z0	q1	0. Have to push on this one
q0	1	z0	1 z0	q1	...or this one
q1	0	0	0 0	q1	1a.Assume not at midpoint
q1	0	1	0 1	q1	Have to push on this one
q1	0	0	ε	q1	1b. Assume at midpoint
q1	1	1	1 1	q1	2a. Assume not at midpoint
q1	1	0	1 0	q1	Have to push on this one
q1	1	1	ε	q1	2b. Assume at midpoint
q1	ε	z0	z0	q2	3. Matched around midpoint

Fig. 13.6. A PDA for the language $L_0 = \{ww^R \mid w \in \{0,1\}^*\}$

A deterministic context-free language (DCFL) is one for which there is a DPDA that accepts the same language. Consider the language

Language $L_0 = \{ww^R \mid w \in \{0,1\}^*\}$.

L_0 is not a DCFL because in any PDA, the use of nondeterminism is essential to "guess" the midpoint. Figure 13.6 presents the δ function of a PDA designed to recognize L_0. This PDA is described by the structure

$$P_{L_0} = (\{q_0, \}, \{0, 1\}, \{0, 1, z_0\}, \delta, q_0, z_0, \{q_0, q_2\}).$$

This PDA begins by stacking 0 or 1, depending on what comes first. The comments **1a** and **1b** describe the nondeterministic selection of assuming *not* being at a midpoint, and being a midpoint, respectively. A similar logic is followed in **2a** and **2b** as well. Chapter 14 describes PDA construction in greater detail.

Let us further our intuitions about PDAs by considering a few languages:

L_1: $\{a^i b^j c^k \mid if\ i = 1\ then\ j = k\}$.

L_2: $\{a^i b^j c^k d^m \mid i, j, k, m \geq 0 \wedge if\ i = 1\ then\ j = k\ else\ k = m\}$

L_3: $\{ww \mid w \in \{0, 1\}^*\}$.

L_4: $\{0, 1\}^* \setminus \{ww \mid w \in \{0, 1\}^*\}$.

L_5: $\{a^i b^j c^k \mid i = j\ or\ i = k\}$.

L_6: $\{a^n b^n c^n \mid n \geq 0\}$.

L_1 is a DCFL, because after seeing whether $i = 1$ or not, a deterministic algorithm can be employed to process the rest of the input. A DPDA can be designed for $reverse(L_1)$ also. Likewise, a DPDA can be designed for L_2. However, as discussed in Section 8.1.4, $reverse(L_2)$ is not a DCFL, as it is impossible to keep *both* decisions – whether $j = k$ or $k = m$ – ready by the time i is encountered. L_3 is not a CFL at all. However, L_4, the complement of L_3, is a CFL. L_5 is a CFL (but not a DCFL) – the guesses of $i = j$ or $i = k$ can be made nondeterministically. Finally, L_6 is not a CFL, as we cannot keep track of the length of three distinct strings using one stack.   □

13.5.3 Some Factoids

Here are a few more factoids that tie together ambiguity (of grammars) and determinism (of PDA):

- If one can obtain a DPDA for a language, then that language is not inherently ambiguous. This is because for an inherently ambiguous language, *every* CFG admits two parses, thus meaning that there cannot be a DPDA for it.
- There are CFLs that are not DCFLs (have no DPDA), and yet they have non-ambiguous grammars. The grammar

```
S -> 0 S 0 | 1 S 1 | e
```

is non-ambiguous, and yet denotes a language that is not a DCFL. In other words, this CFG generates all the strings of the form ww,[R] and these strings have only one parse tree. However, since the mid-point of such strings isn't obvious during a left-to-right scan, a nondeterministic PDA is necessary to parse such strings.

13.6 Right- and Left-Linear CFGs

A right-linear CFG is one where every production rule has exactly one non-terminal and that it also appears rightmost. For example, the following grammar is right-linear:

```
S -> 0 A | 1 B | e
A -> 1 C | 0
B -> 0 C | 1
C -> 1 | 0 C.
```

Recall that S -> 0 A | 1 B | e is actually *three* different production rules S -> 0 A, S -> 1 B, and S -> e, where each rule is right-linear. This grammar can easily be represented by the following NFA obtained almost directly from the grammar:

```
IS - 0 -> A
IS - 1 -> B
IS - e -> F1
A  - 1 -> C
A  - 0 -> F2
B  - 0 -> C
B  - 1 -> F3
C  - 0 -> C
C  - 1 -> F4.
```

A left-linear grammar is defined similar to a right-linear one. An example is as follows:

```
S -> A 0 | B 1 | e
A -> C 1 | 0
B -> C 1 | 1
C -> 1 | C 0.
```

A purely left-linear or a purely right-linear CFG denotes a regular language. However, the converse is not true; that is, if a language is regular, it does not mean that it has to be generated by a purely left-linear or purely right-linear CFG. Even non-linear CFGs are perfectly capable of sometimes generating regular sets, as in

```
S -> T T | e
T -> 0 T | 0.
```

It also must be borne in mind that we cannot "mix up" left- and right-linear productions and expect to obtain a regular language. Consider the productions

```
S -> 0 T  | e
T -> S 1.
```

In this grammar, the productions are linear - left or right. However, since we use *left- and right-linear rules*, the net effect is as if we defined the grammar

```
S -> 0 S 1  | e
```

which generates the non-regular context-free language

$$\{0^n 1^n \mid n \geq 0\}.$$

Conversion of purely left-linear grammars to NFA

Converting a left-linear grammar to an NFA is less straightforward. We first *reverse* the language it represents by reversing the grammar. Grammar reversal is approached as follows: given a production rule

$$S \rightarrow R_1 R_2 \ldots R_n,$$

we obtain a production rule for the reverse of the language represented by S by reversing the production rule to:

$$S^r \rightarrow R_n^r R_{n-1}^r \ldots R_1^r.$$

Applying this to the grammar at hand, we obtain

```
Sr -> 0 Ar | 1 Br | e
Ar -> 1 Cr | 0
Br -> 1 Cr | 1
Cr -> 1 | 0 Cr.
```

Once an NFA for this right-linear grammar is built, it can be reversed to obtain the desired NFA.

13.7 Developing CFGs

Developing CFGs is much like programming; there are no hard-and-fast rules. Here are reasonably general rules of the thumb for arriving at CFGs:

1. *(Use common idioms):* Study and remember many common patterns of CFGs and use what seems to fit in a given context. Example: To get the effect of matched brackets, the common idiom is

   ```
   S -> ( S ) | e.
   ```

2. *Break the problem into simpler problems:*
   Example: $\{a^m b^n \mid m \neq n, \ m, n \geq 0\}$.

   a) So, a's and b's must still come in order.
   b) Their numbers shouldn't match up.
      i. Formulate matched up a's and b's
         ```
         M -> e | a M b
         ```

      ii. Break the match by adding either more A's or more B's
         ```
         S -> A M | M B
         A -> a | a A
         B -> b | b B
         ```

13.8 A Pumping Lemma for CFLs

Consider any CFG $G = (N, \Sigma, P, S)$. A Pumping Lemma for the language of this grammar, $L(G)$, can be derived by noting that a "very long string" $w \in L(G)$ requires a very long derivation sequence to derive it from S. Since we only have a finite number of non-terminals, some non-terminal must repeat in this derivation sequence, and furthermore, the *second* occurrence of the non-terminal must be a result of expanding the first occurrence (it must lie within the parse tree generated by the first occurrence).

For example, consider the CFG

```
S -> ( S ) | T | e

T -> [ T ] | T T | e.
```

Here is an example derivation:

```
S => ( S ) => (( T )) => (( [ T ] )) => (( [ ] ))
                ^                ^
```

```
        Occurrence-1    Occurrence-2
```

```
Occurrence-1 involves Derivation-1: T => [ T ] => [ ]
Occurrence-2 involves Derivation-2: T => e
```

Here, the second T arises because we took T and expanded it into
[T] and then to [ ]. Now, the basic idea is that we can use
Derivation-1 used in the first occurrence in place of Derivation-2, to
obtain a longer string:

```
S => (S) => ((T))  => (( [ T ] )) => (( [[ T ]] )) => (( [[ ]] ))
              ^              ^
```

```
        Occurrence-1 Use Derivation-1 here
```

In the same fashion, we can use Derivation-2 in place of Derivation-1
to obtain a shorter string, as well:

```
S => ( S ) => ( ( T ) ) => ( ( ) )
                    ^
```

```
        Use Derivation-2 here
```

When all this happens, we can find a repeating non-terminal that
can be pumped up or down. In our present example, it is clear that
we can manifest $((([^i \ ]^i))$ for $i \geq 0$ by either applying Derivation-2
directly, or by applying some number of Derivation-1s followed by
Derivation-2. In order to conveniently capture the conditions mentioned so far, it is good to resort to parse trees. Consider a CFG with
$|V|$ non-terminals, and with the right-hand side of each rule containing
at most b syntactic elements (terminals or non-terminals). Consider a
b-ary tree built up to height $|V|+1$, as shown in Figure 13.7. The string
yielded on the frontier of the tree $w = uvxyz$. If there are two such parse
trees for w, pick the one that has the fewest number of nodes. Now,
if we avoid having the same non-terminal used in any path from the
root to a leaf, basically each path will "enjoy" a growth up to height at
most $|V|$ (recall that the leaves are terminals). The string $w = uvxyz$
is, in this case, of length at most $b^{|V|}$. This implies that *if we force
the string to be of length* $b^{|V|+1}$ (called p hereafter), a parse tree for
this string will have some path that repeats a non-terminal. Call the
higher occurrence V_1 and the lower occurrence (contained within V_1)
V_2. Pick the lowest two such repeating pair of non-terminals. Now, we
have these facts:

- $|vxy| \leq p$; if not, we would find two other non-terminals that exist
  lower in the parse tree than V_1 and V_2, thus violating the fact that
  V_1 and V_2 are the lowest two such.

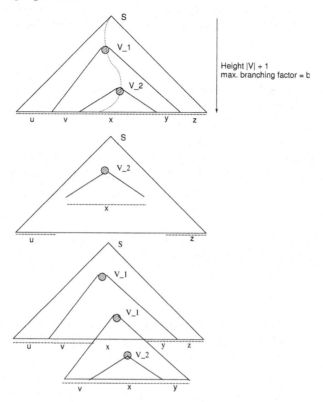

Fig. 13.7. Depiction of a parse tree for the CFL Pumping Lemma. The upper drawing shows a very long path that repeats a non-terminal, with the lowest two repetitions occurring at V_2 and V_1 (similar to Occurrence-1 and Occurrence-2 as in the text). With respect to this drawing: (i) the middle drawing indicates what happens if the derivation for V_2 is applied in lieu of that of V_1, and (ii) the bottom drawing depicts what happens if the derivation for V_2 is replaced by that for V_1, which, in turn, contains a derivation for V_2

- $|vx| \geq 1$; if not, we will in fact have $w = uxz$, for which a shorter parse tree exists (namely, the one where we directly employ V_2).
- Now, by pumping, we can obtain the desired repetitions of v and y, as described in Theorem 13.1.

Theorem 13.1. *Given any CFG $G = (N, \Sigma, P, S)$, there exists a number p such that given a string w in $L(G)$ such that $|w| \geq p$, we can split w into $w = uvxyz$ such that $|vy| > 0$, $|vxy| \leq p$, and for every $i \geq 0$, $uv^i xy^i z \in L(G)$.*

We can apply this Pumping Lemma for CFGs in the same manner as we did for regular sets. For example, let us sketch that L_{ww} of page 230 is not context-free.

Illustration 13.8.1 Suppose L_{ww} were a CFL. Then the CFL Pumping Lemma would apply. Let p be the pumping length associated with a CFG of this language. Consider the string $0^p1^p0^p1^p$ which is in L_{ww}. The segments v and y of the Pumping Lemma are contained within the first 0^p1^p block, in the middle 1^p0^p block or in the last 0^p1^p block, and in each of these cases, it could also have fallen entirely within a 0^p block or a 1^p block. By pumping up or down, we will then obtain a string that is not within L_{ww}.   □

Exercise 13.13 demonstrates another "unusual" application of the CFG Pumping Lemma.

Chapter Summary

This chapter discussed the notion of context-free grammars and context-free languages. We emphasized 'getting a grammar right' by showing that it has two facets—namely *consistency* and *completeness*. Fixed-point theory helps appreciate context-free grammars in terms of recursive equations whose least fixed-point is the "desired" context-free language. We discussed ambiguity and disambiguation—two topics that compiler writers deeply care about. After discussing the Chomsky hierarchy, we discuss the topics of closure properties (or lack thereof under intersection and complementation). We present how CFGs may be simplified. We then move on to push-down automata, which are machines with a finite control and *one* stack. We discuss the fact that NPDAs and DPDAs are not equivalent. We close off with a discussion of an *incomplete* Pumping Lemma for CFLs. Curiously, there is also a complete Pumping Lemma for CFLs ("strong Pumping Lemma" [124]). We do not discuss this lemma (it occupies nearly one page even when stated in a formal mathematical notation).

Exercises

13.1. Draw the parse tree for string

$$a\,a\,a\,b\,b\,a\,b\,b\,b\,b\,b\,b\,a\,a\,b\,a\,b\,a\,a\,a$$

with respect to grammar G_5, thus showing that this string can be derived according to the grammar.

13.2.

1. Parenthesize the following expression according to the rules of standard precedence for arithmetic operators, given that $\sim$ stands for unary minus:

$$\sim 1 * 2 - 3 - 4 / \sim 5.$$

2. Convert the above expression to Reverse Polish Notation (post-fix form).

13.3. Prove by induction that the following grammar generates only strings with an odd number of 1s. Clearly argue the basis case(s), the inductive case(s), and what you prove regarding T and regarding S.

$$S \rightarrow S\,T\,0 \mid 0\,1$$
$$T \rightarrow 1\,1\,T \mid \epsilon$$

13.4. Write the consistency proof pertaining to G_9 in full detail. Then write a proof for the completeness of the above grammar (that it generates *all* well-parenthesized strings).

13.5. Which other solution to the language equation of $L(S_6)$ of page 225 exists?

13.6. *Prove* that $G_{\overline{ww}}$ of Page 231 is a CFG for the language $\overline{L_{ww}}$. *Hint:* The productions S -> A and S -> B generate odd-length strings. Also, S -> AB and S -> BA generate all strings that are *not* of the form ww. This is achieved by generating an even-length string pq where $|p| = |q|$ and if p is put "on top of" q, there will be at least one spot where they both differ.

13.7. Argue that a DPDA satisfying the definition in Section 13.5.1 cannot be designed for the language $\{ww^R \mid w \in \Sigma^*\}$.

13.8. (Adapted from Sipser [111]) Determine whether the context-free language described by the following grammar is regular, showing all the reasoning steps:

```
S -> T T | U
T -> 0 T | T 0 | #
U -> 0 U 0 0 | #.
```

13.9. Answer whether true or false:

1. *There are more regular languages (RLs) than CFLs.*
2. *Every RL is also a CFL.*

3. *Every CFL is also a RL.*
4. *Every CFL has a regular sublanguage* ("sublanguage" means the same as "subset").
5. *Every RL has a CF sublanguage.*

13.10.
1. Obtain one CFG G_1 that employs left-linear and right-linear productions (that cannot be eliminated from G_1) such that $L(G_1)$ is regular.
2. Now obtain another grammar G_2 where $L(G_2)$ is non-regular but is a DCFL.
3. Finally, obtain a G_3 where $L(G_3)$ is not a DCFL.

13.11. Using the Pumping Lemma, show that the language $\{0^n 1^n 2^n \mid n \geq 0\}$ is not context-free.

13.12. Show using the Pumping lemma that the language $\{ww \mid w \in \{0,1\}^*\}$ is not context-free.

13.13. Prove that any CFG with $|\Sigma| = 1$ generates a regular set. *Hint:* use the Pumping Lemma for CFLs together with the ultimate periodicity result for regular sets. Carefully argue and conclude using the PL for CFLs that we are able to generate only periodic sets.

13.14. Argue that the *syntax* of regular expressions is *context-free* while the syntax of *context-free grammars* is *regular*!

14

Push-down Automata and Context-free Grammars

This chapter details the design of push-down automata (PDA) for various languages, the conversion of CFGs to PDAs, and vice versa. In particular, after formally introducing push-down automata in Section 14.1, we introduce two notions of acceptance - by final state and by empty stack - in Sections 14.1.2 and 14.1.3, respectively. In Section 14.2, we show how to prove PDAs correct using the *Inductive Assertions* method of Floyd. We then present an algorithm to convert a CFG to a language-equivalent PDA in Section 14.3, and an algorithm to convert a PDA to a language-equivalent CFG in Section 14.4. This latter algorithm is non-trivial - and so we work out an example entirely, and also show how to simplify the resulting CFG *and* prove it correct. In Section 14.5, we briefly discuss a *normal form* for context-free grammars called the *Chomsky normal form*. We do not discuss other normal forms such as the *Greibach normal form*, which may be found in most other textbooks. We then describe the Cocke-Kasami-Younger (CKY) parsing algorithm for a grammar in the Chomsky normal form. Finally, we briefly discuss closure and decidability properties in Section 14.6.

14.1 Push-down Automata

A push-down automaton (PDA) is a structure $(Q, \Sigma, \Gamma, \delta, q_0, z_0, F)$ where Γ is the stack alphabet (that usually includes the input alphabet Σ), z_0 is the initial stack symbol, and $\delta : Q \times (\Sigma \cup \{\varepsilon\}) \times \Gamma \rightarrow 2^{Q \times \Gamma^*}$ is the transition function that takes a state, an input symbol (or ε), and a stack symbol (or ε) to a set of states and stack contents. In particular, the $2^{Q \times \Gamma^*}$ in the range of the signature indicates that the PDA can nondeterministically assume one of many states and stack contents. Also, as the signature of the δ function points out, in each move, a PDA may or may not read an input symbol (note the ε in the signature), but must read the top of the stack in *every* move (note the absence of a ε associated with Γ).

We must point out that many variations on the above signature are possible. In [111] and in the JFLAP tool [66], for instance, PDAs may also optionally read the top of the stack (in effect, they employ the signature $\delta : Q \times (\Sigma \cup \{\varepsilon\}) \times (\Gamma \cup \{\varepsilon\}) \rightarrow 2^{Q \times \Gamma^*}$). Such variations do not fundamentally change the "power" of PDAs. We adopted our convention—of always reading and popping the stack during every move—because it yields an intuitively clearer algorithm for converting PDAs to CFGs[1] (following [60]).

Notions of Acceptance:

There are two different notions of acceptance of a string by a PDA. According to the first, a PDA accepts a string when, after reading the entire string, the PDA is in a final state. According to the second, a PDA accepts a string when, after reading the entire string, the PDA has emptied its stack. We define these notions in Sections 14.1.2 and 14.1.3. In both these definitions, we employ the notions of *instantaneous descriptions* (ID), and step relations $\vdash$, as well as its reflexive and transitive closure, $\vdash^*$.

Instantaneous Description:

An *instantaneous description* (ID) for a PDA is a triple of the form

  (state, unconsumed input, stack contents)

Formally, the type of the instantaneous description of a PDA is $T_{ID} = Q \times \Sigma^* \times \Gamma^*$. The type of $\vdash$ is $\vdash \subseteq T_{ID} \times T_{ID}$. The $\vdash$ relation is as follows:

$$(q, a\sigma, b\gamma) \vdash (p, \sigma, g\gamma))\rangle \text{ iff}$$
$$a \in \Sigma_\varepsilon \wedge b \in \Gamma \wedge g \in \Gamma^* \wedge \exists (p, g) \in \delta(q, a, b).$$

In other words, if δ allows a move from state q and stack top b to state p via input $a \in \Sigma \cup \{\varepsilon\}$, then $\vdash$ does allow that. In this process, the stack top b is popped, and the new stack contents described by g is pushed on. The first symbol of g ends up at the top of the stack. Sometimes, the last symbol of g is set to b, thus helping restore b (that was popped). In some cases, g is actually made equal to b, thus modeling the fact that the stack did not suffer any changes.

---

[1] In fact, a PDA move that optionally reads the top of the stack may be represented by a PDA move that reads whatever is on top of the stack, but pushes that symbol back.

14.1.1 Conventions for describing PDAs

We prefer to draw tables of PDA moves. Please make the tables detailed. *Do write comments* - after all, you are coding in a pretty low-level language that is *highly error-prone*; therefore, the more details you provide, the better it is for readers to follow your work. A diagram will also be highly desirable, as is included in Figure 14.1.

In Section 14.2, we present a method to *formally prove* the correctness of PDAs using the *inductive assertions* method of Floyd [40]. This technique should convince the reader that arguing the correctness of a PDA is akin to verifying a program; both are activities that can be rendered difficult if comments and clear intuitive explanations are not provided.

The diagramming style we employ for PDAs resembles state diagrams used for NFAs and DFAs, the only difference being that we now annotate moves by $insymb, ssymb \to sstr$ where

- $insymb$ is an input symbol or ε,
- $ssymb$ is a stack symbol, and
- $sstr$ is a string of stack symbols that is pushed onto the stack when the move is executed.

Also, recall that PDAs don't need to specify a behavior for every possible $insymb/ssymb$ combination at every state. If an unspecified combination occurs, the next state of the PDA is undefined. In effect, PDAs are partial functions from inputs and stack contents to new stack contents and new states.

As said earlier, a PDA accepts an input if the input leads it to one of the final states. There is one important difference between DFAs and *D*PDAs: the latter may have undefined input/stack combinations. In other words, one does not have to fully decode inputs and transition to "*black hole*" states upon arrival of illegal inputs, as with a DFA. Finally, recall the difference between NPDAs and DPDAs pointed out in Section 13.5.1. Now we define the different notions of acceptance of PDAs in more detail.

14.1.2 Acceptance by final state

A PDA **accepts a string** w **by final state** if and only if, for some $q_f \in F$, the final set of states of the PDA, $(q_0, w, z0) \vdash^* (q_f, \varepsilon, g)$. For any given PDA, our default assumption will be that of acceptance by final state. The language of the PDA will be defined as follows:

$$\{w \mid \exists g \cdot (q_0, w, z0) \vdash^* (q_f, \varepsilon, g) \; for \; q_f \in F\}.$$

Current State	Input	Stack top	String pushed	New State	Comments
q0	0	z0	0 z0	q1	0. Have to push on this one
q0	1	z0	1 z0	q1	...or this one
q1	0	0	0 0	q1	1a.Assume not at midpoint
q1	0	1	0 1	q1	Have to push on this one
q1	0	0	ε	q1	1b. Assume at midpoint
q1	1	1	1 1	q1	2a. Assume not at midpoint
q1	1	0	1 0	q1	Have to push on this one
q1	1	1	ε	q1	2b. Assume at midpoint
q1	ε	z0	z0	q2	3. Matched around midpoint

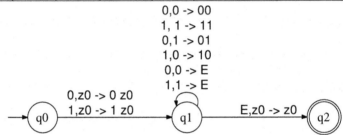

```
                                 0,0 -> 00
                                 1, 1 -> 11
                                 0,1 -> 01
                                 1,0 -> 10
                                 0,0 -> E
                                 1,1 -> E
                 0,z0 -> 0 z0
                 1,z0 -> 1 z0                E,z0 -> z0
    ->( q0 )---------------------->( q1 )---------------->(( q2 ))
```

```
        WINNER TOKEN                           LOSER TOKEN

            (q0,001100,    z0)                     (q0,001100,  z0)
  push  |- (q1, 01100,   0z0)         push  |- (q1, 01100, 0z0)
  push  |- (q1,  1100,  00z0)         pop   |- (q1,  1100,  z0)
  push  |- (q1,   100, 100z0)         stuck! |- can't accept
  pop   |- (q1,    00,  00z0)
  pop   |- (q1,     0,   0z0)
  pop   |- (q1,      ,    z0)
  accept |- (q2,      ,    z0)

        ACCEPT!                               REJECT!
```

Fig. 14.1. Transition table and transition graph of a PDA for the language $L_0 = \{ww^R \mid w \in \{0,1\}^*\}$, and an illustration of the $\vdash$ relation on input 001100

For a PDA P whose acceptance is defined by final states, we employ the notation "$L(P)$" to denote its language. In contrast, for a PDA P whose acceptance is defined by empty stack, discussed next in Section 14.1.3, we employ the notation "$N(P)$" to denote its language. These are, respectively, subsets of Σ^* that lead the PDA into a final state or cause its stack to be emptied.

14.1.3 Acceptance by empty stack

To further highlight PDAs that accept by empty stack, we leave out the F component from their seven-tuple presentation, thus obtaining the six-tuple $P_2 = (Q, \Sigma, \Gamma, \delta, q_0, z_0)$. For such PDAs, a string w is in its language exactly when the following is true:

$$(q_0, w, z0) \vdash^* (q, \varepsilon, \varepsilon).$$

Here, $q \in Q$, i.e., q is *any state*. All that matters is that the input is entirely consumed *and* an empty stack results in doing so.

Consider the PDA for language L_0 defined in Figure 13.6, reproduced in Figure 14.1 for convenience. This figure also shows how IDs evolve. In particular, nondeterminism is clearly shown by the fact that for the same input string, namely 001100, one token (called the "winner") can progress towards acceptance, while another token (called "loser") progresses towards demise. Each token also carries with it the PDA stack. As long as one course of forward progress through $\vdash$ exists, and leads to a final state (q2, in our present example), the given string is accepted. The other tokens "die out.[2]" Such animations are best observed using tools such as JFLAP [66]. In fact, JFLAP allows users to choose the acceptance criterion—through final states, through empty stack, or *both* (when a final state is reached on an empty stack[3]). JFLAP maintains a view of each token as it journeys through the labyrinth of a PDA transition diagram, therefore watching JFLAP animations is a good way to build intuitions about PDAs.

An arbitrarily given PDA may reach a final state without having emptied its stack. A given PDA may also have an empty stack in a state other than its final state. It is, however, possible to modify a given PDA so that it enters a final state or empties its stack only in a controlled

---

[2] Nondeterminism in PDAs is akin to the "fork" operation in operating systems such as Unix: an entire clone of the PDA, including its stack, are created at every nondeterministic choice point, and these clones—or tokens as we have been referring to them—either "win" or "lose."

[3] "...on an empty stomach?!"

manner. Specifically, Section 14.1.4 describes how to convert a PDA that accepts by final state into one that empties its stack exactly when in a final state, and Section 14.1.4 describes how to convert a PDA that accepts by empty stack into one that goes into a final state exactly when its stack is empty.

Start state = q00

Current State	Input	Stack top	String pushed	New State	Comments
q00	ε	z00	z0 z00	q0	Start stack with z00; add z0 here.
q0	ε	z0	z0	qS	q0 is a final state; so jump to qS
qS	ε	any	ε	qS	qS drains the stack regardless of what's on top of the stack.
q0	0	z0	0 z0	q1	1a. Decide to stack a 0
q0	1	z0	1 z0	q1	2a. Decide to stack a 0
q1	0	0	0 0	q1	1a'. Decide to stack a 0
q1	0	1	0 1	q1	Forced to stack
q1	0	0	ε	q1	1b. Decide to match
q1	1	1	1 1	q1	2a'. Decide to stack a 1
q1	1	0	1 0	q1	Forced to stack
q1	1	1	ε	q1	2b. Decide to match
q1	ε	z0	z0	q2	Prepare to drain the stack
q2	ε	z0	z0	qS	Jump to stack-draining state qS

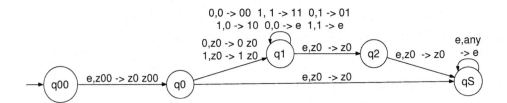

Fig. 14.2. The PDA of Figure 13.6 converted to one that accepts by empty stack. There are some redundancies in this PDA owing to our following a standard construction procedure.

14.1.4 Conversion of P_1 to P_2 ensuring $L(P_1) = N(P_2)$

Given a PDA P_1 that accepts by final state, we can obtain a PDA P_2 that accepts by empty stack such that $N(P_2) = L(P_1)$, simply by ensuring that P_2 has an empty stack exactly when P_1 reaches a final state (for the same input w seen by both these PDAs). The following construction achieves the above condition:

- To avoid the stack of P_2 becoming empty "in between," introduce an extra symbol in P_2's stack alphabet, say z_{00}.
- Start P_2 with its stack containing z_{00}, and then z_0 riding above it.
- The remaining moves of P_2 are similar to that of P_1. However, "final" states are insignificant for P_2. Therefore, whenever P_1 reaches a final state, we introduce in P_2, a move from it to a new *stack-draining state* qS. While in qS, P_2 empties its stack completely.
- No state other than qS tests for the stack-top being z_{00}. Hence, the stack is totally emptied, including z_{00}, only in state qS.

Figure 14.2 illustrates this construction.

14.1.5 Conversion of P_1 to P_2 ensuring $N(P_1) = L(P_2)$

Given a PDA that is defined according to the "accept by empty stack" criterion, how do we convert it to a PDA that accepts by final state? A simple observation tells us that the stack can become empty at any control state. Therefore, the trick is to start the PDA with a new bottom of stack symbol z_{00}. Under normal operation of the PDA, we do not see z_{00} on top of the stack, as it will be occluded by the "real" top of stack z_0. However, in any state, if z_{00} shows up on top of the stack, we add a transition to a newly introduced final state qF. qF is the only final state in the new PDA. Hence, whenever the former PDA drains its stack, the new PDA ends up in state qF.

Illustration 14.1.1 *Develop a push-down automaton for $L_{a^m b^n c^k}$ of Illustration 13.2.1.*

The PDA is shown in Figure 14.3. The PDA will first exercise a nondeterministic option: either I shall decide to match a's and b's, or do b's against c's. Recall that PDAs begin with z0 in the stack, and further we must pop one symbol from the stack in each step. Also, in each move, we can push zero, one, or more (a finite number) symbols back onto the stack.

Here are some facts about this PDA (based on intuitions - no proofs):

Initial state = Q0 Final states = Q0,Qc,Qd

Current State	Input	Stack top	String pushed	New State	Comments
Q0	ε	z0	z0	Qab	Nondeterministically proceed to match a's against b's
Q0	ε	z0	z0	Qbc	..or proceed to match b's against c's
Qab	a	z0	a z0	Qab	Stack the first 'a'
Qab	a	a	a a	Qab	Continue stacking a's
Qab	b	a	ε	Qb	The first b to come; match against an 'a'
Qb	b	a	ε	Qb	One more b came; perhaps more to come; so stay in Qb
Qb	ε	z0	z0	Qc	Go to Qc ("eat c" state), an accept state
Qc	c	z0	z0	Qc	Any number of c's are OK in Qc
Qab	ε	z0	z0	Qc	Enter the "eat c" accept state
Qbc	a	z0	z0	Qbc	Any number of a's can come. Qbc ignores them; it will match b's and c's
Qbc	b	z0	b z0	Qbc1	First b to come; no more a's allowed
Qbc	b	b	b b	Qbc1	Continue stacking b's; no no more a's
Qbc1	c	b	ε	Qm	Continue matching c's; no more b's allowed
Qm	c	b	ε	Qm	Continue matching c's; no more a's or b's
Qm	ε	z0	z0	Qd	A token goes to Qd whenever z0 is on top of the stack

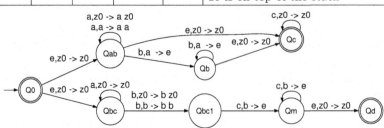

- Nondeterminism is essential. We do not know whether we are going to match a's and b's or b's and c's. In fact, for a string "abc," there must be two different paths that lead to some final state - hence, nondeterminism exists.
- This language is inherently ambiguous. For string "abc" it must be possible to build two distinct parse trees *no matter which grammar is used to parse it.*

14.2 Proving PDAs Correct Using Floyd's Inductive Assertions

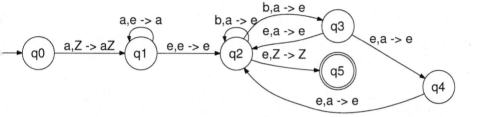

Fig. 14.4. A PDA whose language is being proved correct using Floyd's method

Consider the PDA in Figure 14.4. What is its language? *Think hard before you proceed reading!*   □

Guessing the language and proving its correctness

We guess the language of this PDA to be

$$\{a^{i_a} b^{i_b} \mid i_b \leq i_a \leq 3.i_b\}.$$

How do we prove this claim? We will use Floyd's method which rests on finding *loop invariants*. To simplify the discussion of the method a bit, we assume that the PDA has arrived into state q2, having stacked all the a's (this being the only way this PDA can proceed to accept anything). We seek a loop invariant (explained below) for the loop at state q2.

Let i_a be the number of a's initially in the input; likewise for i_b. Let s_a be the number of a's on the stack (note that b's don't get into the stack, ever). Let n_b be the number of b's yet to be read. Let p_a be the number of a's popped so far.

We explain Floyd's method with respect to a single loop (for more details, see [78]). With this assumption,

- Floyd's method works by pretending that we have arrested the program (PDA in this case) suddenly within its loop, at an arbitrary point during its execution.
- We are then asked to find an accurate description relating all "important" variables used in the loop. This is known as the *loop invariant*.
- The assertion and the variables participating in it must be sufficiently comprehensive so that when we bring the loop to its exit point, the final answer falls out as a special case of the loop invariant.

Considering all this, we come up with these equations:

1. $i_a = s_a + p_a$. This is because all the a's are stacked, and then some are popped, with the rest remaining in the stack.
2. $(i_b - n_b) \leq p_a \leq 3.(i_b - n_b)$. This is because for each 'a' that is popped, we match it against one to three b's. Therefore, the b's read thus far, namely $(i_b - n_b)$, are as per this equation.

Now, these must be *inductive assertions* as far as any q2 to q2 path is concerned. Let us check this:

- In any q2 to q2 traversal, the a's that are popped are the ones that are removed from the stack; hence, the first assertion is inductive.
- Consider the q2 to q3 to q2 traversal (the rest can be similarly argued - see Exercise 14.3). We have n_b going down by 1 while p_a goes up by 2. Thus we have to prove

$$(i_b - n_b) \leq p_a \leq 3.(i_b - n_b) \Rightarrow (i_b - n_b + 1) \leq p_a + 2 \leq 3.(i_b - n_b + 1),$$

which follows from simple arithmetic.

Now, specializing the invariant to the exit point, we observe that exiting occurs when $p_a = i_a$ and $n_b = 0$. This immediately gives us $i_b \leq i_a \leq 3.i_b$. □

14.3 Direct Conversion of CFGs to PDAs

When given the option of capturing a context-free language using a PDA or a CFG, what would one choose? In many cases, a CFG would be easier to first obtain; in that case, there exists a rather *elegant* direct conversion algorithm to convert that CFG into a PDA. This algorithm, in effect, is a *nondeterministic* parsing algorithm. The opposite conversion - a PDA to a CFG - is *much more involved* and is discussed in Section 14.4.

By determinizing the CFG to PDA conversion algorithm, we can obtain an exponential-time parsing algorithm for any CFG. Determinization can be achieved by arranging a *backtracking search*; whenever the NPDA is faced with a nondeterministic choice, we arrange a piece of code that recursively searches for *one* of the paths to accept.[4] In Section 14.5, we discuss the Chomsky Normal Form for a CFG, and in its context, discuss an $O(N^3)$ parsing algorithm attributed to Cocke, Kasami, and Younger (Section 14.5.1).

In the CFG to PDA conversion algorithm, the non-terminals and terminals of the given grammar constitute the stack alphabet of the PDA generated. In addition, the stack alphabet contains z_0. The conversion proceeds as follows:

- Start from state q_0 with z_0 on top of the stack.
- From q_0, jump to state q_M (for "main state") with S, the start symbol of the grammar, on top of the stack, and z_0 below it (restored in the jump).
- In state q_M,
  - If the top of the stack is z_0, jump to state q_F, the only accepting state.
  - If the top of the stack is the terminal x, jump back to state q_M upon input x, without restoring x on top of the stack. Essentially, the parsing goal of x has been fulfilled.
  - If the top of the stack is the non-terminal X, and there is a rule $X \rightarrow R$, where R is a string of terminals and non-terminals, jump to state q_M by popping X and pushing R. Essentially, the parsing goal of X is turned into zero or more parsing subgoals.

Illustration 14.3.1 Let us convert the CFG in Illustration 13.2.1 into a PDA. The resulting PDA is given in Figure 14.5. First set up S to be the *parsing goal*. The PDA can then take a nondeterministic jump to two different states. One state sets up the parsing goals M and C, with M on top of the stack. The other path sets up A and N.

Suppose parsing goal M is on top of the stack. We can then set up the parsing goal a M b, with a on top of the stack. Discharging the

---

[4] In a technical sense, your computer program would then serve as a deterministic Turing machine that simulates your NPDA.

Initial state = Q0 Final states = QF

Current State	Input	Stack top	String pushed	New State	Comments
Q0	ε	z0	S z0	Qmain	Qmain is the main state of this PDA
Qmain	ε	S	M C	Qmain	Create subgoals, ignoring actual input
Qmain	ε	S	A N	Qmain	Create subgoals, ignoring actual input
Qmain	ε	M	a M b	Qmain	Create subgoals, ignoring actual input
Qmain	ε	M	ε	Qmain	Epsilon production for M
Qmain	ε	N	b N c	Qmain	Create subgoals, ignoring actual input
Qmain	ε	N	ε	Qmain	Epsilon production for N
Qmain	ε	C	c C	Qmain	Create subgoals, ignoring actual input
Qmain	ε	C	ε	Qmain	Epsilon production for C
Qmain	ε	A	a A	Qmain	Create subgoals, ignoring actual input
Qmain	ε	A	ε	Qmain	Epsilon production for A
Qmain	a	a	ε	Qmain	Eat 'a' from input - a parsing goal
Qmain	b	b	ε	Qmain	Eat 'b' from input - a parsing goal
Qmain	c	c	ε	Qmain	Eat 'c' from input - a parsing goal
Qmain	ε	z0	z0	QF	Accept when z0 surfaces (parsing goals met)

Fig. 14.5. CFG to PDA conversion for the CFG of Illustration 13.2.1

parsing goal a is easy: just match a with the input. On the other hand, with parsing goal M on top of the stack, we could also have set up the parsing goal "ε" which means – *we could be done*! Hence, another move can simply empty M from the stack. Then, finally, when z0 shows up on top of stack, we accept, as there are no parsing goals left.

14.4 Direct Conversion of PDAs to CFGs

We first illustrate the PDA to CFG conversion algorithm with an example. As soon as we present an example, we write the corresponding general rule in *slant* fonts. Further details, should you need them, may be found in the textbook of Hopcroft, Motwani, and Ullman [60] whose algorithm we adopt. A slightly different algorithm appears in Sipser's book [111].

```
delta              contains        Productions
-----------------  ------------    ------------------------------
                                   S -> [p,Z0,x] for x in {p,q}

<p,(,Z0>           <p,(Z0>         [p,Z0,r_2] -> ( [p,(,r_1] [r_1,Z0,r_2]

                                       for r_i in {p,q}

<p,(,(>            <p,((>          [p,(,r_2] -> ( [p,(,r_1] [r_1,(,r_2]

                                       for r_i in {p,q}

<p,),(>            <p,e>           [p,(,p]  -> )

<p,e,Z0>           <q,e>           [p,Z0,q] -> e
```

Fig. 14.6. PDA to CFG conversion. Note that e means the same as ε.

Consider the PDA that *accepts by empty stack*,

$$(\{p,q\}, \{(,)\}, \{(,), Z0\}, \delta, p, Z0)$$

with δ given in Figure 14.6. Recall that since this is a PDA that accepts by empty stack, we do not specify the F component in the PDA structure. The above six-tuple corresponds to $(Q, \Sigma, \Gamma, \delta, q_0, z_0)$. This figure also shows the CFG productions generated following the PDA moves. The method used to generate each production is the following. Each step is explained with a suitable section heading.

14.4.1 Name non-terminals to match stack-emptying possibilities

Notice that the non-terminals of this grammar have names of the form [a,b,c]. Essentially, such a name carries the following significance:

> It represents the language that can be generated by starting in state a of the PDA with b on top of the stack, and being able to go to state c of the PDA *with the same stack contents* as was present while in state a.

This is top-down recursive programming at its best: we set up top-level goals, represented by non-terminals such as [a,b,c], without immediately worrying about how to achieve such complicated goals. As it turns out, these non-terminals achieve what they seek through subsequent recursive invocations to other non-terminals - letting the *magic of recursion* make things work out!

> *General rule: For all states* $q_1, q_2 \in Q$ *and all stack symbols* $g \in \Gamma$, *introduce a non-terminal* $[q_1, g, q_2]$ *(most of these non-terminals will prove to be useless later).*

14.4.2 Let start symbol S set up all stack-draining options

All the CFG productions are obtained systematically from the PDA transitions. The only exception is the first production, which, for our PDA, is as follows:

```
S -> [p,Z0,x] for x in {p,q}.
```

In other words, two productions are introduced, they being:

```
S -> [p,Z0,p]
S -> [p,Z0,q].
```

Here is how to understand these productions. S, the start symbol of the CFG, generates a certain language. This is the *entire* language of our PDA. The entire language of our PDA is nothing but *the set of all those strings* that take the PDA from its start state p to some state, *having gotten rid of everything in the stack.* In our PDA, since it starts with Z0 on top of stack, that's the only thing to be emptied from the stack. Since the PDA could be either in p or q after emptying the stack (and since we don't care where it ends up), we introduce both these possibilities in the productions for S.

> *General rule: For all states* $q \in Q$, *introduce one production* $S \rightarrow [q_0, z_0, q]$.

14.4.3 Capture how each PDA transition helps drain the stack

A PDA transition may either get rid of the top symbol on the stack *or* may end up *adding* several new symbols onto the stack. Therefore, many PDA transitions do *not* help achieve the goal of draining the stack. However, we can set up recursive invocations to clear the extra symbols placed on top of the stack, thus still achieving the overall goals.

To see all this clearly, consider the fact that δ contains a move, as shown below:

```
delta           contains
----------------------
<p,(,Z0>        <p,(Z0>
```

In this PDA, when in state p, upon seeing (in the input and Z0 on top of the stack, the PDA will jump to state p, having momentarily gotten rid of Z0, but promptly restoring (as well as Z0. Then the PDA has to "further struggle" and get rid of (as well as Z0, reaching some states after these acts. It is only *then* that the PDA has successfully drained the Z0 from its stack. Said differently, to drain Z0 on the stack while in state p, read (, invite more symbols onto the stack, and then recursively get rid of them as well. All this is fine, except we don't know rightaway where the PDA will be after getting rid of (, and subsequently getting rid of Z0. However, this is no problem, as we can *enumerate all possible states*, thus obtaining as many "catch all" rules as possible. This is *precisely* what the set of context-free grammar rules generated for this grammar says:

```
[p,Z0,r_2] -> ( [p,(,r_1] [r_1,Z0,r_2] for r_i in {p,q}
```

The rule says: "if you start from state p with a view to *completely* drain Z0 from the stack, you will end up in some state r_2. That, in turn, is a three step process:

- Read (and, for sure, we will be in state p.
- From state p, get rid of (recursively, ending up in some state r_1.
- From state r_1, get rid of Z0, thus ending up in the very same state r_2!

Fortunately, this is *precisely* what the above production rule says, according to the significance we assigned to all the non-terminals. We will have *sixteen* possible rules even for this single PDA rule!! Many of these rules will prove to be useless.

General rule: If $\delta(p, a, g)$ contains $\langle q, g_1, \ldots, g_n$, introduce one generic rule

$$[p, g, q_0] \to a \, [q, a, q_1][q_1, g_1, q_2] \ldots [q_n, g_n, q_0]$$

and create one instance of the rule for each $q_i \in Q$ chosen in all possible ways.

14.4.4 Final result from Figure 14.6

We apply this algorithm to the PDA in Figure 14.6, obtaining an extremely large CFG. We hand simplify, by throwing away rules as well as non-terminals that are never used. We further neaten the rules by assigning shorter names to non-terminals as shown below:

```
Let A=[p,Z0,p], B=[p,Z0,q], C=[q,Z0,p], D=[q,Z0,q],
    W=[p,(,p], X=[p,(,q], Y=[q,(,p], Z=[q,(,q].
```

Then we have the following rather bizzare looking CFG:

```
S -> A | B

A -> ( W A  |  ( X C
B -> ( W B  |  ( X D

W -> ( W W  |  ( X Y
X -> ( W X  |  ( X Z

W -> )
B -> e
```

How are we sure that this CFG is even close to being correct?

We simplify the grammar based on the notions of *generating* and *reachable* from the previous chapter. This process proceeds as follows:

1. Notice that C,D,Y,Z are *not* generating symbols (they can never generate any terminal string). Hence we can eliminate production RHS using them.
2. W and B are generating (W ->) and B -> e).
3. X is not generating. Look at X -> (W X. While (is generating and W is generating, X on the RHS isn't generating – we are doing a "bottom-up marking." The same style of reasoning applies also to X -> (X Z.
4. Even A is not generating!

Therefore, in the end, we obtain a short (but still 'bizzare looking') grammar:

```
S -> ( W S | e
W -> ( W W | )
```

Fortunately, this grammar is now small enough to apply our verification methods based on *consistency* and *completeness*:

Consistency: Any string s generated by S must be such that it has an equal number of (and). Further, in any of its proper prefixes, the number of (is greater than or equal to the number of).

Completeness: All such strings must be generated by S.

Proof outline for consistency:

Let us establish the 'same number of (and) part. Clearly, e (ε) satisfies this part. How about (W S? For this, we must state and prove a lemma about W:

- Conjecture: W has one more) than (.
- True for both arms of W, by induction.
- Hence, this conjecture about W is true.

Therefore, s has an equal number of (and).

Now, to argue that in any of the proper prefixes of s, the number of (is greater than or equal to the number of), we again need a lemma about W:

- Conjecture: In any prefix of a string generated by W, number of) is at most one more than the number of (.
- This has to be proved by induction on W.

Hence, S satisfies consistency.

Completeness

To argue completeness with respect to S, we state and prove a completeness property for W.

All the following kinds of strings are generated by W: In any prefix of a string generated by W, number of) is at most one more than the number of (.

The proof would proceed as illustrated in Figure 13.3. Now, the completeness of S may be similarly argued, as Exercise 14.1 requests.[5]

---

[5] In fact, the plot will be simpler for this grammar, as there will be no zero-crossings. There could be occasions where the plot touches the x-axis, and if it continues, it promptly takes off in the positive direction once again.

14.5 The Chomsky Normal Form

Given a context-free grammar G, there is a standard algorithm (described in most textbooks) to obtain a context-free grammar G' in the *Chomsky normal form* such that $L(G') = L(G) - \{\varepsilon\}$. A grammar in the Chomsky normal form has two kinds of productions: $A \to BC$, as well as $A \to a$. If ε is required to be in the new grammar, it is explicitly added at the top level via a production of the form $S \to \varepsilon$. There is another well-known normal form called the *Greibach normal form* (GNF) which may be found discussed in various textbooks. In the GNF, all the production rules are of the form $A \to aB_1 B_2 \ldots B_k$ where a is a terminal and $A, B_1, \ldots, B_k$, for $k \geq 0$, are non-terminals (with $k = 0$, we obtain $A \to a$). Obtaining grammars in these normal forms facilitates proofs, as well as the description of algorithms. In this chapter, we will skip the actual algorithms to obtain these normal forms, focusing instead on the advantages of obtaining grammars in these normal forms.

A grammar G in the Chomsky normal form has the property that *any* string of length n generated by G must be derived through exactly $2n - 1$ derivation steps. This is because all derivations involve a binary production $A \to BC$ or an unary production $A \to a$. For example, given the following grammar in the Chomsky normal form,

```
S -> A B | S S    B -> b    A -> a,
```

a string `abab` can be derived through a seven step $(2 \times 4 - 1)$ derivation

```
S => SS => ABS => ABAB => aBAB => abAB => abaB => abab.
```

In the next section, we discuss a parsing algorithm for CFGs, assuming that the grammar is given in the Chomsky normal form.

14.5.1 Cocke-Kasami-Younger (CKY) parsing algorithm

The CKY parsing algorithm uses *dynamic programming* in a rather elegant manner. Basically, given any string, such as 0 0 1, and a Chomsky normal form grammar such as

$S \to S T \mid 0$
$T \to S T \mid 1,$

the following steps describe how we "parse the string" (check that the string is a member of the language of the grammar):

- Consider *all possible* substrings of the given string of length 1, and determine all non-terminals which can generate them.

- Now, consider *all possible* substrings of the given string of length 2, and determine all pairs of non-terminals in juxtaposition which can generate them.
- Repeat this for strings of lengths 3, 4, ..., until the full length of the string has been examined.

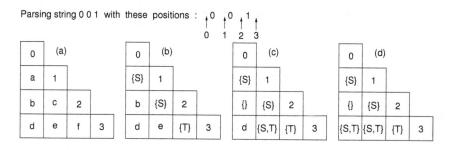

Fig. 14.7. Steps of the CKY parsing algorithm on input 001

To capture all this information, we choose a convenient tabular representation as in Figure 14.7(a). The given string 001 has *four* positions (marked 0 through 3) in it. Position a in the table represents the portion of the string between positions 0 and 1, i.e., the first "0" in the string. Likewise, positions c and f represent 0 and 1, respectively. Let us fill these positions with the set of all non-terminals that can generate these strings. We know that S can generate a 0, and nothing else. Therefore, the *set* of non-terminals that generates 0 happens to be $\{S\}$. Likewise, $\{T\}$ is the set of non-terminals that generate a 1. Filling the table with these, we obtain Figure 14.7(b).

What can we say about position b in this table? It represents the region in the string between positions 0 and 2. *Which non-terminal can generate the region of the string between positions 0 and 2?* The answer depends on which non-terminals generate the region of the string between positions 0 and 1, and which non-terminals generate the region of the string between positions 1 and 2. We know these to be $\{S\}$ and $\{S\}$. The set of non-terminals that generate the substring 02 are then *those non-terminals that yield SS*. Since *no* non-terminals yield SS, we fill position b with $\{\}$. By a similar reasoning, we fill position e with $\{S, T\}$. The table now becomes as shown in Figure 14.7(c).

Finally, position d remains to be filled. Substring 03 can be generated in two distinct ways:

- Concatenating substring 01 and 13, or

- Concatenating substring 02 and 23.

Substring 01 is generated by {S} and substring 13 by {S,T}. Therefore, substring 03 is generated by all the non-terminals that generate {S}{S,T}, i.e., those that generate {SS,ST, i.e., {S,T}. *No* non-terminal generates substring 02, hence we don't pursue that possibility anymore. Hence, we fill position d with {S,T} as in Figure 14.7(d).

The parsing succeeds because we managed to write an S in position d—the *start* symbol can indeed yield the substring 03.

14.6 Closure and Decidability

In this section, we catalog the main results you should remember, plus some justifications. Details are omitted for now.

1. Given a CFG, it *is* decidable whether its language is empty. Basically, if you find that S is not generating, the language of the grammar is empty! It is the bottom-up marking algorithm discussed above.
2. Given a CFG, it is *not decidable* whether its language is Σ^*.
3. The equivalence between two CFGs is not decidable. This follows from the previous result, because one of the CFGs could easily be encoding Σ^*.
4. Given a CFG, whether the CFG is ambiguous is not decidable.
5. Given a CFG, whether the CFG generates a *regular* language is not decidable.
6. CFLs are closed under union, concatenation, and starring because these constructs are readily available in the CFG notation.
7. CFLs are closed under reversal because we know how to "reverse a CFG."
8. CFLs are *not* closed under complementation, and hence also not closed under intersection.
9. CFLs are closed under intersection with a *regular language*. This is because we can perform the product state construction between a PDA and a DFA.
10. CFLs are closed under homomorphism.

14.7 Some Important Points Visited

We know that if L is a regular language, then L is a context-free language, but not vice versa. Therefore, the *space* of regular languages is *properly contained* in the space of context-free languages. We note some facts below:

- It **does not** follow from the above that the union of two CFLs is always a *non-regular* CFL; it is *not* so, in general. Think of $\{0^n 1^n \mid n \geq 0\}$ and the complement of this language, both of which are context-free, and yet, their union is Σ^* which is context-free, *but also regular*.
- The union of a context-sensitive language and a context-free language can be a regular language. Consider the languages L_{ww} and $\overline{L_{ww}}$ of Section 13.4.1.

All this is made clear using a real-world analogy:

- In the real world, we classify music (compared to context-free languages) to be "superior" to white noise (compared to the regular language Σ^*) because music exhibits superior patterns than white noise.
- By a stretch of imagination, it is possible to regard white noise as music, but usually not vice versa.
- By the same stretch of imagination, *utter silence* (similar to the regular language $\emptyset$) can also be regarded as music.
- If we mix music and white noise in the air (they are simultaneously played), the result is white noise. This is similar to taking $\{0^n 1^n \mid n \geq 0\} \cup \Sigma^*$ which yields Σ^*.
- However, if we mix music and silence in the air, the result is still music (similar to taking $\{0^n 1^n \mid n \geq 0\} \cup \emptyset$).
- Regular languages other than $\emptyset$ and Σ^* 'sound different.' For instance, $\{(01)^n \mid n \geq 0\}$ 'sounds like' a square wave played through a speaker. Therefore, the result of taking the union of a context-free language and a regular language is either context-free or is regular, depending on whether the strings of the regular language manage to destroy the delicate patterns erected by the strings of the CFL.

It must also be clear that there are $\aleph_0$ regular languages and the same number of context-free languages, even though not all context-free languages are regular. This is similar to saying that not all natural numbers are prime numbers, and yet both have cardinality $\aleph_0$.

Illustration 14.7.1 Consider $\{a^m b^m c^m \mid m \geq 0\}$. This is not a CFL. Suppose it is a CFL. Let us derive a contradiction using the CFL Pumping Lemma. According to this lemma, there exists a number n such that given a string w in this language such that $|w| \geq n$, we can split w into $w = uvwxy$ such that $|vx| > 0$, $|vwx| \leq n$, and for every $i \geq 0$, $uv^iwx^iy \in L(G)$.

Select the string $a^nb^nc^n$ which is in this language. These are the cases to be considered:

- v, w, and x fall exclusively in the region "a".
- v, w, and x fall exclusively in the region "b".
- v, w, and x fall exclusively in the region "c".
- v and x fall in the region "a" and "b", respectively.
- v and x fall in the region "b" and "c", respectively.

In all of these cases, "pumping" takes the string outside of the given language. Hence, the given language is not a CFL.

Illustration 14.7.2 We illustrate the CKY parsing algorithm on string aabbab with respect to the following grammar:

```
S -> AB | BA | SS | AC | BD
A -> a  B -> b  C -> SB  D -> SA
```

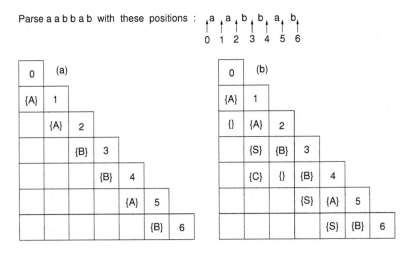

Fig. 14.8. Steps of the CKY parsing algorithm on input aabbab

The basic idea is to subdivide the string into "regions" and apply dynamic programming to "solve" all the shorter regions first, and use that information to solve the "larger" regions. Let us build our table now. The region 01 is generated by the set of non-terminals A. We just write A below. We write likewise the other non-terminals (Figure 14.8(a)).

The next table is obtained as follows: There is *no* non-terminal that has a right-hand side of the production as **AA**. So we put a $\emptyset$ (**phi**) at 02. Since **S -> AB**, **A** marks 12, and **B** marks 23, we put an **S** at 13. We proceed in the same manner for the remaining entries that are similar. Last but not least, we write a **C** at 14, because 14 is understood to be representing 12–24 or 13–34. 12–24 is **A phi** (**A** concatenated with the empty language $\emptyset$), and so we ignore it. 13–34 is **SB**, and **C -> SB**; therefore, we write "C" there. We fill the remaining table entries similarly. The results are shown in Figure 14.8(b). The parsing is successful if, in position "06", you manage to write a set of non-terminals that contain "S". Otherwise, the parsing fails.

Illustration 14.7.3 *Prove that any context-free grammar over a singleton alphabet generates a regular language.*

We provide a proof sketch, leaving details to the reader (see [45, page 86] for a full proof). To solve this problem, we can actually use the Pumping Lemma for context-free languages in an unusual way! The CFL Pumping Lemma says that for a long w (longer than some "k"), we can regard $w = uvxyz$ such that $uv^i x y^i z \in L$. Each pump up via i increases the length by $v + y$. However, since $|vxy| \leq k$, there are only a finite number of $v + y$'s we can get. These are the *periodicities* (in the ultimate periodicity sense). If a set is described by a finite number of periods $p_1, p_2, \ldots$, it is easily described by the *product* of these periods. This was the argument illustrated in Section 12.2 when we tackled a regular language Pumping Lemma problem, and chose $0^n 1^{n+n!}$ to be the initial string. In that problem, the $n!$ we chose served as the product of all the values possible for $|y|$. For instance, if a set S is such that

- it has strings of a's in it, and
- S is infinite, and for a sufficiently large i,
    - if $a^i \in S$ then $a^{i+4} \in S$ as well as $a^{i-4} \in S$,
    - if $a^i \in S$ then $a^{i+7} \in S$ as well as $a^{i-7} \in S$,

then S is ultimately periodic with period 28.

Therefore, we conclude that any CFL over a singleton alphabet has its strings obeying lengths that form an ultimately periodic set. This means that the language is regular.

14.7.1 Chapter Summary – Lost Venus Probe

In this chapter, we examined many topics pertaining to PDAs and CFGs: notions of acceptance, interconversion, and proofs of correctness. We also examined simple parsing algorithms based on the Chomsky normal form of CFGs. The theory of context-free languages is one of the pinnacles of achievement by computer scientists. The theoretical rigor employed has truly made the difference between "winging" parsing algorithms—which are highly likely to be erroneous—versus producing highly reliable parsing algorithms that silently work inside programs. For instance, Hoare [55] cites the story of a Venus probe that was lost in the 1960's due to a FORTRAN programming error. The error was quite simple - in hindsight. Paraphrased, instead of typing a "DO loop" as

```
DO 137 I=1,1000
...
137 CONTINUE,
```

the programmer typed

```
DO 137 I=1 1000
...
137 CONTINUE.
```

The missed comma caused FORTRAN to treat the first line as the assignment statement DO137I=11000 — meaning, an assignment to a newly introduced variable DO137I, the value 11000. The DO statement essentially did not loop 1000 times as was originally intended! FORTRAN's permissiveness was quickly dispensed with when the theory of context-free languages lead the development of "Algol-like" block-structured languages.

Sarcastically viewed, progress in context-free languages has helped us leapfrog into the era of deep semantic errors in programs, as opposed to unintended simple syntactic errors that caused programs to crash. The computation engineering methods discussed in later chapters in this book do help weed out semantic errors, which are even more notoriously difficult to pin down. We hope for the day when even these errors appear to be as shallow and simpleminded as the forgotten comma.

Exercises

14.1. Argue the consistency and completeness of S and W.

14.2. The following "optimization" is proposed for the PDA of Figure 14.1: merge states q0 and q1 into a new state q01; thus, (i) q01 will now be the start state, and (ii) for any move between q0 and q1 or from q1 to itself, now there will be a q01 to q01 move. Formally argue whether this optimization is correct with respect to the language L_0; if not, write down the language now accepted by the PDA.

14.3. Argue the remaining cases of the proof in Section 14.2, namely the direct q2 to q2 traversal, and the q2 to q3 to q4 to q2 traversal.

14.4. Prove one more case not covered in Exercise 14.3; prove that the language of this PDA cannot go outside the language of regular expressions $a^* b^*$.

14.5. Consider the following list of languages, and answer the questions given below the list:

- $L_{a^i b^j c^k} =$

$$\{a^i b^j c^k \mid i, j, k \geq 0 \text{ and if } odd(i) \text{ then } j = k\}.$$

  In other words, if an odd number of a's are seen at first, then an equal number of j's and k's must be seen later.
- $L_{b^j c^k a^i} =$

$$\{b^j c^k a^i \mid i, j, k \geq 0 \text{ and if } odd(i) \text{ then } j = k\}.$$

- $L_{a^i b^j c^k d^l} =$

$$\{a^i b^j c^k d^l \mid i, j, k, l \geq 0 \text{ and if } odd(i) \text{ then } j = k \text{ else } k = l\}.$$

- $L_{b^j c^k d^l a^i} =$

$$\{b^j c^k d^l a^i \mid i, j, k, l \geq 0 \text{ and if } odd(i) \text{ then } j = k \text{ else } k = l\}.$$

1. Which of these languages are deterministic context-free?
2. Which are context-free?
3. Write the pseudocode of a parsing algorithm for the strings in this language. Express the pseudocode in a tabular notation similar to that in Figure 14.1.
4. For each language that is context-free, please design a PDA and express it in a tabular or graphical notation.
5. For each language that is context-free, please design a CFG.
6. For each of these CFGs, convert each to a PDA using the CFG to PDA conversion algorithm.

14.6. Prove using Floyd's method that the PDA of Figure 14.1 is correct.

14.7. Convert the following PDA to a CFG:

```
delta              contains
----------------------------
<p,(,z0>           <q,(z0>
<q,(,(>            <q,((>
<q,),(>            <q,e>
<p,e,z0>           <r,e>
<q,e,z0>           <r,e>
```

14.8.
1. Develop a PDA for the language

$$w \mid w \in \{0,1\}^* \wedge \#_0(w) = 2 \times \#_1(w)\}$$

In other words, w has twice as many 0's as 1's.
2. Prove this PDA correct using Floyd's method
3. Convert this PDA into a CFG
4. Simplify the CFG
5. Prove the CFG to be correct (consistent and complete)

15

Turing Machines

We live in a digital society "steeped" in computers. Computers are employed in everyday devices ranging from toys and shoes, telephones, automobiles, and spacecraft. Such was not the world in the early 20th century when no computers were around, and logicians and philosophers such as Hilbert were discussing the possibility of automating computation. The extent to which computation can be automated was the main subject of discussions. In that era, a simple machine called the Turing machine was proposed as a *formal model* of computers. Turing machines embodied the notion of *effective* (algorithmic, mechanical) computability in such an unambiguous and elementary fashion that it was taken to be *the* canonical device that defined the limits of mechanical computability. Many alternative mechanisms such as Thue systems, the Lambda calculus, and the combinatory logic [31] have also been shown to be capable of defining the notion of computation. In fact, in combinatory logic, merely *two letters*, S and K, representing two specific Lambda calculus terms, and their reduction rules,[1] have been shown to be complete. This means that the computation ensuing from any arbitrary computer program running on arbitrary input data can be modeled through a sequence of rewrites performed on a string comprised of S and K. These notations have been shown to be equivalent in power to Turing machines. Yet, Turing machines are "king," in the sense that they most closely resemble the kinds of stored-program computers that we are most familiar with.

This chapter begins in Section 15.1 with historical accounts of early work by Church and Turing from Andrew Hodges's web site [118] and the Stanford Encyclopedia of Philosophy web site [119]. We then define Turing machines formally in Section 15.2, touching on TM variants as

---

[1] $S = \lambda xyz.\, xz(yz)$ and $K = \lambda xy.\, x$.

well as related machines. We then define notions such as *acceptance* and
Halting in Section 15.3, and provide examples of deterministic Turing
machines in Section 15.4. Finally, we provide an account of NDTMs in
Section 15.5.

15.1 *Computation:* Church/Turing Thesis

The Encyclopedia of Philosophy web site [119] gives the following account of the evolution of the Church Turing thesis. We now provide
direct excerpts from this web site.
(Begin excerpts)

The Princeton logician Alonzo Church had slightly outpaced Turing
in finding a satisfactory definition of what he called *effective calculability.* Church's definition required the logical formalism of the Lambda
calculus. This meant that from the outset Turing's achievement merged
with and superseded the formulation of Church's Thesis, namely the
assertion that the Lambda calculus formalism correctly embodied the
concept of effective process or method. Very rapidly it was shown
that the mathematical scope of Turing's computability coincided with
Church's definition (and also with the scope of the general recursive
functions defined by Gödel). Turing wrote his own statement (Turing
1939, p. 166) of the conclusions that had been reached in 1938; it is *in
the Ph.D. thesis that he wrote under Church's supervision,* and so this
statement is the nearest we have to a joint statement of the Church
Turing thesis:

> A function is said to be *effectively calculable* if its values can be
> found by some purely mechanical process. Although it is fairly
> easy to get an intuitive grasp of this idea, it is nevertheless de-
> sirable to have some more definite, mathematically expressible
> definition. Such a definition was first given by Gödel at Prince-
> ton in 1934. These functions were described as *general recursive*
> by Gödel. Another definition of effective calculability has been
> given by Church who identifies it with lambda-definability. The
> author [i.e. Turing] has recently suggested a definition corre-
> sponding more closely to the intuitive idea. It was stated above
> that a function is effectively calculable if its values can be found
> by a purely mechanical process. We may take this statement lit-
> erally, understanding by a purely mechanical process one which
> could be carried out by a machine. It is possible to give a math-
> ematical description, in a certain normal form, of the structures

of these machines. The development of these ideas leads to the author's definition of a computable function, and to an identification of computability with effective calculability. It is not difficult, though somewhat laborious, to prove that these three definitions are equivalent.

Church accepted that Turing's definition gave a compelling, intuitive reason for why Church's thesis was true. The recent exposition by Davis (2000) emphasizes that Gödel also was convinced by Turing's argument that an absolute concept had been identified (Gödel 1946). The situation has not changed since 1937.

15.1.1 "Turing machines" according to Turing

Two excerpts from Turing's own paper, [117, 118] "On Computable Numbers, with an Application to the Entscheidungsproblem"—which, in German, stands for *the decision problem*—are given below:

1. Computing is normally done by writing certain symbols on paper. We may suppose this paper is divided into squares like a child's arithmetic book. In elementary arithmetic the two-dimensional character of the paper is sometimes used. But such a use is always avoidable, and I think that it will be agreed that the two-dimensional character of paper is no essential of computation. I assume then that the computation is carried out on one-dimensional paper, i.e. on a tape divided into squares. I shall also suppose that the number of symbols which may be printed is finite ...
2. The behavior of the [human] computer at any moment is determined by the symbols which he is observing, and his state of mind at that moment ....

(End of excerpts)

Turing argued that his formalism was sufficiently general to encompass anything that a human being could do when carrying out a definite method. Turing also proposed the notion of universal Turing machines capable of simulating the operation of any Turing machine.

15.2 Formal Definition of a Turing machine

The Turing machine is a conceptual machine with an infinite sequential access store called a "tape." It serves the purpose of modeling actual computers as well as computations occurring within them. In a sense, the use of Turing machines with infinite tape storage to model computations that can occur in finite-memory real computers is analogous to the use of *real numbers* that have infinite precision to model *rational numbers* that only have finite precision; in both cases, only the finite counterparts can be directly represented and manipulated within a computer. Given that all "activities of interest" must either transpire in a finite amount of time *or* be subject to a finitary description, it is clear that one will never be able to use an infinite amount of memory or a description with infinite precision within a finite amount of time. On the other hand, placing an arbitrary upper bound on storage requirements, or the precision allowed, will a priori rule out many computations/rational numbers as the case may be; this is not desirable. Hence, conceptual devices with an infinite capacity become essential.

Formally, a Turing machine M is a structure $(Q, \Sigma, \Gamma, \delta, q_0, B, F)$. The finite state control of M ranges over the control states in Q, beginning at the initial control state $q_0 \in Q$. States $F \subseteq Q$ of M are called *final* states, and are used to define when a machine *accepts* a string. The input on which M operates is initially presented on the input tape. It is a string over the input alphabet Σ. Once started, it is possible that a Turing machine may *never* halt; it may keep zigzagging on the tape, writing symbols all over, and running amok, much like many tricky programs do in real life. A Turing machine cannot manufacture new symbols ad infinitum - so all the symbols written by a Turing machine on its tape do belong to a *finite* tape alphabet, Γ. Notice that $\Sigma \subset \Gamma$, since Γ includes the *blank* symbol B that is not allowed within Σ. We assume that a Turing machine begins its operation scanning cell number 0 of a doubly-infinite tape (meaning that there are tape cells numbered $+x$ or $-x$ for any $x \in Nat$); more on this is in the following section. A fact to remember is this: in order to feed the string ε to a TM, one must present to the Turing machine a tape *filled* with blanks (B). However, some authors alter this convention slightly, allowing ε to be fed to a Turing machine by ensuring that, in the initial state, the symbol under the tape head is blank (B) (*i.e.,* the rest of the tape could contain non-blank symbols). In any case, a normal Turing machine input is such that for every $i \in length(w)$, $w[i] \neq B$ is presented on tape cell i, with all remaining tape cells containing B, and the head of the Turing machine faces $w[1]$ at the beginning of a computation.

A TM may be deterministic or nondeterministic. The signature of δ for a deterministic Turing machine (DTM) is

$$\delta : Q \times \Gamma \to Q \times \Gamma \times \{L, R\}.$$

This signature captures the fact that a TM can be in a certain state $q \in Q$ and looking at $a \in \Gamma$. It can then write a' on the tape in lieu of a, move to a state q', and move its head left (L), or right (R), depending on whether $\delta(q, a) = \langle q', a', L \rangle$ or $\delta(q, a) = \langle q', a', R \rangle$, respectively.

For an NDTM, $\delta(q, a)$ returns a set of next control states, tape symbol replacements, and head move directions. The signature of δ for a nondeterministic Turing machine (NDTM) is

$$\delta : Q \times \Gamma \to 2^{Q \times \Gamma \times \{L,R\}}.$$

Think of a nondeterministic Turing machine as a C program where instead of the standard if-then-else construct, we have an `if/fi` construct of the following form (as in the Promela language of Chapter 21):

```
if :: condition_1 -> action_1
   :: condition_2 -> action_2
   :: ...
   :: condition_n -> action_n
fi
```

The intended semantics is as follows. Each `condition` may be a Boolean expression such as `(x > 0)`. There may be more than one condition becoming true at any given time. In that case, one of the conditions is *nondeterministically* chosen. The `actions` can be `goto`, `assignment`, `while-loops`, etc., as in a normal C program. With *just* this change to the C syntax, we have a class of nondeterministic C programs that are equivalent to nondeterministic Turing machines.

Nondeterministic Turing machines may be regarded as "fictitious," since real-world computers do not behave nondeterministically in the above manner.[2] They are, however, conceptual devices that play a fundamental role in the study of *complexity theory*. Chapter 19 fully delves into this topic; an example also appears in Section 15.5.2.

> *In this program, we shall treat Turing machines and programs synonymously. Therefore, "build a program" will also mean "build a Turing machine."*[3]

15.2.1 Singly- or doubly-infinite tape?

Some authors assume that the tape of a TM is a singly-infinite list of cells "going to the right," while others allow a doubly-infinite tape

---

[2] Unless they have a circuit board that is loose inside them, making erratic electrical contact!

[3] This is done, suspecting that most students will understand "programming" far more readily than "building a Turing machine."

(going to the left and the right). Those using a singly-infinite tape are taking the view that, after all, a description of the contents of the tape as well as the position of the TM head can be given in terms of a singly-infinite sequence of characters. However, this view forces one to answer the question, "what happens when the head attempts to move towards the left of the leftmost cell?" The most common answer is, "the head cannot move to the left of the leftmost cell." Hence, the TM would behave as if the "belt that moves it head slips," making the head stay at the leftmost cell. Those working with a doubly-infinite tape may, on the other hand, assume that there is a special "left-end marker" symbol. This allows one to answer the question "what happens when the TM attempts to move to the left of the leftmost cell" more uniformly: the head will then face the left-end marker symbol. If we assume that the transition function δ of the TM will then specify a move to the right, it will end up restoring the head to the left-most position of the "working region" of the tape. In this book, we will work with both these representations; that is, we will not explicitly show the δ move on a left-end marker; sometimes (e.g., when we discuss experiments with the JFLAP tool), we may even ignore the notion of there being a leftmost cell, and truly allow the machine to span any range going left and right with respect to the initial cell.

These and many other variations of the TM specification can be shown to be equivalent in the sense that any computation that one type of TM performs can also be performed by another TM type. Such TMs are termed *universal* in the sense they can perform all possible computations. The extreme amount by which one can vary the specification of a TM and retain universality is tribute to the high degree of *robustness* exhibited by TMs.

To sum up, when a TM with a doubly-infinite tape begins operation, its head is scanning the first character of the input string. The entire input string lies to the right of the head. The portion of the tape towards the left side of the head contains only blanks (B).

The reader is urged to download and experiment with the JFLAP tool at this point. It provides very intuitive animations of TMs, especially NDTMs.

15.2.2 Two stacks+control = Turing machine

In previous chapters, we discussed how DFAs can be used to model finite-state C programs (C programs that do not contain recursive calls and have only a finite number of variables). We also discussed how PDAs can be used to model C programs in which the only infinite-state component appears in the form of a push-down stack. One can view the DFA as just a finite-state control device (*i.e.*, a finite-state control device coupled with *zero* stacks). Likewise, one can view the PDA as finite-state control coupled with *one* stack. Along the same lines, a TM can be viewed as finite-state control coupled with *two* stacks, by modeling the infinite tape using two stacks:

- When a TM goes one step left from its current position, the tape segment to the left of the head shrinks by one cell while the segment to the right of the head grows by one segment. This can be viewed in terms of popping the left-hand stack and pushing the right-hand stack of the finite-state control.
- A TM taking one step to the right can be viewed in terms of popping the right-hand stack and pushing onto the left-hand stack.

In the next chapter, we will discuss the fact that adding more stacks does not increase the power of the machine; and hence, "*two stacks are necessary and sufficient.*"

15.2.3 Linear bounded automata

Linear Bounded Automata (LBA) (see Figure 13.5) are machines that have the same overall structure as a TM. The only difference between a TM and an LBA is in the input tape and the δ function. For an LBA, each input string is presented on its tape bracketed by a left-end marker and a right-end marker. Without loss of generality, we can assume that the right-end marker is at least two cells to the right of the left-end marker (recall how ε is presented to a TM through a string of blanks; the same convention is used for the LBA). The δ of an LBA is defined such that when faced with the left-end marker in *any* state, the head moves one step to the right, and similarly, when faced with the right-end marker, the head moves one step left. Hence, we obtain a special TM that cannot go beyond the extent of the initial input. Such a machine is termed an LBA.

An LBA properly subsumes a PDA in power, and can be used to recognize context-sensitive languages such as L_{ww} of Section 13.4.1. Unlike with other machine types, it is not known whether or not *non-deterministic* LBAs (NLBAs) and deterministic LBAs are equivalent in power or not. This has been an open problem for quite some time now.

15.2.4 Tape vs. random access memory

The most familiar hardware view of a computer is as a stored-program computer running machine-language instructions. Each machine-language instruction describes what the machine does in its current "step," as well as which machine-language instruction is eligible to be executed in the next step; in fact, this is what the abstraction of finite-state control ends up being in a random access machine. However, unlike in a TM, each machine-language instruction specifies the addresses of zero or more memory cells to be loaded from and/or stored into. *The key advantage of a TM is that the current head position is never recorded anywhere explicitly.* In a TM, updates happen at the cell facing the current head position, and the TM then moves one step to the left or right *relative* to the current head position. These conventions differ in a fundamental way from those in a random access memory based computer, where addresses are needed to address a memory cell, and all addresses belong to a *finite* range. Hence, in a random access memory based computer, only a finite number of locations can be accessed to fetch or store data, and to fetch instructions from. Therefore, every hardware realization of a computer is a DFA. However, we derive the most insights by viewing these computations as occurring within a single uniform device called a TM, as opposed to viewing them in terms of one DFA for every machine-language instruction set and machine-level program.

15.3 Acceptance, Halting, Rejection

Given a Turing machine $M = (Q, \Sigma, \Gamma, \delta, q_0, B, F)$, the transition function δ is *partial* - it need not be defined for all inputs and states. When not defined, the machine becomes "stuck" - and is said to *halt*. Therefore, note that **a TM can halt in any state**. Also, by convention, no moves are defined out of the control states within F. Therefore, **a TM always halts when it reaches a state $f \in F$**. Here are further definitions:

- A TM **accepts** when halting at $f \in F$.
- A TM **rejects** when halting in any other state outside F.
- A TM **loops** when not halting.

15.3.1 "Acceptance" of a TM closely examined

Compared to DFAs and PDAs, the notion of acceptance for a TM is unusual in the following sense: a TM can accept an input without fully reading the input—or, in an extreme situation, *not even reading one cell of the input* (e.g., if $q_0 \in F$)! Hence, curiously, any TM with $q_0 \in F$ has language Σ^*. On the other hand, if a TM never accepts, its language is $\emptyset$. This can be the result of the TM looping on every input, or the TM getting stuck in a reject state for every input. As a final illustration, given an arbitrary language L_1, a TM that accepts any string within the set L_1 and does not accept (loops or rejects) strings outside L_1, has L_1 as its language.

15.3.2 Instantaneous descriptions

While capturing the 'snapshot' of a TM in operation, we need to record the control state of the machine, the contents of the tape, and what the head is scanning. All these are elegantly captured using an *instantaneous description* containing the tape contents, w, with a single state letter q placed somewhere within it. Specifically, suppose that the string lqr represents that the tape contents is lr, the finite-state control is in state q, and the head is scanning $r[0]$. The initial ID is $q_0 w$, for initial input string w. We define $\vdash$, the 'step' function that takes IDs to IDS, as follows:

$l_1 p a r_1 \vdash l_1 b q r_1$ if and only if $\delta(p, a) = (q, b, R)$. The TM changes the tape cell contents a to b and moves right one step, facing $r_1[0]$, the first character of r_1.

Similarly, $l_1 c p a r_1 \vdash l_1 q c b r_1$ if and only if $\delta(p, a) = (q, b, L)$. The TM changes an a to a b and moves left one step to now face c, the character that was to the left of the tape cell prior to the move.

We define the language of a TM M using $\vdash^*$:

$$L(M) = \{w \mid q_0 w \vdash^* l q_f r, \ for \ q_f \in F, \ and \ l, r \in \Sigma^*\}.$$

In other words, a TM that starts from the initial ID $q_0 w$ and attains an ID containing q_f (namely $l q_f r$, for some l and r in Σ^*) ends up accepting w.

15.4 Examples

We now present short examples that illustrate various concepts about Turing machines (Section 15.4.1). This is followed by a deterministic Turing machine that accepts strings of the form $w\#w$ for $w \in \Sigma^*$ (Section 15.4.2). Section 15.5 introduces NDTMs, and Section 15.5.2 presents an NDTM that accepts strings of the form ww.

Turing machines are specified fully at the *implementation level* by completely specifying their δ function (or δ relation for NDTMs) in a tabular form as in Figure 15.1 or diagrammatic form as in Figure 15.3. After gaining sufficient expertise with Turing machines, we will allow them to be specified at the *high level* through pseudo-code or precise English narratives. When resorting to high-level descriptions, the specification writer must strive to ensure sufficient clarity so that a reader can, in principle, reconstruct an implementation level description if necessary.

15.4.1 Examples illustrating TM concepts and conventions

Illustration 15.4.1 Consider the Turing machine with a doubly-infinite tape $M = (Q, \Sigma, \Gamma, \delta, q0, B, F)$, where $Q = \{q0, qa\}$, $\Sigma = \{0, 1\}$, $\Gamma = \{0, 1, B\}$, $F = \{qa\}$, and δ is as below:

$\delta(q0, 0) = (q0, 0, L)$
$\delta(q0, 1) = (q0, 1, L)$
$\delta(q0, B) = (q0, B, L)$.

This TM will not stop running - it will keep moving left (even this is called "looping"). Its language is $\emptyset$.

Illustration 15.4.2 Now consider changing δ to the following:

$\delta(q0, 0) = (qa, 0, L)$
$\delta(q0, 1) = (qa, 1, L)$
$\delta(q0, B) = (qa, B, L)$.

Now the language of this Turing machine is Σ^*. Notice that we need not specify moves for state **qa** for input B. In other words, this Turing machine will move one step to the left and get "stuck" in state q_a which is accepting.

Illustration 15.4.3 Suppose the entries for (qa,1,L) as well as (qa,B,L) are removed from the transition table given in Illustration 15.4.2. The language of this Turing machine will then be $0(0+1)^*$. This is because:

- Neither $\delta(q0, 1)$ nor $\delta(q0, B)$ is specified—hence, the input has to begin with a 0.
- State q_a accepts everything. Hence, $(0 + 1)^*$ can follow.

	q0	q1	q2	q3	q4	q5	q6
q0	-	a; X, R	-	-	Y; Y, R	-	B; B, R
q1	-	a; a, R	b; Y, R	-	-	-	-
	-	Y; Y, R	-	-	-	-	-
q2	-	-	Z; Z, R	c; Z, L	-	-	-
	-	-	b; b, R	-	-	-	-
q3	X; X, R	-	-	b; b, L	-	-	-
	-	-	-	Y; Y, L	-	-	-
	-	-	-	a; a, L	-	-	-
	-	-	-	Z; Z, L	-	-	-
	-	-	-	c; c, L	-	-	-
q4	-	-	-	-	Y; Y, R	Z; Z, R	-
q5	-	-	-	-	-	Z; Z, R	B; B, R
q6	-	-	-	-	-	-	-

Fig. 15.1. A TM for $\{a^n b^n c^n \mid n \geq 0\}$, with start state q0, final state q6, and moves occurring from the row-states to column-states

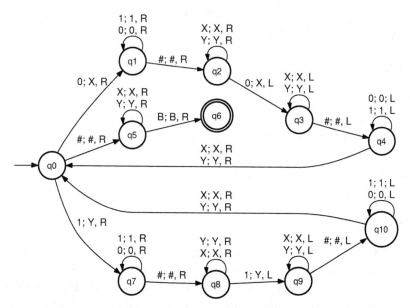

Fig. 15.2. A Turing machine for $w\#w$

A Turing machine for $a^n b^n c^n$

A Turing machine that recognizes $L_{anbncn} = \{a^n b^n c^n \mid n \geq 0\}$ is given in Figure 15.1 in the JFLAP tool's saved-table syntax. The algorithm implemented is one of turning a's into X's, b's into Y's, and c's into Z's. Here is an execution on aabbcc:

- We start in state q0 where we seek an a, changing it to an X when we see one, and at the same time entering state q1.
- In q1, we skip over a's going right, while staying in q1. Upon encountering a b, we change it to a Y, and move over to state q2.
- In q2, we move over b's going right, until we encounter a c, turning it into a Z, and then move over to state q3, and start moving in the left (L) direction.
- Notice that in q3, we keep moving left upon seeing any one of b, Y, a, Z, or c, and stop only when we see an X. Then we sweep over the input once again.

In q0, encountering an X is impossible (why?). Also notice that we explicitly decode the B (blank) input case in the q0 state. While in q0, encountering a Z is impossible. We may, however, encounter a Y (e.g., consider the input 'abc' which would be changed to XYZ; and then in state q3, we would skip over the X and be in state q0 facing a Y). The reader is invited to argue that this DTM is correct with respect to the advertised language $L_{a^n b^n c^n}$.

15.4.2 A DTM for $w \# w$

In Figure 15.2, we provide a deterministic Turing machine for the language of strings of the form $w \# w$, where $w \in \Sigma^*$. Notice how the presence of # allows the midpoint to be deterministically located. The Turing machine basically hovers to either side of #, scoring off matching characters.

15.5 NDTMs

An NDTM is a Turing machine with nondeterminism in its finite-state control, much like we have seen for earlier machine types such as PDAs. To motivate the incorporation of nondeterminism into Turing machines, consider a problem such as determining whether an undirected graph

G has a clique (a completely connected subgraph) of k nodes.[4] No efficient algorithm for this problem is known: all known algorithms have a worst-case *exponential* time complexity. Computer scientists have found that there exist thousands of problems such as this that arise in many practical contexts. They have not yet found a method to construct a *polynomial* algorithm for these problems. However, they have discovered another promising approach:

> They have found a way to formally define a class called "NP-complete" such that finding a polynomial algorithm for even *one* of the problems in the NP-complete class will allow one to find a polynomial algorithm for *all* of the problems in the NP-complete class. Furthermore, most of these thousands of problems that have confounded scientists have been shown to belong to the NP-complete class.

In short, scientists now have the strong hope of resorting to the maxim introduced in Chapter 1, namely: "solving one implies solving all," meaning solving even one NP-complete problem using a polynomial-time algorithm will provide a polynomial-time algorithm for the thousands of known NP-complete problems.

The aforesaid techniques rely on measuring the runtime of nondeterministic algorithms in a certain way which will be made precise in Chapter 20, but briefly consists of the following approach:

> If an NDTM can solve a certain problem P in polynomial-time, then the problem belongs to the class "NP." If, in addition, problem P belongs to the "NP-hard" class, then this combination (being in NP and NP-hard) ensures that P is in the NP-complete class.

15.5.1 Guess and check

While all this may sound bizarre, the fundamental ideas are quite simple. The crux of the matter is that many problems can be solved by the "guess and check" approach. For instance, finding the prime factors of very large numbers is hard. As [98] summarizes, it was conjectured

---

[4] If you walk into a room full of people, and imagine drawing a graph of *who knows each other mutually*, then a k-clique exists wherever all pairs within a group of k people know each other. Whether there exists a group of such "tight-knit" people is, essentially, the clique problem.

by Mersenne[5] that $2^{67} - 1$ is prime. This conjecture remained open
for two centuries until Frank Cole showed, in 1903, that it wasn't; in
fact, $2^{67} - 1 = 193707721 \times 761838257287$. Therefore, if we could somehow have guessed that 193707721 and 761838257287 are the factors of
$2^{67} - 1$, then checking the result would have been extremely easy! The
surprising thing is that there are two gradations of "difficulty" among
problems for which only exponential algorithms seem to exist:

- those for which short guesses exist, *and* checking the guesses is easy,
  and
- those for which the existence of short guesses is, as yet, *unknown*.

To sum up:

- If, for a problem p, we can generate a "short guess" and check the
  guess efficiently, then p belongs to the class NP. *Clique* is in NP (as
  we will see in more detail in the next chapter) because a "guess" will
  be short (simply write out k of the graph nodes) and the "check" is
  easy (see if these nodes include a k-clique).
- If a problem is NP-complete, it is believed to be unlikely that it will
  have a polynomial algorithm, although this issue is open.
- Problems for which the guesses are not short, and also checking
  guesses is not easy, do not belong to NP. Hence, these problems are
  thought to be much harder to solve.

For instance, $\overline{Clique}$ is the problem: "the given graph does not have
a k-clique." There is no known way to produce a succinct guess of a
solution for this problem, let alone check the guess efficiently. Every
purported solution that a graph *does not* have a k-clique seems to
warrant providing a guess of a solution of the form, "this set of k nodes
does not span a clique; neither does this other set of k nodes; etc. etc."
This may, however, end up enumerating all k-node combinations, which
are exponential in number.

NDTMs are machines that make the study of complexity theory
in the above-listed manner possible. Their use in defining complexity-
classes such as NP-hard and NP-complete forms the main hope for
finding efficient algorithms for thousands of naturally occurring NP-
complete problems—or to prove that such algorithms cannot exist.

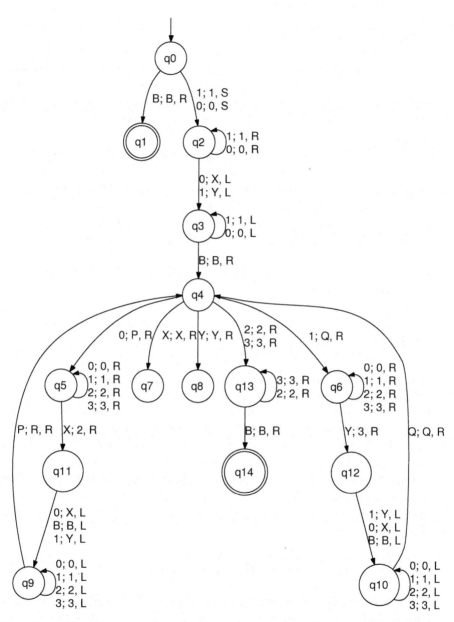

Fig. 15.3. A Nondeterministic Turing machine for *ww*. Letter 'S' means that the head stays where it is

15.5.2 An NDTM for ww

In Figure 15.3, we provide a nondeterministic Turing machine for the language of strings of the form ww, where $w \in \Sigma^*$. Letter 'S' in the edge from q0 to q2 means that the head stays where it is (can be simulated by an R followed by an L). This TM has to "guess" the midpoint; this happens in the initial nondeterministic loop situated at state q2. Notice that the Turing machine can stay in q2, skipping over the 0s and 1s or exit to state q3, replacing a 0 with an X or a 1 with a Y. This is how the Turing machine decides the midpoint; after this step, the Turing machine zigzags and tries to match and score off around the assumed midpoint. Any wrong guess causes this check phase to fail. One guess is guaranteed to win if, indeed, the input is of the form ww.

The animation of this NDTM in action using JFLAP would be highly intuitive, and the reader is strongly urged to do so.

15.6 Simulations

We show that having multiple tapes or having nondeterminism does not change the inherent power of a Turing machine.

15.6.1 Multi-tape vs. single-tape Turing machines

A k-tape Turing machine simply maintains k instantaneous descriptions. In each step, its δ function specifies how each ID evolves. One can simulate this behavior on a single tape Turing machine by placing the k logical tapes as segments, end-to-end, on a single actual tape, and also remembering where the k tape heads are through "dots" kept on each segment. One step of a k-tape Turing machine now becomes a series of k activities conducted one after the other on the k tape segments.

15.6.2 Nondeterministic Turing machines

A nondeterministic Turing machine can be conveniently simulated on a single tape deterministic Turing machine. However, it is much more convenient to explain how a nondeterministic Turing machine can be simulated on a multi-tape deterministic Turing machine. For the ease

---

[5] Prime numbers of the form $2^p - 1$ are known as Mersenne primes. The 42nd known Mersenne prime was discovered on February 18, 2005 and is $2^{25964951} - 1$, having 7,816,230 decimal digits.

of exposition, we will carry this explanation out with respect to the example given in Figure 15.3.

We will employ a 3-tape deterministic Turing machine to simulate the NDTM in Figure 15.3 (hereafter called *ww_ndtm*). The first tape maintains a read-only copy of the initial input string given to *ww_ndtm*. We call it the input tape. The second tape maintains a tree path in the nondeterministic computation tree of *ww_ndtm*. We call it the tree path tape. The third tape is the "working tape."

Tree path conventions

Notice that state q2 is the only nondeterministic state in *ww_ndtm*; and hence, the computation tree of *ww_ndtm* will have a binary nondeterministic split every so often—whenever the nondeterminism in q2 is invoked. The tree paths in *ww_ndtm* can be specified as follows, with the associated computations shown next to it:

ε—the empty computation beginning at $q0$,
0—the computation $q0 \to q1$ of length 1,
1—the computation $q0 \to q2$,
1, 0—the computation $q0 \to q2 \to q2$, and
1, 1—the computation $q0 \to q2 \to q3$.

We will uniformly use the 0 path for a "self" loop (if any), and 1 for an exit path; example: 1, 0 and 1, 1 above. The only exception to this convention is at state $q4$ where there are three exits, and we can number them 0, 1, and 2, going left to right. Therefore, note that the tree path 1,1,1,4,0 refers to the march q0, q2, q3, q4, q13, and back to q13. Notice that we do not have a path 0, 0 or 0, 1, as state $q1$ has no successor. When we require the next path in numeric order to be generated below, we skip over such paths which do not exist in the computation tree, and instead go to the next one in numeric order.

15.6.3 The Simulation itself

Here is how the simulation proceeds, with the second tape (tree path tape) containing ε:

- If the machine has accepted, accept and halt.
- Copy the first (input) tape to the third (working) tape.
- Generate the next tree path in numeric order on the tree path tape.

- Pursue the δ function of the NDTM according to the tree path specified in the tree path tape, making the required changes to the working tape. If, at any particular point, the tree path does not exist or the symbol under the TM tape head does not match what the δ function is forced to look, move on to the next option in the tree path enumeration.
- Repeat the above steps.

For example, if the input $0, 1, 0, 1$ is given on the input tape, the simulation will take several choices, all of which will fail except the one that picks the correct midpoint and checks around it. In particular, note that in state $q2$, for input string $0, 1, 0, 1$, the self-loop option can be exercised at most three times; after four self-loop turns, the Turing machine is faced with a blank on the tape and gets stuck (rejects). The outcome of this simulation is that

> the given NDTM accepts ww if and only if the DTM that simulates the NDTM as described above accepts.

The astute reader will, of course, have noticed that we are walking the exponential computation tree of the nondeterministic Turing machine *repeatedly*. In fact, we are not even performing a breadth-first search (which would have been the first algorithm thought of by the reader) because

- Performing breadth-first search (BFS) requires maintaining the frontier
- Even after the extra effort towards maintaining a BFS frontier, the overall gain is not worth it: we essentially might achieve an $O(2^n)$ computation instead of an $O(2^{n+1})$ computation.

Chapter Summary

This chapter provided a historical account of TMs. It presented some examples of DTMs and NDTMs. It was shown that NDTMs and DTMs have the same expressive power. We hope to provide enough intuitions about these topics to permit users to appreciate the benefits of formal methods to system construction. The next chapter introduces languages defined by TMs more formally, and also discusses several known undecidability results as well as proof techniques for showing undecidability.

Exercises

15.1. Consider a *Queue Machine* (QM) - a variant of a Turing machine which uses an unbounded queue instead of two stacks or an infinite tape. Assume that the head and tail of this queue are available for manipulation. QM may, in one step, read what is at the head of its queue, dequeue this item, change the item to something belonging to the "tape alphabet" Γ, and enqueue this item back. Is QM equivalent in power to a regular Turing machine? Justify your answer.

15.2. Argue that performing a BFS — as opposed to walking the computation tree repeatedly as in Section 15.6.3 — does not reduce the asymptotic worst-case complexity (the "big $O()$ complexity").

15.3. Argue that we cannot perform a DFS search in the simulation discussed in Section 15.6.3, and still claim the equivalence between the NDTM and the DTM that simulates it.

In all the JFLAP experiments requiring simulation to check various machines, conduct a sufficient number of simulation runs to cover a reasonable number of corner cases, choosing strings inside the language of interest as well as strings outside.

15.4. Using JFLAP, develop a TM for $rev(L_1)$ where

$$L_1 = \{a^i b^j c^k d^l \mid i,j,k,l \geq 0 \wedge \text{ if } i = 1 \text{ then } (j = k) \text{ else } (k = l)\}.$$

Obtain a DTM if possible; if not, explain why a DTM is not possible to find, and obtain an NDTM. Simulate and check.

15.5. Build a deterministic Turing machine in JFLAP that recognizes the language

$$L_{subseq} = \{x \# y \mid x, y \in \{0,1\}^* \wedge x \text{ is a subsequence of } y\}.$$

Simulate and check.

15.6. Build a nondeterministic Turing machine in JFLAP for L_{subseq}. Simulate and check.

And thus the Turing Machine was born.

16

Basic Undecidability Proofs

In this chapter, we define the notion of *decidability, semi-decidability,* and *undecidability.* These notions pertain to *degrees of solvability* of problems by Turing machines. We will present three proof methods in this chapter: (i) through contradiction (Section 16.2.3), (ii) through reductions from languages unknown to be decidable (Section 16.2.4), and (iii) through mapping reductions (Section 16.2.5). Chapter 17 will discuss two additional proof methods: (iv) Rice's theorem, and (v) computational history method.

Methods (ii), (iii), and (iv) are strongly related to each other, in the following sense:

- Applications of method (ii), namely reduction *from* a known undecidable language, A, *to* the language in question, B, employs an 'if and only if' argument of the form $x \in A \Leftrightarrow f(x) \in B$.
- Method (iii), namely *mapping reductions*, isolates this 'if and only if' argument into a *mapping reduction* principle which is quite powerful, and also applicable in other contexts (e.g., in our study of NP-completeness in Chapter 19).
- Method (iv), namely *Rice's Theorem*, capitalizes on the core argument underlying all these proofs. It tends to make the connection that

    Hidden in every undecidability proof is a proof of the undecidability of the Halting problem.

    Rice's Theorem makes it even more convenient to carry out undecidability proofs.

The parallels between this chapter and Chapter 19 are worth reiterating. In this chapter, we present mapping reductions of the form $A \leq_m B$ where there exists a function f such that $x \in A \Leftrightarrow f(x) \in B$.

All we require of f is that it be a total computable function. In Chapter 19, we will define $A \leq_P B$ where the function in question will be one that has polynomial runtime. Also, in Chapter 19, we will point out the analogous fact that

> *Hidden in every NP-complete problem is a proof of the NP-completeness of Boolean satisfiability.*

With these introductions, we now proceed to study the three reduction methods.

At this juncture, it pays to recall the facts introduced in Chapter 1, namely:

- We can deem two computers to be equivalent if they can solve the same class of problems — *ignoring the actual amount of time taken* (see discussions on page 5).
- Problem solving can be modeled in terms of deciding membership in languages (page 5).
- Some problems are *unsolvable*. Formally stated, there exists a class of problems P such that for any $p \in P$, one can model p using a language L_p such that L_p admits no membership deciders.
- Other problems have deciders, but these may take different amounts of runtime to decide language membership.

In Section 16.1.1, we examine a list of *decidable* problems with a view to sharpen our intuitions in this area. The *existence* of deciders is shown by presenting the pseudocode of an algorithm. In Section 16.1.2, we examine a list of *un*decidable problems.

16.1 Some Decidable and Undecidable Problems

16.1.1 An assortment of decidable problems

In all descriptions below, we use $\langle \rangle$ to indicate the *code* or *description*; for instance, $\langle G \rangle$ stands for a grammar G's description as a character- or bit-string. Also, for a Turing machine M, $\langle M \rangle$ will mean its description, say in the form of a table such as in Figure 15.1. Sometimes, we omit $\langle \ldots \rangle$ if the intent is clear from the context.

$\diamond$ ALL_{DFA}: Given a DFA, is its language Σ^*?
This problem is modeled as a language membership question in the language:

$$ALL_{DFA} = \{\langle A \rangle \mid A \text{ is a DFA that recognizes } \Sigma^*\}.$$

ALL_{DFA} can be shown to be decidable by minimizing the given DFA and examining the result.

◇ $A\varepsilon_{CFG}$: Given a CFG, does it generate ε?

$$A\varepsilon_{CFG} = \{\langle G \rangle \mid G \text{ is a CFG that generates } \varepsilon\}.$$

One approach is to trace all ε-generating productions using a bottom-up marking algorithm similar to how we found generating non-terminals in Section 13.4.2, except we focus on which non-terminals generate ε. Call this notion ε-*generating*. Now, for any $a \in \Sigma$, a is not ε-generating. If $A \to \varepsilon$, then A is ε-generating. If $A \to B_1 \ldots B_n$ and if all of B_i are ε-generating, then so is A. Finally, check whether the start symbol S is ε-generating.

◇ $INFINITE_{DFA}$: Given a DFA, does it have an infinite language?

$$INFINITE_{DFA} = \{\langle A \rangle \mid A \text{ is a DFA and } L(A) \text{ is Infinite}\}.$$

Exercise 16.2 asks you to show that this is a decidable language.

◇ $NOODD_{DFA} =$

$$\{\langle A \rangle \mid A \text{ is a DFA that does not accept any string with odd 1s}\}.$$

Exercise 16.3 asks you to show that this is a decidable language.

◇ $ONESTAR_{CFG}$: Given a CFG, does it include *some* strings from 1^*?

$$ONESTAR_{CFG} = \{\langle G \rangle \mid G \text{ is a CFG over } \{0,1\} \text{ and } 1^* \cap L(G) \neq \emptyset\}$$

One algorithm is to build the product machine of a DFA for 1^* and a PDA for G with a view to obtain the intersection of their languages. The result will be a PDA. We can then check the language emptiness of this PDA, which is decidable (e.g., by converting the resulting PDA to a CFG and running the bottom-up marking algorithm on the CFG to see if the start symbol, S, of the CFG, is generating). Details of this product construction algorithm are left to the reader (Exercise 16.4, which offers some hints also). Exercise 16.5 requests an alternative algorithm.

◇ $ALLSTAR_{CFG}$: Given a CFG, does it include *all* strings from 1^*?

$$ALLSTAR_{CFG} = \{\langle G \rangle \mid G \text{ is a CFG over } \{0,1\} \text{ and } 1^* \subseteq L(G)\}.$$

Exercise 16.6 requests an algorithm (also offers hints).

$\diamond$ $EMPTY_{CFG}$:

$$EMPTY_{CFG} = \{\langle G\rangle \mid \text{ is a CFG and } L(G) = \emptyset\}.$$

One can employ the marking algorithm of Section 13.4.2 to decide whether starting from the initial non-terminal, S, one can generate a terminal-only string.

16.1.2 Assorted undecidable problems

We now present a list of undecidable problems and sketch reasons for them to be undecidable. In Section 16.2, we present the actual undecidability proofs, after introducing basic notions such as recursive enumerability.

- The universality of the language of a CFG is undecidable. A language L is universal if $L = \Sigma^*$.
- The equivalence of two CFGs is undecidable.
- In-equivalence of two CFGs is undecidable.
- Whether a given Turing machine accepts string w is undecidable.
- Whether a given Turing machine halts on string w is undecidable.
- The emptiness of the language of a Turing machine is undecidable.
- Whether a given Turing machine's language is context-free is undecidable.

Here are intuitive arguments that support the above claims:

- $L_G = \Sigma^*$: Informally, given a CFG G, it seems one must find a string that G cannot generate. One can, of course, keep checking the strings within Σ^* in an ascending order of lengths, with each check taking a finite amount of time; however, this process does not have a definite stopping criterion. *A formal proof will be given later,* but intuitively it does appear that this is undecidable – and let us assume so for the purpose of supporting the following discussions.
- $L_{G_1} = L_{G_2}$: If this were to be decidable, we would be able to decide $L_G = \Sigma^*$.
- $L_{G_1} \neq L_{G_2}$: We observe that we must examine whether every string generated by G_1 is generated by G_2, and vice versa. This appears to be a search with no definite stopping criterion. However, since $L_{G_1} = L_{G_2}$ is not decidable, $L_{G_1} \neq L_{G_2}$ cannot be decidable (if set S is decidable, so is $\overline{S}$; otherwise, we can employ the algorithm to decide $\overline{S}$ as an algorithm to decide S. Please think why).

- We will state and prove results about the halting and acceptance of Turing machines. It also will turn out that we can find similar proofs for the emptiness of the language of a Turing machine being undecidable, and for whether a given Turing machine's language is context-free being undecidable. This series of undecidability results about Turing machines will be captured by one "master theorem" called Rice's Theorem.

In Section 16.2, we will motivate the important concept of *recursive enumerability*. We will show that all pairs $\langle G_1, G_2 \rangle$ such that $L_{G_1} \neq L_{G_2}$ are enumerable, in the sense that every such pair can be found in a finite amount of time and printed out. This will mean that the language of pairs $\langle G_1, G_2 \rangle$ such that $L_{G_1} = L_{G_2}$ is *not* enumerable (if a set S and its complement are enumerable, then membership in S becomes decidable, as will be re-iterated soon). This is another reason why the language of pairs $\langle G_1, G_2 \rangle$ such that $L_{G_1} \neq L_{G_2}$ is not decidable.

16.2 Undecidability Proofs

16.2.1 Turing recognizable (or recursively enumerable) sets

The terms Turing recognizable (TR) and recursively enumerable (RE) will be used interchangeably as they essentially mean the same thing, but from two different perspectives. A language L is TR if it is the language of *some* Turing machine M. We write L_M for emphasis. A language L is RE if there exists a Turing machine M that can enumerate the strings in L (say, on an "output tape") such that any member $x \in L$ is guaranteed to appear in a finite amount of time.

> Notice that if L is L_M for one TM M, then it is the language of an infinite number ($\aleph_0$) of other machines, M', as we can simply pad M with i "no op" instructions, for every $i \in Nat$.

For instance,

$$NEQ_{CFG} = \{\langle G_1, G_2 \rangle \mid G_1, G_2 \text{ are CFGs and } L(G_1) \neq L(G_2)\}$$

is TR, with a candidate Turing machine being the following:

- Input: $\langle G_1, G_2 \rangle$.
- Output: If $\langle G_1, G_2 \rangle \in NEQ_{CFG}$, then "accept," else "loop."
- Method:
  - Generate strings $x \in \Sigma^*$ in numeric order, feeding x to a parser (PDA) for G_1 and another for G_2.

 – For each such x, if x can be parsed by G_1 and not by G_2 (or vice versa), go to *accept*.

Notice that we do not say "or else, keep continuing." This is typical of *semi-algorithms* which may keep continuing *ad infinitum* – essentially going into an "infinite loop."

A recursively enumerable (RE) language is the language that an *enumerator* Turing machine can enumerate. An enumerator Turing machine is a Turing machine that has no input tape, but has an output tape. In addition, it may employ a working tape. It keeps generating (finite) strings, and appends each generated string to the output tape. Here is an enumerator for NEQ_{CFG}:

- Keep enumerating all possible pairs of grammars $\langle G_1, G_2 \rangle$ over the given alphabet, on a working tape. This is possible (see Exercise 13.14) because the *syntax* of any context-free grammar over a given set of terminals and non-terminals is expressible as a regular expression, and one can generate random strings and filter those passing the regular expression as a legal CFG.
- Keep enumerating strings $x \in \Sigma^*$ in enumeration order, also on the working tape.
- Run one additional step of a parsing algorithm for *all* the grammar pairs $\langle G_1, G_2 \rangle$ generated so far acting on *all* the inputs x generated so far (these are called the 'simulations in progress').
- If one of the simulations in progress reports that the parser for grammar G_1 accepted an x while the parser for the corresponding grammar G_2 rejected x (or vice versa), then write $\langle G_1, G_2 \rangle$ on the output tape.

The above enumerator guarantees that every pair of nonequivalent grammars will, eventually, be enumerated. From the above constructions, it is an easy exercise to conclude the following theorem.

Theorem 16.1. A language is TR if and only if it is RE.

The main argument is that given an enumerator (in the sense of RE), we can build a recognizer (in the sense of TR), and vice versa.

Dovetailing and systematic enumeration methods

In many of our Turing machine constructions, we face the situation of enumerating the Cartesian product of a collection of sets S_i, $i \in k$ for some $k \in N$. For example, we may have to enumerate pairs of grammars and strings, thus effectively enumerating triples from sets that, individually, have $\aleph_0$ cardinality. There are many systematic approaches for achieving this end; we now summarize a standard approach, taking triples of Nat as an example:

- Enumerate all the triples that add up to i before enumerating any triple that adds up to $i+1$. Within each group that adds up to i, employ a lexicographic order.
- As an example, enumerate $\langle 0,0,0\rangle$, followed by $\langle 0,0,1\rangle$, $\langle 0,1,0\rangle$, $\langle 1,0,0\rangle$, followed by $\langle 0,1,1\rangle$, $\langle 1,0,1\rangle$, $\langle 1,1,0\rangle$, $\langle 0,0,2\rangle$, $\langle 0,2,0\rangle$, $\langle 2,0,0\rangle$, etc.

16.2.2 Recursive (or decidable) languages

A recursive language L_M is the language of a Turing machine M that, given any $x \in L_M$, accepts x, and given any $y \notin L_M$, rejects y. In other words, M does not loop on any input. Note that it is possible to have another machine N such that $L(M) = L(N)$ and N loops on inputs $y \notin L_M$; however, so long as there exists *one* decider M, we can conclude that L_M is recursive (or decidable).

Another *very important* characterization of recursive languages is this:

Theorem 16.2. L is recursive if and only if L and $\overline{L}$ are RE (equivalently, are TR).

Imagine an enumerator enumerating L and another enumerating $\overline{L}$. To decide whether some $x \in \Sigma^*$ is in L, all we need to do is watch which enumerator outputs x.[1] Decidable languages correspond to algorithmically solvable problems.

Non-RE languages

The cardinality of the set of all Turing machine descriptions is $\aleph_0$. This is because a Turing machine can be described through a finite number of bits that model its states and its transitions, and such a description can be read as a natural number (these numbers are known as Gödel numbers). On the other hand, there are $\aleph_1$ languages over Σ. Therefore, there are non-TR (non-RE) languages.

> This means that there exist languages in which membership testing cannot be carried out by *any* Turing machine.

Exercise 16.8 asks you to construct such a language.

---

[1] Imagine two big 'spigots,' one pouring out the contents of L and another pouring out the contents of $\overline{L}$. One can decide membership of $x \in L$ by watching which spigot emits x.

16.2.3 Acceptance (A_{TM}) is undecidable (important!)

This is one of the most fundamental results that we shall encounter in our study of Turing machines and computation. Define

$$A_{TM} = \{\langle M, w \rangle \mid M \text{ is a Turing machine that accepts string } w\}.$$

Deciding membership in A_{TM} is tantamount to asking "does a given Turing machine M accept a string w?" We prove this set to be *undecidable* through contradiction, as follows:

- **Suppose** there exists a decider H for A_{TM}. H expects to be given a Turing machine M and a string w. Notice that "giving H a Turing machine" means "giving it a character string representing a Turing machine program." Hence, in reality, we will be feeding it $\langle M, w \rangle$ as mentioned in Section 16.1.1.
- Build a program called D as follows:
  - D takes a single argument M.
  - As its first step, D invokes H on $\langle M, M \rangle$.[2]
  - If $H(\langle M, M \rangle)$ rejects, $D(\langle M \rangle)$ accepts.
  - If $H(\langle M, M \rangle)$ accepts, $D(\langle M \rangle)$ rejects.
- Now we can ask what $D(\langle D \rangle)$ will result in (to preserve the clarity of our arguments, the reader is invited to suppress any occurrence of $\langle \ldots \rangle$ in the text below):
  - The $D(\langle D \rangle)$ "call" turns into an $H(\langle D, D \rangle)$ call.

  - Suppose $H(\langle D, D \rangle)$ rejects. In that case, $\boxed{D(\langle D \rangle) \text{ accepts.}}$
  - But, according to the advertised behavior of H — which is that it is a decider for A_{TM} — the fact that $H(\langle D, D \rangle)$ rejects means that D is *not* a Turing machine that will accept $\langle D \rangle$, or that $\boxed{D(\langle D \rangle) \text{ rejects!}}$

  - Suppose $H(\langle D, D \rangle)$ accepts. In that case, $\boxed{D(\langle D \rangle) \text{ rejects.}}$
  - But, according to the advertised behavior of H — which is that it is a decider for A_{TM} — the fact that $H(\langle D, D \rangle)$ accepts means that D *is* a Turing machine that accepts $\langle D \rangle$, or that $\boxed{D(\langle D \rangle) \text{ accepts!}}$

Therefore, we obtain a contradiction *in both cases*.[3] Hence, the claimed decider H for A_{TM} cannot exist, or

---

[2] Basically, we feed $\langle M \rangle$ twice over, just to 'please' H that expects two arguments.
[3] Please recall our discussions of Section 2.3.2.

The *acceptance problem* for Turing machines is undecidable. □

16.2.4 Halting ($Halt_{TM}$) is undecidable (important!)

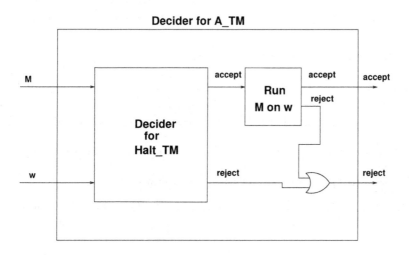

Fig. 16.1. A_{TM} to $Halt_{TM}$ reduction. Notice that *if* we assume that the inner components – namely the OR-gate, the ability to run M on w, and $D_{Halt_{TM}}$ exist, then D_{ATM} can be constructed; and hence, $D_{Halt_{TM}}$ cannot exist!

In Chapter 1, page 11, we presented the idea of "solving one implies solving all." This idea is captured by the central concept of *reduction*.

The golden rule of reduction is: **Reduce an existing ("old") undecidable problem to the given ("new") problem.** This way, if we assume that the new problem is decidable, we would be forced to conclude that the existing undecidable problem is decidable — a contradiction. See Figure 16.1 where the "*new problem*" is $Halt_{TM}$, and by assuming that it is decidable, we can assume the existence of the decider $D_{Halt_{TM}}$, and using it, build a decider for A_{TM}, the "*old problem*" already shown to be undecidable; this obtains a contradiction. Hence, $D_{Halt_{TM}}$ cannot exist.

Don't get reduction backwards!

We should not go by the English language meaning of the term "reduction" that can lead us astray. Getting the meaning of 'reduction' backwards means trying to reduce the new problem to an existing ("old") problem. This does not help us. To see why, assume this to be the right direction. Then we will be trying to employ proof by contradiction (following reduction) in the following futile way: "IF the existing undecidable problem is decidable, THEN we would have shown the new problem to be decidable." However, this is a statement of the form "IF false, THEN assert X." Clearly, in this case, we cannot assert X.

An introduction to mapping reduction

Reduction is an abbreviation for *mapping reduction* — a concept fully explored in Section 16.2.5. In this section, we informally apply the (mapping) reduction idea to show that $Halt_{TM}$ is undecidable. Define

$$Halt_{TM} = \{\langle M, w \rangle \mid M \text{ is a Turing machine that halts on string } w\}.$$

We show $Halt_{TM}$ to be undecidable as follows:

> Suppose not; i.e., there is a decider for $Halt_{TM}$. Let's now build a decider for A_{TM} (call it $D_{A_{TM}}$). $D_{A_{TM}}$'s design will be as follows:
> - $D_{A_{TM}}$ will first feed M and w to $D_{Halt_{TM}}$, the claimed decider for $Halt_{TM}$.
> - If $D_{Halt_{TM}}$ goes to $accept_{D_{Halt_{TM}}}$, $D_{A_{TM}}$ knows that it can safely run M on w, which it does.
> - If M goes to $accept_M$, $D_{A_{TM}}$ will go to $accept_{D_{A_{TM}}}$.
> - If M goes to $reject_M$, or if $D_{Halt_{TM}}$ goes to $reject_{D_{Halt_{TM}}}$, $D_{A_{TM}}$ will go to $reject_{D_{A_{TM}}}$.

Notice that we have labeled the accept and reject states of the two machines $D_{Halt_{TM}}$ and $D_{A_{TM}}$. After one becomes familiar with these kinds of proofs, higher-level proof sketches are preferred. Here is such a higher-level proof sketch:

> Build a decider for A_{TM}. This decider accepts input $\langle M, w \rangle$ and runs *Halt_decider* (if it exists) on it. If this run accepts, then we can safely (without the fear of looping) run M on w, and return the accept/reject result that this run returns; else return "reject."

A diagram that illustrates this construction is in Figure 16.1. Therefore, we conclude that

> the *Halting problem* for Turing machines is undecidable.   □

Two observations that the reader can make after seeing many such proofs (to follow) are the following:

- One *cannot* write statements of the form "if $f(x)$ loops, then ..." in any algorithm, because termination is not detectable. Of course, one *can* write "if $f(x)$ halts, then ... ." This asymmetry is quite fundamental, and underlies all the results pertaining to halting / acceptance.
- One cannot examine the code ("program") of a Turing machine and decide what its language is. More precisely, one cannot build a classifier program Q that, given access only to Turing machine programs P_m (which encode Turing machines m), classify the m's into two bins (say "good" and "bad") according to the language of m. Any such classifier will have to classify all Turing machines as "good" or all as "bad, " or itself be incapable of handling all Turing machine codes (not be *total*). This result will be proved in Chapter 17 as *Rice's Theorem*.

16.2.5 Mapping reductions

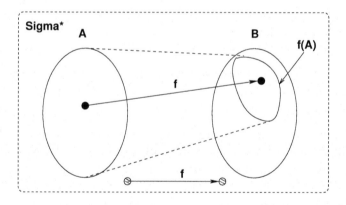

Fig. 16.2. Illustration of mapping reduction $A \leq_M B$

Definition 16.3. A *computable* function $f : \Sigma^* \to \Sigma^*$ is a mapping reduction from $A \subseteq \Sigma^*$ into $B \subseteq \Sigma^*$ if for all $x \in \Sigma^*$, $x \in A \Leftrightarrow f(x) \in B$.

In Chapter 19, we will employ polynomial-time mapping reductions, which are denoted by $\leq_P$.

```
How a decider for A_TM is obtained:

Step 1: Here is the initial tape.
-----------------------------------------------------------------
| M | w |
-----------------------------------------------------------------

Step 2. Build M' and put it on the tape
-----------------------------------------------------------------
| M | w | ..build M' that incorporates M here.. |
-----------------------------------------------------------------

Step 3. Put w on the tape.
-----------------------------------------------------------------
| M | w | ..build M' that incorporates M here.. | ..put w here.. |
-----------------------------------------------------------------

Step 4. Run Halt_TM_decider on M' and w  and return its decision
-----------------------------------------------------------------
| M | w | ..build M' that incorporates M here.. | ..put w here.. |
-----------------------------------------------------------------
```

$$D_{Halt_{TM}}(M',w) = \begin{cases} accepts \Rightarrow M' \ halts \ on \ w \Rightarrow & M \ accepts \ w \\ rejects \Rightarrow M' \ doesn't \ halt \ on \ w \Rightarrow & M \ doesn't \ accept \ w \end{cases}$$

Fig. 16.3. How the mapping reduction from A_{TM} to $Halt_{TM}$ works

Definition 16.4. A polynomial-time mapping reduction $\leq_P$ is a mapping reduction where the reduction function f has polynomial-time asymptotic upper-bound time complexity.[4]

See Figure 16.2 which illustrates the general situation that A maps into a subset denoted by $f(A)$ of B, and members of A map into $f(A)$ while non-members of A map outside of B (that means they map outside of even $B \setminus f(A)$). Also note that A and B need not be disjoint sets, although they often are. A mapping reduction can be (and usually is) a non-injection and non-surjection; *i.e.*, it can be many to one and not necessarily onto. It is denoted by $\leq_m$. By asserting $A \leq_m B$, the existence of an f as described above is also being asserted.
Typically mapping reductions are used as follows:

- Let A be a language known to be undecidable ("old" or "existing" language).

---

[4] Using the familiar notation $\mathcal{O}(\dots)$ for asymptotic upper-bounds, polynomial-time means $\mathcal{O}(n^k)$ for an input of length n, and $k > 1$. See [5, 88, 29] for details.

```
M'(x) {
    if x <> w then loop ; // could also goto reject_M' here
    Run M on w ;
    If M accepts w, goto accept_M' ;
    If M rejects w, goto reject_M' ; }
```

How a decider for E_TM is obtained:

Step 1: Build above M' and put it on the tape

| M | w | ..build M' that incorporates M and w here.. |

Step 2: Run E_TM_decider on M' and return its decision

| M | w | ..build M' that incorporates M and w here.. |

$$Decider_{E_{TM}}(M') = \begin{cases} accepts \Rightarrow L(M') \ is \ empty \Rightarrow & M \ does \ not \ accept \ w \\ rejects \Rightarrow L(M') \ is \ not \ empty \Rightarrow & M \ accepts \ w \end{cases}$$

Fig. 16.4. Mapping reduction from A_{TM} to E_{TM}

```
M'(x) {
    if x is of the form 0^n 1^n then goto accept_M' ;
    Run M on w ;
    If M accepts w, goto accept_M' ;
    If M rejects w, goto reject_M' ; }
```

$$Decider_{Regular_{TM}}(M') = \begin{cases} accepts & \Rightarrow L(M') \ is \ regular \\ \quad \Rightarrow Language \ is \Sigma' & \Rightarrow M \ accepts \ w \\ rejects & \Rightarrow L(M') \ is \ not \ regular \\ \quad \Rightarrow Language \ is \ 0^n1^n & \Rightarrow M \ does \ not \ accept \ w \end{cases}$$

Fig. 16.5. Mapping reduction from A_{TM} to $Regular_{TM}$

- Let B be the language that must be shown to be undecidable ("new" language).
- Find a mapping reduction f from A into B.
- Now, if B has a decider D_B, then we can decide membership in A as follows:
  - On input z, in order to check if $z \in A$, find out if $D_B(f(z))$ accepts or not. If it accepts, then $z \in A$, and if it rejects, then $z \notin A$.

Mapping Reduction From A_{TM} to $Halt_{TM}$

We first illustrate mapping reductions by taking $A = A_{TM}$ and $B = Halt_{TM}$ with respect to Figure 16.2. Function f takes a member of A_{TM}, namely a pair $\langle M, w \rangle$, as input, and prints out $\langle M', w \rangle$ on the tape as its output. Function f, in effect, generates the *text* of the program M' from the text of the program M. Here is the makeup of M':

$M'(x) =$
    Run M on x
    If the result is "accept," then "accept"
    If the result is "reject," then loop

Notice that the text of M' has "spliced" within itself a copy of the text of program M that was input. Mapping reductions such as f illustrated here need not "run" the program they manufacture; they simply accept a program such as M, and a possible second input, such as w, and manufacture another program M' (and also copy over w) and then consider their task done! The reason such a process turns out to be useful is for the following reasons:

> Suppose someone were to provide a decider for $Halt_{TM}$. The mapping reduction f then makes it possible to obtain a decider for A_{TM}. When given $\langle M, w \rangle$, this decider will obtain $\langle M', w \rangle = f(\langle M, w \rangle)$, and then feed it to the decider for $Halt_{TM}$.

We have to carefully argue that f is a mapping reduction. We will be quite loose about the argument types of f (namely that it maps Σ^* to Σ^*; we will assume that any $\langle M, w \rangle$ pair can be thought to be a string, and hence a member of a suitable Σ^*. The proof itself is depicted in Figure 16.3.

Mapping reduction From A_{TM} to E_{TM}

We show that

$$E_{TM} = \{\langle M \rangle \mid \text{M is a TM and } L(M) = \emptyset\}$$

is undecidable through a mapping reduction that maps $\langle M, w \rangle$ into $\langle M' \rangle$, as explained in Figure 16.4.

Mapping reduction from A_{TM} to $Regular_{TM}$

Similarly, we can prove $Regular_{TM}$ to be undecidable by building the M' shown in Figure 16.5.

16.2.6 Undecidable problems are "A_{TM} in disguise"

This chapter covered many important notions including decidability, semi-decidability (TR, RE), and mapping reductions. The techniques discussed here lie at the core of the notion of "problem solving" in that they help identify which problems possess algorithms and which do not.

A closing thought to summarize the proofs in this chapter is the slogan that undecidable problems are A_{TM} in disguise. We leave you with this thought, hoping that it will provide you with useful intuitions. Section 19.3.2 proposes a similar slogan with respect to NP-complete problems.

Exercises

16.1. Present an alternate proof for $A\varepsilon_{CFG}$ being decidable by converting the given grammar to the Chomsky normal form and then looking for something special about the resulting grammar.

16.2. Show that $INFINITE_{DFA} =$

$$\{\langle A\rangle \mid A \text{ is a DFA and L(A) is Infinite}\}$$

is decidable.

16.3. Describe an algorithm to decide $NOODD_{DFA}$.

16.4. Describe an algorithm to obtain a PDA whose language is the intersection of the language of a given DFA and a PDA.

Hint: The procedure similar to that described in Section 10.2.2. The main difference occurs when the PDA has an ε move while the DFA, quite naturally, does not have such a move. In this case, the product construction proceeds by advancing the PDA control state while arresting the DFA's control state in its current state. Prove this algorithm to be correct.

16.5. Describe another algorithm for the decidability of $ONESTAR_{CFG}$.

16.6. Describe an algorithm to decide $ALLSTAR_{CFG}$. *Hint:* Consider all production rules to be *simple* – no | on the right-hand side. Next, eliminate any production whose right-hand side sentential form includes a 0 (the only symbol other than 1 in Σ). We now have a CFG over a singleton alphabet which is regular, from Illustration 14.7.3. Complete all these steps.

16.7. Argue that the set of all pairs of equivalent C programs is not recursively enumerable. Argue that the set of all pairs of inequivalent C programs is also not recursively enumerable.

16.8. Describe a non-TR language. *Hint:* Employ diagonalization to specify such a language with respect to an enumeration of all possible Turing machines, M.

16.9. Prove that if $A \leq_m B$, then $\overline{A} \leq_m \overline{B}$. Also prove that $\leq_m$ is a preorder.

16.10. Describe a mapping reduction from A_{TM} to CFL_{TM}, the set of Turing machines that have a context-free language.

16.11. Show that the language

$$EQL_{TM} = \{\langle M_1, M_2 \rangle \mid M_1 \text{ and } M_2 \text{ are TMs and } L(M_1) = L(M_2)\}$$

is neither Turing recognizable nor Co-Turing recognizable. A language L is Co-Turing recognizable if $\overline{L}$ is Turing recognizable.

16.12. Suppose $A \leq_m B$. Identify those assertions below that are true and those that are false. Then rigorously establish your results, providing proof-sketches or counterexamples:

(a) If A is TR then B is TR
(b) If B is TR then A is TR
(c) If A is not TR then B is not TR
(d) If B is not TR then A is not TR
(e) Variations of (a), (b), (c), and (d) with A and B replaced by $\overline{A}$ and $\overline{B}$ in all combinations (derive a sufficient number of the 255 variations, and test your understanding sufficiently)
(f) If A is decidable then B is decidable
(g) If B is decidable then A is decidable

17

Advanced Undecidability Proofs

In this chapter, we will discuss Rice's Theorem in Section 17.1, and the computational history method in Section 17.3. As discussed in Chapter 16, these are two additional important methods to approach the question of decidability.

17.1 Rice's Theorem

Rice's Theorem asserts:

Theorem 17.1. Every non-trivial partitioning of the space of Turing machine codes based on the languages recognized by these Turing machines is undecidable.

Rice's Theorem is, basically, a *general* statement about the undecidability of non-trivial partitions one can erect on Turing machine codes based on the language that the corresponding Turing machines accept. Stated another way, Rice's Theorem asserts the impossibility of building an *algorithmic* classifier for Turing machine codes ("programs") based on the *language* recognized by these Turing machines, if the classification attempted is anything but trivial (a trivial partition puts all the Turing machines into one or the other bin).

Relating these notions more to real-world programs, consider the language L consisting of all ASCII character sequences $s_1, s_2, \ldots$ such that each s_i is a C program c_i. Now suppose that each c_i, when run on inputs from $\{0,1\}$,* accepts only those sequences that describe a regular set. *Rice's Theorem says that languages such as L are not decidable.* In other words, it is impossible to classify all C programs (or equivalently TMs) into those *whose* languages are regular and those whose languages are non-regular. Mathematically, given a property $\mathcal{P}$, consider the set

$$L = \{\langle M \rangle \mid M \text{ is a Turing machine and } \mathcal{P}(Lang(M))\}.$$

Furthermore, let $\mathcal{P}$ be non-trivial — meaning, it is neither $\emptyset$ nor Σ^*. For example, $\mathcal{P}$ could be *"Regular"*; since there *are* TMs that encode regular sets and there are TMs that do not encode regular sets, $\mathcal{P}$ represents a non-trivial partition over TR languages. Rice's Theorem asserts that sets such as L above are undecidable.

17.2 Failing proof attempt

```
M'(x)
    Run M on w ;
    IF this run ends at state reject_M , THEN loop ;
    Manifest N;
    RUN N on x ;
    IF this run accepts, THEN goto accept_M' ;
    IF this run rejects, THEN goto reject_M' ;
```

Fig. 17.1. Machine M' in the proof of Rice's Theorem

For the ease of exposition, we present, as a special case of Rice's Theorem, the proof of Rice's Theorem for $\mathcal{P} = Regular$. Our first proof attempt will fail because of a small technical glitch. The glitch is caused by $\emptyset \in \mathcal{P}$, or in other words, allowing $\mathcal{P}(\emptyset)$. In our special case proof, this glitch manifests in the form of $\emptyset \in Regular$, as we are proving for the special case of $\mathcal{P} = Regular$. We fix this glitch in Section 17.2.1.

Proof, with a small glitch, of a special case of Rice's Theorem:

By contradiction, assume that L has a decider, namely D_L. The Turing machine D_L is capable of classifying Turing machine codes into those whose languages pass the predicate test *Regular*. As noted earlier, *Regular* is a non-trivial property. Therefore, given D_L, it should be possible to find at least *one* Turing machine— say N — whose language is regular (it does not matter which Turing machine this is). In particular, algorithm *Manifest N* is:

Find the first (in numeric order) string from Σ^* accepted by D_L.

Now, we use *Manifest N* in defining a machine M' shown in Figure 17.1. Then, using M' (built with respect to arbitrary M and x), we try to derive a contradiction:

- If M accepts w, $L(M') = Lang(N)$, which is regular. Hence, M' is in $Regular_{TM}$.
- If M does not accept w, $L(M') = \emptyset$. Unfortunately, $\emptyset$ is also regular! Hence, *in this case too, M' is in $Regular_{TM}$. Alas, this is the place where we ought to have obtained a contradiction.* Somehow, we "blew our proof."
- In other words, if we feed M' to D_L, we will get *true* regardless of whether or not M accepts w. Therefore, no contradiction results.

17.2.1 Corrected proof

Surprisingly, the proof goes through if we take $\mathcal{P} = non\text{-}regular$; in this case, we *will* obtain a contradiction. We redo our proof steps as follows:

- Now define L and D_L for the case of $\mathcal{P} = non\text{-}regular$.
- Define *Manifest N* using D_L, which manifests a Turing machine N whose language is not regular.
- Define M' in terms of N. In other words, in M''s body, we will manifest N and also use it.

Let us analyze the language of M':

- If M accepts w, $L(M') = Lang(N)$, which is *non-regular*. Hence, $L(M')$ is in *non-regular*.
- If M does not accept w, $L(M') = \emptyset$. $\emptyset$ is regular, and in this case, $L(M')$ is NOT in *non-regular*.
- Therefore, if the decider D_L exists, we can generate M' and feed it to D_L. Then we will see that if M accepts w, $D_L(M') = true$, and if M does not accept w, $D_L(M') = false$. Hence, if D_L exists, we can decide A_{TM}. Since this is a contradiction, D_L cannot exist; and hence, we obtain a contradiction!

Summary of the proof of Rice's Theorem: We observe that if we go with $\emptyset \notin \mathcal{P}$, then the proof succeeds. The general proof of Rice's Theorem *assumes* that $\mathcal{P}$ does not contain $\emptyset$. This approach is justified; *there is no loss of generality.* This is because if $\mathcal{P}$ indeed contains $\emptyset$, then we can always proceed with $\neg \mathcal{P}$ as our "new $\mathcal{P}$" and finish the proof. When we fail to obtain a contradiction with respect to $\mathcal{P}$ but obtain a contradiction with respect to $\neg \mathcal{P}$, we would have shown that D_L — the decider of Turing machine codes whose language is $\neg \mathcal{P}$ — is undecidable. But, this is tantamount to showing that D_L — the decider of Turing machine codes whose language is $\mathcal{P}$ — is undecidable.

Proof for a general non-trivial property $\mathcal{P}$

The reader can easily verify that substituting $\mathcal{P}$ for *Regular* in the previous proof makes everything work out. In particular, M''s language either would be $\emptyset$ or would be N. If $\emptyset \notin \mathcal{P}$, this would result in a full contradiction when D_L is fed M'. If $\emptyset \in \mathcal{P}$, then M' would be constructed with respect to $M_{has \neg \mathcal{P}}$'s language, and even here full contradiction will result. Therefore, we would end up showing that either L defined with respect to an arbitrary non-trivial property $\mathcal{P}$ has no decider, or that L defined with respect to $\neg \mathcal{P}$ has no decider.

17.2.2 Greibach's Theorem

There is a theorem analogous to Rice's Theorem for PDAs. Known as *Greibach's Theorem*, the high-level statement of the theorem (in terms of its practical usage) is as follows:

> It is impossible to algorithmically classify (using, of course, a Turing machine) context-free grammars on the basis of whether their languages are regular or not.

For details, please see [61, page 205].

17.3 The Computation History Method

As we mentioned in Chapter 1, it is possible to "teach" LBAs (and NPDAs) to answer certain difficult questions about Turing machines. This idea is detailed in Section 17.3.2 through 17.3.4.

We first recap basic facts about linear bounded automata (LBA) and present a decidability result about them (Section 17.3.1). Thereafter, we present three undecidability results based on the computation history method: (i) emptiness of LBA languages (Section 17.3.2), (ii) universality of the language of a CFG (Section 17.3.3), and (iii) Post's correspondence problem (Section 17.3.4). In Chapter 18, Section 18.2.3, we emphasize the importance of the undecidability of Post's correspondence problem (PCP) by presenting a classic proof due to Robert Floyd: we reduce PCP to the validity problem for first-order logic. This proof then establishes that the validity problem for first-order logic is undecidable.

17.3.1 Decidability of LBA acceptance

LBAs are Turing machines that are allowed to access (read or write) only that region of the input tape where the input was originally presented. To enforce such a restriction, one may place two distinguished symbols, say ¢ and $, around the original input string.[1] With these conventions, it can easily be seen that instantaneous descriptions of an LBA that begin as $q_0 w$ will change to the general form lqr, where $|lr| = |w|$. Therefore, there are a finite number, say N, of these IDs (see Exercise 17.1). A decider can simulate an LBA starting from ID $q_0 w$, and see if it accepts within this many IDs; if not, the LBA will not accept its input. Hence, LBA acceptance is decidable.

17.3.2 Undecidability of LBA language emptiness

Suppose a Turing machine M accepts input w; it will then have an accepting computational history of IDs starting with $q_0 w$, going through intermediate IDs, and ending with an ID of the form $aq_f b$ where $q_f \in F$. With respect to a given $\langle M, w \rangle$ pair, it is possible to generate an LBA $LB_{\langle M,w \rangle}$ that accepts a string s exactly when s is a sequence of IDs representing an accepting computational history of M running on w.[2] All $LB_{\langle M,w \rangle}$ need to do is this: check that the first ID is $q_0 w$; check that the $i + 1$st ID follows from the ith ID through a legal transition rule of the Turing machine M; and check that the final ID is of the form $aq_f b$. Hence, if the emptiness of an LBA's language were decidable through a decider D_{E_LBA}, one could apply it to $LB_{\langle M,w \rangle}$.

> By virtue of its design, $LB_{\langle M,w \rangle}$ has an empty language exactly when M does not accept w.

Therefore, the decision of D_{E_LBA} would be tantamount to whether or not M accepts w — a known undecidable problem; and hence, D_{E_LBA} cannot exist.

17.3.3 Undecidability of PDA language universality

The question of whether an NPDA over alphabet Γ has a universal language (language equal to Γ^*) is undecidable. The proof proceeds almost exactly like the proof that D_{E_LBA} cannot exist.
- We will define an NPDA $P_{\langle M,w \rangle}$ (created with respect to Turing machine M and input string w), such that

---

[1] If the input to an LBA is ε, ¢ and $ will lie in adjacent cells.

[2] It is possible to separate these IDs using some fixed separator character that is not in the original Γ.

- The language of $P_{\langle M,w \rangle}$ is Γ^* if M does not accept w.
- The language of $P_{\langle M,w \rangle}$ is $\Gamma^* \setminus \{s\}$ if M accepts w through an accepting computational history s.

If we manage to define such an NPDA, *then* simply feeding it to a claimed decider for universality will allow us to solve the acceptance problem of Turing machines, which is known to be undecidable (Section 16.2.3).

- Here is how $P_{\langle M,w \rangle}$ is designed:

  - $P_{\langle M,w \rangle}$ is designed to examine the computation history of M running on w. In other words, what is fed to $P_{\langle M,w \rangle}$ is a sequence of instantaneous descriptions (ID) of some Turing machine M (for reasons to be made clear soon, we require odd-numbered IDs to be reversed in the input).
  - If what is fed to $P_{\langle M,w \rangle}$ is an *accepting* computational history of M on w, then $P_{\langle M,w \rangle}$ rejects the input.
  - If there is some i such that ID $i+1$ does *not* follow from ID i through a rule of M, then the given sequence of IDs is *not* an accepting computational history of M on w; in this case, $P_{\langle M,w \rangle}$ accepts.

From this design, it is clear that if M does not accept w, then $P_{\langle M,w \rangle}$ has a universal language. This is because *no* string will be an accepting computational history in this situation! On the other hand, if M accepts w, $P_{\langle M,w \rangle}$'s language will precisely miss the accepting computational history of M on w.

- Now, all we need to present is *how* $P_{\langle M,w \rangle}$ can do the said checks. This is easy:

  - An NPDA can be made to nondeterministically pick the ith ID on the tape; it nondeterministically decides to either move over an ID or actually read and stack that ID.
  - Once the ith ID has been picked and stacked, the $i+1$st ID can be compared against it by popping the ith ID from the stack each time one character of the $i+1$st ID is read.
  - The only twist is that the NPDA will have to detect how ID i changed over to ID $i+1$. Fortunately, this comparison can be done around the head of the TM in the ith ID and the head of the TM in the $i+1$st ID. This much (finite) information can be recorded within the finite-state control of the NPDA.

17.3.4 Post's correspondence problem (PCP)

At first glance, a PCP *instance* is a simple puzzle about finite sequences[3] of pairs of strings of the form

$$\langle 01, 1\rangle\langle 01, \varepsilon\rangle\langle 01, 0\rangle\langle 1, 101\rangle.$$

It is customary to think of the above as "tiles" (or "dominoes"), with each tile at the respective index portrayed thus:

```
Index  0    1    2    3

Tile  [01] [01] [01] [  1]
      [ 1] [  ] [ 0] [101]
```

The question is: is there an arrangement of one or more of the above tiles, with repetitions allowed, so that the top and bottom rows read the same? Here is such an arrangement:

```
[  1] [01] [01] [1  ] [01] --> This row reads 10101101
[101] [ 0] [ 1] [101] [  ] --> This row reads 10101101
```

In obtaining this solution, the tiles were picked, *with repetition*, according to the sequence given by the indices 3,2,0,3,1:

```
Index  3    2    0    3    1    --> Solution is
                                     3,2,0,3,1
      [  1] [01] [01] [1  ] [01] --> reads 10101101
      [101] [ 0] [ 1] [101] [  ] --> reads 10101101
```

Given a PCP instance S of length N ($N = 4$ in our example), a *solution* is a sequence of numbers $i_1, i_2, \ldots, i_k$ where $k \geq 1$ and each $i_j \in \{0 \ldots N - 1\}$ for $j \in \{1 \ldots k\}$ such that $S[i_1]S[i_2]\ldots S[i_k]$ has the property of the top and bottom rows reading the same. By the term "solution" we will mean either the above sequence of integers or a sequential arrangement of the corresponding tiles.

Note that 3,2,0 is another solution, as is 3,1. The solution 3,1 is:

```
Index  3    1    --> Solution is 3,1
      [  1] [01] --> reads 101
      [101] [  ] --> reads 101
```

---

[3] Recall from Chapter 8 that sequences and strings are synonymous terms.

17.3.5 PCP is undecidable

The PCP is a fascinating undecidable problem! Since its presentation by Emil L. Post in 1946, scores of theoretical and practical problems have been shown to be undecidable by reduction *from* the PCP.[4] A partial list includes the following problems:

- Undecidability of the ambiguity of a given CFG (see Exercise 17.2).
- Undecidability of aliasing in programming languages [101].
- Undecidability of the validity of an arbitrary sentence of first-order logic (see Section 18.2.3 for a proof).

The impressive diversity of these problems indicates the commonality possessed by this variety of problems that Post's correspondence problems embody.

The main undecidability result pertaining to PCPs can be phrased as follows. Given any alphabet Σ such that $|\Sigma| > 1$, consider the *tile alphabet* $T \subseteq \Sigma^* \times \Sigma^*$. Now consider the language

$$PCP = \{S \mid S \text{ is a finite sequence over } T \text{ that has a solution}\}.$$

Theorem 17.2. PCP is undecidable (a proof-sketch is provided in Section 17.3.6).

In [125], PCP is studied at a detailed level, and a software tool PCPSolver is made available to experiment with the PCP. The following terminology is first defined:

- A *PCP* instance is a member of the language *PCP*.
- If any member of T is of the form $\langle w, w \rangle$, then that *PCP* instance is *trivial* (has a trivial solution).
- The number of pairs in a *PCP* instance is its *size*. The length of the longest string in either position of a pair ("upper or lower string of a tile") in a *PCP* instance is the *width* of the instance.
- An *optimal* solution is the shortest solution sequence. The example given on page 315 has an optimal solution of length 2.

With respect to the above definitions, here are some *fascinating* results cited in [125], that reveal the depth of this simple problem:

- Bounded PCP is NP-complete (finding solutions of length less than an a priori given constant $K \in Nat$). Basically, checking whether solutions below a given length is decidable, but has, in all likelihood, exponential running time (see Chapter 19 for an in-depth discussion of NP-completeness).

---

[4] PCP is taken as the existing undecidable problem, and a mapping reduction to a new problem P is found, thus showing P to be undecidable.

- PCP instances of size 2 are decidable, while PCP instances of size 7 are undecidable (note: no restriction on the width is being imposed). Currently, decidability of sizes 3 through 6 are unknown.

Here are more unusual results that pertain to the shortest solutions that exist for two innocent looking PCP instances:

- The PCP instance below has an optimal solution of length 206:

$$[1000] \ [01] \ [1 \ \ ] \ [00 \ ]$$
$$[\ \ \ 0] \ [\ 0] \ [101] \ [001]$$

- The PCP instance below has two optimal solutions of length 75:

$$[100] \ [0 \ \ ] \ [1]$$
$$[\ \ 1] \ [100] \ [0]$$

These discussions help build our intuitions towards the proof we are about to sketch:

the undecidability of PCP indicates the *inherent inability to bound search while solving a general PCP instance.*

17.3.6 Proof sketch of the undecidability of PCP

The basic idea behind the proof of undecidability of PCP is to use A_{TM} and the computation history method. In particular,

- Given a Turing machine M, we systematically go through the transition function δ of M as well as the elements of its tape alphabet, Γ, and generate a finite set of tiles, $Tiles_M$.
- Now we turn around and ask for an input string w that is in the initial tape of M. Then, with respect to M and w, we generate one additional tile, $tile_{Mw}$. We define $Tiles = Tiles_M \cup \{tile_{Mw}\}$.
- We arrange it so that any solution to $Tiles$ must begin with $tile_{Mw}$. This is achieved by putting special additional characters around the top and bottom rows of each tile, as will soon be detailed.
- We then prove that $Tiles$ has one solution, namely $Soln_{Tiles}$, exactly when M accepts w. If M does not accept w, $Tiles$ will have no solutions, by construction. Furthermore, $Soln_{Tiles}$ would end up being a sequence $t_1, t_2, \ldots, t_k$ such that when the tiles are lined up according to this solution, the top and bottom rows of the tiles would, essentially,[5] be the accepting computation history of M on w.

---

[5] We say "essentially" because there would be extraneous characters introduced to "align" various tiles. In addition, at the very end of the solution sequence, the

- Hence, given a solver for PCP, we can obtain a solver for A_{TM}.

```
      change 0 to 1 move right        change 1 to 0 move left
p01          --->              1q1           --->              A10

   T1            T2     T3    T4      T5       T6     T7     T8
------------- ------ ---- ---- ------- ---- ------ ----

[*#          ] [*p*0] [*1] [*#] [*1*q*1] [*#] [*A*1] [*0]
[*#*p*0*1*#*] [1*q*] [1*] [#*] [A*1*0*] [#*] [A*  ] [0*]

                                T9   T10    T11   T12      T13
                               ---- ------ ---- ------- ---

                               [*#] [*A*0] [*#] [*A*#*#] [*<>]
                               [#*] [A*  ] [#*] [#*     ] [<> ]
```

Fig. 17.2. An accepting computation history and its encoding as per the PCP encoding rules

The crux of achieving the above reduction is to generate each set of tiles carefully; here is how we proceed to generate the members of *Tiles*.[6] Here, the following notations are used: if u is the string of characters $u_1 u_2 \ldots u_n$,

- $*u = *u_1 * u_2 \ldots * u_n$,
- $u* = u_1 * u_2 * \ldots u_n*$, and
- $*u* = *u_1 * u_2 * \ldots u_n*$.

In Figure 17.2, we illustrate the encoding ideas behind the PCP undecidability proof through the example of a Turing machine starting from state p with its tape containing string 0100 — *i.e.*, ID $p01$. This Turing machine first moves one step right to ID $1q1$, and in the process changes the 0 it was initially facing to a 1, as shown in Figure 17.2. Then the Turing machine moves one step left, and in the process changes the 1 it was facing to a 0, as also shown in Figure 17.2. At this point, it enters the accepting state A, as shown by the ID of $A10$ attained by this Turing machine, and hence halts. The general rules below are illustrated on specific tiles mentioned by the annotation "**Tn**" below:

top and bottom rows will shrink from being an accepting ID to an ID that simply contains q_a. Ignoring these characters, we would have the computational history of M on w on the top and bottom.

[6] We base our explanations quite heavily on those provided in [111].

$$[*\#]$$

T1: $tile_{Mw} = [*\#q_0w\#*]$. As said earlier, any solution to the PCP will start with this tile. This is ensured, as we shall soon see, by having a $*$ begin the top and bottom row of this tile.

$$[*\diamond]$$

T13: $Tiles_M$ must include $[\diamond]$. This tile will end any solution to the PCP. The extra $*$ at the top will, as we shall soon see, supply the last needed star in a run of tiles. The $\diamond$ at the top and bottom forces this to be the last tile.

$$[*a]$$

T3,T4,T6,T8,T9,T11: For every $a \in \Gamma$, $[a*]$ is a tile in $Tiles_M$. In

$$[*\#]$$

addition, include $[\#*]$ as a tile in $Tiles_M$.

$$[*qa]$$

T2: For every move $\delta(q, a) = (r, b, R)$, $[br*]$ is a tile in $Tiles_M$. Notice that this pattern in a 2×2 window captures the Turing machine head, changing the 'a' character of the ID qa into a b, and moving right, attaining state r in the process.

$$[*cqa]$$

T5: For every move $\delta(q, a) = (r, b, L)$ and for every $c \in \Gamma$, $[rcb*]$ is a tile in $Tiles_M$. This pattern in a 3×2 window captures the tape head moving left.

$$[*aq_a] \qquad [*q_a a]$$

T10: For every $a \in \Gamma$, $[q_a*]$ and $[q_a*]$ are tiles in $Tiles_M$. These tiles help shrink the top and bottom rows of the solution sequence from being the accepting ID to being the ID q_a.

$$[*q_a\#\#]$$

T12: For $q_a \in F$, $[\#*]$ is in $Tiles_M$. This tile helps finish off the accepting computation history that will form on the top row of a solution sequence.

The reader may verify that the top and bottom rows in Figure 17.2 essentially have the accepting computation history for M on w. Ignoring [,], and $*$, what we have is

p01 # 1q1 # A10 # A0 # A ## $\diamond$.

The crux of the PCP proof was that A_{TM} can be decided if and only if the PCP instance generated by these tile generation rules *can be decided to possess a solution*.

Chapter Summary

This chapter discussed Rice's Theorem, the computation history method, and Post's correspondence problem. We took a semi-formal approach, but highlighting many details and intuitions often lost in highly theoretical presentations of these ideas. The basic techniques are all quite simple, and boil down to the undecidability of the acceptance problem, A_{TM}.

Exercises

17.1. Calculate N, referred to in Section 17.3.1, in terms of $|Q|$ and $|\Gamma|$.

17.2. Show that it is undecidable whether an arbitrary CFG is ambiguous. *Hint:* Let

$$A = w_1, w_2, \ldots, w_n$$

and

$$B = x_1, x_2, \ldots, x_n$$

be two lists of words over a finite alphabet Σ. Let $a_1, a_2, \ldots, a_n$ be symbols that do not appear in any of the w_i or x_i. Let G be a CFG

$$(\{S, S_A, S_B\}, \ \Sigma \cup \{a_1, \ldots, a_n\}, \ P, \ S),$$

where P contains the productions

$S \to S_A$,
$S \to S_B$,
For $1 \le i \le n$, $S_A \to w_i S_A a_i$,
For $1 \le i \le n$, $S_A \to w_i a_i$,
For $1 \le i \le n$, $S_B \to x_i S_B a_i$, and
For $1 \le i \le n$, $S_B \to x_i a_i$.

Now, argue that G is ambiguous if and only if the PCP instance (A, B) has a solution (thus, we may view the process of going from (A, B) to G as a mapping reduction).

17.3. Show that the unary PCP problem — PCP over a singleton alphabet ($|\Sigma| = 1$) — is decidable.

17.4. Modify the Turing machine for which Figure fig:pcp-complete-example is drawn, as follows: in state q, when faced with a 0, it changes 0 to a 1 and moves right, and rejects (gets stuck) in state R. Now go through the entire PCP "tile construction" exercise and show that the PCP instance that emerges out of starting this Turing machine on input string 00 in state p has no solution.

17.5. Someone provides this "proof:" There are PDAs that recognize the language $L = \{0^n 1^n \mid n \geq 0\}$. This PDA erects a non-trivial partitioning of $\{0,1\}^*$. Both the partitions are recursively enumerable. Hence, L is the language of infinitely many Turing machines M_L^i, for i ranging over some infinite index set. The same is true for $\bar{L}$ and $M_{\bar{L}}^j$. However this is a non-trivial partitioning of the space of Turing machines. Hence, L is undecidable.

Describe the main flaw in such a "proof."

Basic Notions in Logic including SAT

This chapter is on propositional logic, first-order logic, and modern Boolean satisfiability methods. In particular, we will prove the undecidability of the validity of first-order logic sentences. Boolean satisfiability is discussed in sufficient detail to ensure that the reader is well motivated to study the theory of NP-completeness in Chapter 19.

Mathematical logic is central to formalizing the notion of assertions and proofs. Classical logics include *propositional* (zeroth-order), *first-order*, and *higher-order* logics. Most logics separate the notion of *proof* and *truth*. A proof is a sequence of *theorems* where each theorem is either an *axiom* or is obtained from previous theorems by applying a *rule of inference.*

Informally speaking, it is intended that all theorems are "true." This is formalized through the notion of a *meaning function* or *interpretation* which maps a *well-formed formula* (a syntactically correct formula, often abbreviated *wff*) to its truth value. If a *wff* is true under all interpretations, it is called a *tautology* (or equivalently, it is called *valid*). Therefore, when we later mention the term, "the validity problem of first-order logic," we refer to the problem of deciding whether a given first-order logic formula is valid (this problem is later shown to be only *semi-decidable*, or equivalently TR or RE).

A logical system consisting of a collection of axioms[1] and rules of inference is *sound* if every theorem is a tautology. A logical system is *complete* if every tautology is a theorem. A logical system has *independent* rules of inference if omitting any rule prevents certain theorems from being derived.

---

[1] Note that axioms can be regarded as rules of inference that have an empty premise.

One uses the terminology of "order" when talking about logics. Propositional (zeroth-order) logic includes Boolean variables that range over true and false (or 1 and 0), Boolean connectives, and Boolean rules of inference. Predicate (first-order) logic additionally allows the use of *individuals* such as integers and strings that constitute infinite sets, variables that are quantified over such sets, as well as predicate and function *constants*. Second (and higher) order logics allow quantifications to occur over function and predicate spaces also. For example, in Chapter 5, Section 5.3, we presented two principles of induction, namely *arithmetic* and *complete*; the formal statement of both these induction principles constitutes examples of higher order logic. There are also special classes of logics that vary according to the kinds of *models* (interpretations) possible: these include *temporal logic*, which is discussed in Chapters 21 and 22. In these logics, notions such as *validity* tend to acquire specialized connotations, such as validity under a certain *model*. In this chapter, however, we will stick to the classical view of these notions.

Section 18.1 discusses a *Hilbert style* axiomatization of propositional logic due to Church. Section 18.2 begins with an example involving "quacks and doctors," presents examples of interpretations for formulas, and closes off with a proof for the fact that the validity problem of first-order logic is undecidable, but semi-decidable (or equivalently, RE or TR).

We then turn our attention to the topic of satisfiability in the setting of propositional logic in Section 18.3. We approach the subject based on notions (and notation) popular in hardware verification - following how "Boolean logic" is treated in those settings (a model theoretic approach). We first examine two normal forms, namely the *conjunctive* and the *disjunctive* normal forms, and discuss converting one to the other. We then examine related topics such as $\neq$-satisfiability, 2-satisfiability, and satisfiability-preserving transformations.

18.1 Axiomatization of Propositional Logic

We now present one axiomatization of propositional calculus following the Hilbert style. Let the syntax of well-formed formulas (wff) be as follows:

$$Fmla ::= Pvar \mid \neg Fmla \mid Fmla \Rightarrow Fmla \mid (Fmla).$$

In other words, a formula is a propositional variable, the negation of a propositional formula, an implication formula, or a parenthesized formula. This grammar defines a complete set of formulas in the sense that all Boolean propositions can be expressed in it (prove this assertion). Following Church [17], a Hilbert calculus for propositional logic can be set up based on three axioms (A1-A3) and two rules of inference (R1-R2) shown below. Here, p and q stand for propositional formulas.

A1: $p \Rightarrow (q \Rightarrow p)$
A2: $(s \Rightarrow (p \Rightarrow q)) \;\Rightarrow\; (s \Rightarrow p) \Rightarrow (s \Rightarrow q)$
A3: $(\neg q \Rightarrow \neg p) \;\Rightarrow\; (p \Rightarrow q)$

R1 (Modus Ponens): If P is a theorem and $P \Rightarrow Q$ is a theorem, conclude that Q is a theorem.
 Example: From p and $p \Rightarrow (p \Rightarrow q)$, infer $p \Rightarrow q$.
R2 (Substitution): The substitution of wffs for propositional variables in a theorem results in a theorem. A substitution is a "parallel assignment" in the sense that the newly introduced formulas themselves are not affected by the substitution (as would happen if, for instance, the substitutions are made serially).
 Example: Substituting $(p \Rightarrow q)$ for p and $(r \Rightarrow p)$ for q in formula $p \Rightarrow q$, results in $(p \Rightarrow q) \Rightarrow (r \Rightarrow p)$. It is as if p and q are replaced by fresh and distinct variables first, which, in turn, are replaced by $(p \Rightarrow q)$ and $(r \Rightarrow p)$ respectively.
 We do *not* perform the substitution of $r \Rightarrow p$ for q first, and then affect the p introduced in this process by the substitution of $(p \Rightarrow q)$ for p.

Given all this, a proof for a simple theorem such as $p \Rightarrow p$ can be carried out – but it can be quite involved:

P1: From A1, through substitution of $p \Rightarrow p$ for q, we obtain

$$p \Rightarrow ((p \Rightarrow p) \Rightarrow p).$$

P2: From A2, substituting p for s, $p \Rightarrow p$ for p, and p for q, we obtain

$$(p \Rightarrow ((p \Rightarrow p) \Rightarrow p)) \;\Rightarrow\; (p \Rightarrow (p \Rightarrow p)) \Rightarrow (p \Rightarrow p).$$

P3: Modus ponens between **P1** and **P2** yields

$$(p \Rightarrow (p \Rightarrow p)) \Rightarrow (p \Rightarrow p).$$

P4: From A1, substituting p for q, we obtain

$$p \Rightarrow (p \Rightarrow p).$$

Modus ponens between **P4** and **P3** results in $(p \Rightarrow p)$.   □

It is straightforward to verify that the above axiomatization is sound. This is because the axioms are true, and every rule of inference is truth-preserving. The axiomatization is also complete, as can be shown via a proof by induction. The take away message from these discussions is that it pays to hone the axiomatization of a formal logic into something that is parsimonious as well as enjoys the attributes of being sound, complete, and independent, as explained earlier.

18.2 First-order Logic (FOL) and Validity

Below, we will introduce many notions of first-order logic intuitively, through examples. We refer the reader to one of many excellent books in first-order logic for details. One step called *skolemization* merits some explanation. Basically, skolemization finds a witness to model the existential variable. In general, from a formula of the form $\exists X.P(X)$, we can infer $P(c)$ where c is a constant in the domain. Likewise, from $P(c)$, we can infer $\exists X.P(X)$; in other words, for an unspecified constant c in the domain,

$$\exists X.P(X) \iff P(c).$$

We will use this equivalence in the following proofs. There is another use of skolemization illustrated by the following theorem (which we won't have occasion to use):

$$\forall X.\exists Y.P(X,Y) \iff P(X, f(X)).$$

Here, f is an unspecified (but fixed) function. This equivalence is valid because of two reasons:

- The right-hand side leaves X as a free variable, achieving the same effect as $\forall X$ goes, as far as validity goes (must be true for all X).
- The right-hand side employs $f(X)$ to model the fact that the selection of $\exists Y$ may depend on X.

18.2.1 A warm-up exercise

Suppose we are given the following proof challenge:

> *Some patients like every doctor. No patient likes a quack. Therefore, prove that no doctor is a quack.*

It is best to approach problems such as this using predicates to model the various classes of individuals[2] instead of employing 'arbitrary constants' such as p and d to denote *patients, doctors*, etc. Introduce predicate p to carve out a subset of people (P) to be patients, and similarly d for doctors. Let l (for "likes") be a binary relation over P. Our proof will consist of instantiation of formulas, followed by Modus ponens, and finally proof by contradiction.

A1: The statement, "Some patients like every doctor:"

$$\exists x \in P : (p(x) \wedge \forall y : (d(y) \Rightarrow l(x, y))).$$

A2: The statement, "No patient likes a quack:"

$$\forall x, y \in P : (p(x) \wedge q(y) \Rightarrow \neg l(x, y)).$$

Proof goal: "No doctor is a quack."

$$\forall x \in P : (d(x) \Rightarrow \neg q(x)).$$

Negate the proof goal: $\exists x \in P : (d(x) \wedge q(x))$.

Skolemize negated proof goal: $(x_0 \in P \wedge d(x_0) \wedge q(x_0))$.

Skolemize A1: $c \in P \wedge (p(c) \wedge \forall y : (d(y) \Rightarrow l(c, y)))$ (we will suppress domain membership assertions such as $c \in P$ from now on).

Specialize: From A1, we specialize y to x_0 to get:
$p(c) \wedge (d(x_0) \Rightarrow l(c, x_0))$.

But since $d(x_0)$ is true in the negated proof goal, we get:
$p(c) \wedge l(c, x_0)$ (more irrelevant facts suppressed).

Since $q(x_0)$ is true in the negated proof goal, we also get:
$p(c) \wedge l(c, x_0) \wedge q(x_0)$.

Use A2 and specialize x to c and y to x_0 to get $\neg l(c, x_0)$.

Contradiction: Since we have $l(c, x_0)$ and $\neg l(c, x_0)$, we get a contradiction.

18.2.2 Examples of interpretations

Manna's book [78] provides many insightful examples, some of which are summarized below. We focus on the notion of interpretations, which, in case of first-order logic: (i) chooses domains for constants, predicates, and function symbols to range over, and (ii) *assigns* to them. Examples below will clarify.

---

[2] Pun intended; in FOL, members of domains other than Booleans are called individuals.

Example 1

Consider the formula

$$Fmla1 = \exists F.F(a) = b$$
$$\wedge (\forall x).[p(x) \Rightarrow F(x) = g(x, F(f(x)))]$$

We will now provide *three distinct* interpretations for it.
Interpretation 1.

D = Nat
a = 0
b = 1
f = $\lambda x.(x = 0 \rightarrow 0, x - 1)$
g = *
p = $\lambda x.x > 0$

Interpretation 2.

D = Σ^*
a = ε
b = ε
f = $\lambda x.(tail(x))$
g(x,y) = concat(y,head(x))
p = $\lambda x.x \neq \varepsilon$

Interpretation 3.

D = Nat
a = 0
b = 1
f = $\lambda x.x$
g(x,y) = y+1
p = $\lambda x.x > 0$

It is clear that under Interpretation 1, *Fmla*1 is true, because there indeed exists a function F, namely the factorial function, that makes the assertion true. It is also true under Interpretation 2, while it is false under Interpretation 3 (Exercise 18.4 asks for proofs). Hence, this formula is *not* valid — because it is not true under all interpretations.

Example 2

Consider the formula

$Fmla2 = \forall P.P(a)$
$\quad \wedge(\forall x).[(x \neq a) \wedge P(f(x)) \Rightarrow P(x)]$
$\Rightarrow (\forall x.P(x))$

We can interpret this formula suitably to obtain the principle of induction over *many* domains: for example, *Nat*, *strings*, etc.

18.2.3 Validity of first-order logic is undecidable

Valid formulas are those that are true under *all* interpretations. For example,

$$\forall x.f(x) = g(x) \Rightarrow \exists a.f(a) = g(a).$$

Validity stems from the innate structure of the formula, as it must remain true under *every conceivable* interpretation. We will now summarize Floyd's proof (given in [78]) that the validity problem for first-order logic is undecidable. First, an abbreviation: for $\sigma_i \in \{0,1\}$, use the abbreviation

$$f_{\sigma_1,\sigma_2,\dots,\sigma_n}(a) = f_{\sigma_n}(f_{\sigma_{n-1}}(\dots f_{\sigma_1}(a))\dots).$$

The proof proceeds by building a FOL formula for a given "Post system" (an instance of Post's correspondence problem).
Given a Post system S $= \{(\alpha_1, \beta_1), (\alpha_2, \beta_2), \dots, (\alpha_n, \beta_n), \ n \geq 1$ over $\Sigma = \{0,1\}$, construct the wff W_S (we will refer to the two antecedents of W_S as A1 and A2, and its consequent as C1):

$\wedge_{i=1}^{n} p(f_{\alpha_i}(a), f_{\beta_i}(a))$ (A1)
$\wedge \ \forall x \forall y \ [p(x, y) \Rightarrow \wedge_{i=1}^{n} p(f_{\alpha_i}(x), f_{\beta_i}(y))]$ (A2)
$\Rightarrow \exists z \ p(z, z)$ (C1)

We now prove that S has a solution iff W_S is valid.
Part 1. (W_S valid) $\Rightarrow$ (S has a solution).
If valid, it is true for all interpretations. Pick the following interpretation:

$a = \varepsilon$
$f_0(x) = x0$ (string 'x' and string '0' concatenated)
$f_1(x) = x1$ (similar to the above)
$p(x, y) =$ There exists a non-empty sequence $i_1 i_2 \dots i_m$ such that
$\quad x = \alpha_{i_1} \alpha_{i_2} \dots \alpha_{i_m}$ and $y = \beta_{i_1} \beta_{i_2} \dots \beta_{i_m}$

Under this interpretation, parts A1 and A2 of W_S are true. Here is why:

- Under the above interpretation, $f_{\alpha_i}(a) = \varepsilon\alpha_i = \alpha_i$ and similarly $f_{\beta_i}(a) = \beta_i$.
- Thus A1 becomes $\wedge_{i=1}^{n} p(\alpha_i, \beta_i)$. Each conjunct in this formula is true by p's interpretation; hence A1 is true.
- The part $[p(x,y) \Rightarrow \wedge_{i=1}^{n} p(f_{\alpha_i}(x), f_{\beta_i}(y))]$ reduces to the following claim: $p(x,y)$ is true means that x and y can be written in the form $x = \alpha_{i_1}\alpha_{i_2}\ldots\alpha_{i_m}$ and $y = \beta_{i_1}\beta_{i_2}\ldots\beta_{i_m}$; the consequent of this implication then says that we can append some α_i and the corresponding β_i to x and y, respectively. The consequent is also true by p's interpretation. Thus A2 is also true.
- Since W_S is valid (true), C1 must also be true. C1 asserts that the Post system S has a solution, namely some string z that lends itself to being interpreted as some sequence $\alpha_{i_1}\alpha_{i_2}\ldots\alpha_{i_m}$ as well as $\beta_{i_1}\beta_{i_2}\ldots\beta_{i_m}$. That is,

$$\alpha_{i_1}\alpha_{i_2}\ldots\alpha_{i_m} = z = \beta_{i_1}\beta_{i_2}\ldots\beta_{i_m}.$$

Part 2. $(W_S \text{ valid}) \Leftarrow (S \text{ has a solution}).$

If S has a solution, let it be the sequence $i_1 i_2 \ldots i_m$. In other words, $\alpha_{i_1}\alpha_{i_2}\ldots\alpha_{i_m} = \beta_{i_1}\beta_{i_2}\ldots\beta_{i_m} = Soln$. Now, in order to show that W_S is valid, we must show that for *every* interpretation it is true. We approach this goal by showing that under every interpretation where the antecedents of W_S, namely A1 and A2, are true, the consequent, namely C1, is also true (if any antecedent is false, W_S is true, so this case is not considered).

From A1, we conclude that

$$p(f_{\alpha_{i_1}}(a), f_{\beta_{i_1}}(a))$$

is true. Now using A2 as a rule of inference, we can conclude through Modus ponens, that

$$p(f_{\alpha_{i_2}}(f_{\alpha_{i_1}}(a)), f_{\beta_{i_2}}(f_{\beta_{i_1}}(a)))$$

is true. In other words,

$$p(f_{\alpha_{i_1}\alpha_{i_2}}(a), f_{\beta_{i_1}\beta_{i_2}}(a))$$

is true. We continue this way, applying the functions in the order dictated by the assumed solution for S; in other words, we arrive at the assertion that the following is true (notice that the subscripts of f describe the order in which the solution to S considers the α's and β's):

$$p(f_{\alpha_{i_1}\alpha_{i_2}\ldots\alpha_{i_m}}(a), f_{\beta_{i_1}\beta_{i_2}\ldots\beta_{i_m}}(a)).$$

However, since

$$\alpha_{i_1}\alpha_{i_2}\ldots\alpha_{i_m} = \beta_{i_1}\beta_{i_2}\ldots\beta_{i_m} = Soln,$$

we have essentially shown that

$$p(f_{Soln}(a), f_{Soln}(a)).$$

Now, $p(f_{Soln}(a), f_{Soln}(a))$ means that there exists a z such that $p(z, z)$, namely $z = f_{Soln}(a)$.   □

18.2.4 Valid FOL formulas are enumerable

It was proved by Gödel in his dissertation that first-order predicate calculus is complete. In other words, there are axiomatizations of first-order logic in which every valid formula has a proof. Hence, by enumerating proofs, one can enumerate all valid first-order logic formulas. Thus the set of valid FOL formulas is recursively enumerable (or is Turing recognizable).

18.3 Properties of Boolean Formulas

18.3.1 Boolean satisfiability: an overview

Research on Boolean satisfiability methods (SAT) is one of the "hottest" areas in formal methods, owing to the fact that BDDs are often known to become exponentially sized, as already hinted in Chapter 11. In [11], the idea of model checking without BDDs was introduced. This work also coincided with the arrival on the scene, of efficient Boolean satisfiability methods. These modern SAT solvers (e.g., [81, 90]) followed the basic "DPLL" procedure presented in [33, 32], but made considerable improvements, including intelligent methods for backtracking search over the space of satisfying assignments, and improving the efficiency of computer memory system (e.g., cache) utilization in carrying out these algorithms. Easily modifiable versions of these tools are now freely available (e.g., [87]). Boolean SAT methods are now able to often handle extremely large system models, establish correctness properties, and provide explanations when the properties fail. Andersson [7] presents many of the basic complexity results, including the inevitability of exponential blow up, even in the domain of SAT. We discuss a summary of these issues, and offer a glimpse at the underlying ideas behind modern SAT tools. Unfortunately, space does not permit our detailed presentation of the use of SAT techniques or some of their unusual applications in system verification (e.g., [46]). The reader may find many tutorials on the Internet or web sites such as [105].

18.3.2 Normal forms

We now discuss several properties of Boolean (or propositional) formulas to set the stage for discussions about the theory of NP-completeness. There are two commonly used normal forms for Boolean formulas: the conjunctive normal form (CNF) and the disjunctive normal form (DNF).[3]   Given a Boolean expression (function) over n variables $x_1 \ldots x_n$, a sum-of-products (SOP) or disjunctive normal form (DNF) expression for it is one where products of literals are disjoined (OR-ed) – a literal being a variable or its negation. For example, $nand(x,y)$ is expressed in SOP (DNF) as

$$nand(x,y) = \neg x \neg y + \neg x y + x \neg y.$$

A POS (CNF) expression for $nand$ would be $(\neg x + \neg y)$. A systematic way to obtain DNF and CNF expressions from truth tables is illustrated below: basically, the DNF form for a Boolean function is obtained by disjoining (OR-ing) the min terms at which the function is a 1, while the CNF form for a Boolean function is obtained by conjoining (AND-ing) the max terms at which the function is a 0. All these are illustrated in Figure 18.1, and details may be found in any standard digital system design text book (e.g., [12]).

Row	x	y	nand(x,y)	Minterm	Maxterm
0	0	0	1	$m_0 = \neg x \neg y$	$M_0 = (x + y)$
1	0	1	1	$m_1 = \neg x y$	$M_1 = (x + \neg y)$
2	1	0	1	$m_2 = x \neg y$	$M_2 = (\neg x + y)$
3	1	1	0	$m_3 = x y$	$M_3 = (\neg x + \neg y)$

Fig. 18.1. Min/Max terms for $nand$, whose DNF form is $m_0 + m_1 + m_2$ and CNF form is M_3 (the only max term where the function is 0)

Here are important points pertaining to these normal forms:

- Conversion between the CNF and the DNF forms incurs an exponential cost. This is best illustrated by a conversion program included in Figures 18.2 and Figures 18.3.[4] You are encouraged to run this code

---

[3] Strictly speaking, we should call the normal forms being discussed here the *canonical sum-of-products form* and the *canonical product-of-sums form*, since each product term or sum term consists of n literals.

[4] These programs *do not prove* that the conversion is exponential, but provide strong intuitions as to why.

under an Ocaml system. You will see that an exponential growth in formula size can occur, as demonstrated in the tests at the end of Figure 18.3. While these tests indicate the results on DNF to CNF conversion, the same complexity growth exists even for CNF to DNF conversion (Exercise 18.8 asks you to modify this program to one that obtains the DNF form of a given formula. The modifications are based on the duality between ∨ and ∧, and hence the complexity follows).

- CNF satisfiability will be shown to be NP-complete in the next chapter, whereas DNF satisfiability is linear-time (see exercise below). However, due to the exponential size blow up, one cannot "win" against NP-completeness by converting CNF satisfiability to DNF satisfiability.

18.3.3 Overview of direct DNF to CNF conversion

Consider Figure 18.3 in which some terms representing DNF formulas are given. In particular, note that formula f5 is a 4-DNF formula with eight product terms. The expression List.length (gencnf(f5) 1 []).cnf converts this formula to CNF, and measures its length, which is found to be 65536, showing the exponential growth in length. Here is a brief explanation of this code.[5]

In Figure 18.2, the types for literals, clauses, and CNF formulas are given. The Ocaml List of lists [[1; 3; -2]; [1; -1]] stands for the CNF formula $(x_1 \lor x_3 \lor \neg x_2) \land (x_1 \lor \neg x_1)$. The syntax of formulas, fmla, declares what formula terms are permissible. According to these conventions, *true* is represented by [], *i.e.*, an empty list of clauses, while *false* is represented by [[]], *i.e.*, one empty clause.

The program proceeds driven by pattern matching. Suppose fmla matches *true* (Tt); this results in an Ocaml record structure as shown in the following line:

```
Tt    -> {cnf=[ ]; nvgen=0; algen=[]}
```

Note that [] is how *true* is represented. This is a list of lists with the outer list empty. Since the outer list is a list of conjunctions, an empty outer list corresponds to the basis case of a list of conjunctions, namely *true*. Likewise, [[]] represents *false* because the inner list is empty, and its basis case (for a list of disjunctions) is *false* (Ff). We also record the number of new variables generated in nvgen, and an association list of

---

[5] This code can convert arbitrary Boolean expressions to CNF - we choose to illustrate it on DNF.

```
type literal = int         (* 1, 3, -2, etc         *)
type clause = literal list (* [1; 3; -2] -> [ ] means TRUE and [ [ ] ] means FALSE *)
type cnffmla = clause list (* [[1; 3; -2]; [1; -1]] -> (x1 \/ x3 \/ ~x2)/\(x1 \/ ~x1)*)

type fmla =
    Ff  | Tt  | Pv of string    | And of fmla * fmla | Or  of fmla * fmla
              | Eq of fmla * fmla | Imp of fmla * fmla | Not of fmla

type cnf_str = {cnf:cnffmla;                 (* CNF formula as a list *)
                nvgen:int;                   (* Variable allocation index *)
                algen: (string * int) list}  (* Association list between variable-names
                                                and integers representing them  *)
let rec gencnf(fmla)(next_var_int)(var_al) =
  match fmla with
      Tt    -> {cnf=[ ]; nvgen=0; algen=[]}
    | Ff    -> {cnf=[[]]; nvgen=0; algen=[]}
    | Pv(s) ->
        if (List.mem_assoc s var_al)
        then {cnf = [ [ (List.assoc s var_al) ] ];
              nvgen = 0; algen = []}
        else {cnf = [ [ next_var_int ] ]; nvgen = 1; (* 1 new var generated *)
              algen = [(s,next_var_int)]}

    | And(q1,q2) ->
        let {cnf=cnf1; nvgen=nvgen1; algen=algen1}
          = gencnf(q1)(next_var_int)(var_al) in
        let {cnf=cnf2; nvgen=nvgen2; algen=algen2}
          = gencnf(q2)(next_var_int + nvgen1)(algen1 @ var_al) in
        {cnf=doAnd(cnf1)(cnf2); nvgen=nvgen1+nvgen2;
         algen=algen1 @ algen2}

    | Or(q1,q2) ->
        let {cnf=cnf1; nvgen=nvgen1; algen=algen1}
          = gencnf(q1)(next_var_int)(var_al) in
        let {cnf=cnf2; nvgen=nvgen2; algen=algen2}
          = gencnf(q2)(next_var_int + nvgen1)(algen1 @ var_al) in
        {cnf=doOr(cnf1)(cnf2); nvgen=nvgen1+nvgen2;
         algen=algen1 @ algen2}

    | Imp(q1,q2) -> gencnf(Or(Not(q1),q2))(next_var_int)(var_al)
    | Eq(q1,q2)  -> gencnf(And(Or(Not(q1),q2), Or(Not(q2),q1)))(next_var_int)(var_al)
    | Not(Pv(s)) ->
        if (List.mem_assoc s var_al)
        then {cnf = [ [ (0-(List.assoc s var_al)) ] ] ;
              nvgen = 0; algen = []}
        else {cnf = [ [ (0-next_var_int) ] ]; nvgen = 1;  (* 1 new var generated *)
              algen = [(s,next_var_int)]}
    | Not(q1) ->
        let {cnf=cnf1; nvgen=nvgen1; algen=algen1}
          = gencnf(q1)(next_var_int)(var_al) in
          {cnf=doNot(cnf1); nvgen=nvgen1;
           algen=algen1}

and
  doAnd(cnf1)(cnf2) =
  (match (cnf1,cnf2) with
     | ([],cnf2')       -> cnf2'
     | (cnf1',[])       -> cnf1'
     | ([[]],_)         -> [[]]
     | (_,[[]])         -> [[]]
     | _                -> List.append(cnf1)(cnf2) )

  (* See PART-2 ... *)
```

Fig. 18.2. A CNF generator, Part-1 (continued in Figure 18.3)

```
and
  doOr(cnf1)(cnf2) =
    (match (cnf1,cnf2) with
        | ([],_)          -> []
        | (_,[])          -> []
        | ([[]],cnf2')    -> cnf2'
        | (cnf1',[[]])    -> cnf1'
        | ([cl1],[cl2])   -> [List.append(cl1)(cl2)]
        | (cnf1',[cl2])   -> doOr([cl2])(cnf1')
        | (cnf1', (cl2::cls)) ->
            let door1 = doOr(cnf1')([cl2]) in
            let door2 = doOr(cnf1')(cls) in
                doAnd(door1)(door2) )
and
  doNot(cnf1) =
    (match cnf1 with
        | []   -> [[]]
        | [[]] -> []
        | (cl1::cls) ->
            let compclause = comp_clause(cl1) in
            let donot' = doNot(cls) in
                doOr(compclause)(donot') )
and
  comp_clause(clause) =
    (match clause with
        | []                -> []
        | (lit::lits) ->
            let cl = 0-lit in              (* complement literal *)
            let cl_cnf = [[cl]] in         (* turn literal into CNF *)
            let rest = comp_clause(lits) in
                doAnd(cl_cnf)(rest) )
;;
(* To run tests, load these definitions, and type, e.g., gencnf(f5) 1 [];;
let f1 = And(And(Pv "x", Pv "y"), And(Pv "z", Pv "w"));;
let f2 = And(And(Pv "p", Pv "q"), And(Pv "r", Pv "s"));;
let f3 = Or(f1,f2);;
let f4 = Or(f3,f3);;
let f5 = Or(f4,f4);;

f5 is really (xyzw + pqrs + xyzw + pqrs + xyzw + pqrs + xyzw + pqrs)

List.length ( gencnf( f5 ) 1 [] ).cnf;;
- : int = 65536
```

Fig. 18.3. A CNF generator, Part-2 (continued from Figure 18.2)

variable names to their integer values in **algen**. These are respectively 0 and [] for the case of **Tt**. Formula **Ff** is handled similarly.

A propositional variable is converted by the code under **Pv(s)**. We look up s in the association list **var_al**, and if found, generate its CNF with respect to the looked-up result. Else, we generate a CNF clause containing **next_var_int**, the next variable to be allocated, and also return **nvgen=1** and **algen** suitably records the association between s and **next_var_int**.

The remaining cases with propositional connectives are handled as shown: for **And**, we recursively convert its arguments, and call **doAnd** (Figure 18.3). The first four cases of **doAnd** deal with one of the arguments being *false* or *true*. The last case simply appends `cnf1` and `cnf2`, as they are already in CNF form. In case of `doOr`, we perform a more elaborate case analysis. The case

```
| ([cl1],[cl2])    -> [List.append(cl1)(cl2)]
```

corresponds to two CNF clauses; here, we simply append the list of disjuncts to form a longer list of disjuncts. Given the case

```
| (cnf1',[cl2])    -> doOr([cl2])(cnf1')
```

This helps bring the first argument towards a single clause, at which time we can apply the **append** rule described earlier.

Let us discuss the last case:

```
| (cnf1', (cl2::cls)) ->
     let door1 = doOr(cnf1')([cl2]) in
     let door2 = doOr(cnf1')(cls) in
        doAnd(door1)(door2) )
```

We recursively OR `cnf1'` with the head of the second list, namely `cl2`, and the tail of the list, namely `cls`, and call `doAnd` on the result.

These are the places which cause an exponential size growth of formulas.

18.3.4 CNF-conversion using gates

To contain the size explosion during the conversion of Boolean formulas to CNF or DNF, one can resort to a circuit-based representation of clauses. This keeps the sizes of formulas *linear*, but introduces a linear number of *new* intermediate variables. Theoretically, therefore, the exponential cost remains hidden. This is because during Boolean satisfiability, an exponential number of assignments can be sought for these new variables. In practice, however, we can often avoid suffering this cost.

We now illustrate the idea of CNF conversion using gates through an example. Consider converting the DNF expression `xyzw + pqrs` to CNF. We build a circuit net-list as follows, by introducing temporary nodes `t1` through `t7`:

```
t1 = and(x,y)   ; t2 = and(z,w)   ; t3 = and(p,q)   ; t4 = and(r,s)
t5 = and(t1,t2) ; t6 = and(t3,t4) ; t7 = or(t5,t6)
```

Now, we convert each gate to its own CNF representation. We now illustrate one conversion in detail:

- $t1 = x \wedge y.$

- Treat it as $t1 \Rightarrow (x \wedge y)$ and $(x \wedge y) \Rightarrow t1$.
- The former is equivalent to $(\neg t1) \vee (x \wedge y)$ which is equivalent to $(\neg t1 \vee x) \wedge (\neg t1 \vee y)$.
- The latter is equivalent to $(\neg x \vee \neg y \vee t1)$.
- Hence, $t1 = x \wedge y$ can be represented through *three* clauses.

The following ML functions accomplish the conversion of AND gates and OR gates into clausal form ($t1 = x \wedge y$ was an example of the former):

```
let gen_andG_clauses({gatetype = Andg; inputs = (i1,i2); output = output} as andG)
  = [ [-i1; -i2; output]; [-output; i1]; [-output; i2] ]

let gen_orG_clauses ({gatetype = Org;  inputs = (i1,i2); output = output} as orG)
  = [ [i1; i2; -output]; [output; -i1]; [output; -i2] ]
```

Using these ideas, the final result for the example xyzw + pqrs is the following 15 variables (8 variables were in the original expression, and seven more – namely, t1 through t7 – were introduced) and 22 clauses (7 gates, each represented using 3 clauses each; the last clause is the *unit* clause t7 — encoded as variable number 15).

```
x=1,    y=2,    z=3,    w=4,    p=5, q=6, r=7, s=8, t1=9, t2=10,
t3=11, t4=12, t5=13, t6=14, t7=15
```

```
[ [-1; -2; 9]    ; [-9; 1]  ; [-9; 2]   ]          t1 = and(x,y)
[ [-3; -4; 10]   ; [-10; 3] ; [-10; 4]  ]          t2 = and(z,w)
[ [-5; -6; 11]   ; [-11; 5] ; [-11; 6]  ]          t3 = and(p,q)
[ [-7; -8; 12]   ; [-12; 7] ; [-12; 8]  ]          t4 = and(r,s)
[ [-9; -10; 13]  ; [-13; 9] ; [-13; 10] ]          t5 = and(t1,t2)
[ [-11; -12; 14] ; [-14; 11]; [-14; 12] ]          t6 = and(t3,t4)
[ [13; 14; -15]  ; [15; -13]; [15; -14] ]          t7 = or(t5,t6)
[ 15 ]                                             t7
```

18.3.5 DIMACS file encoding

At this point, having generated the gate net-list and their clauses, we now need to generate a file format representing the conjunction of these clauses. The standard format employed is the so called DIMACS format, where the literals in each clause are listed on separate lines, terminated by 0.

Given such a DIMACS file, one can feed the file to a Boolean satisfiability solver. Three examples of SAT solvers are [87, 81, 90], with reference [105] listing some of the latest news in the area of Boolean satisfiability.

```
p cnf 15 22  # Problem CNF format : Nvars and Nclauses.
-1 -2 9      0 # Each line lists the literals of each clause.
-9 1         0 # All lines end with a 0.
-9 2         0
-3 -4 10     0
-10 3        0
-10 4        0
-5 -6 11     0
-11 5        0
-11 6        0
-7 -8 12     0
-12 7        0
-12 8        0
-9 -10 13    0
-13 9        0
-13 10       0
-11 -12 14   0
-14 11       0
-14 12       0
13 14 -15    0
15 -13       0
15 -14       0
15           0
```

```
Type command ''zchaff CNF-File'' - here the file is "cnf"
```

The result of feeding the above file to **zchaff** is shown below:

```
[ganesh@localhost CH19]$ zchaff cnf
Z-Chaff Version: ZChaff 2003.11.04
Solving cnf ......
22 Clauses are true, Verify Solution successful.
Instance satisfiable

1 2 3 4 -5 -6 -7 -8 9 10 -11 -12 13 -14 15
Max Decision Level 7
Num. of Decisions 8
Num. of Variables 15
Original Num Clauses 22
Original Num Literals 50
Added Conflict Clauses 0
Added Conflict Literals 0
Deleted Unrelevant clause 0
Deleted Unrelevant literals 0
Number of Implication 15
Total Run Time 0

SAT
```

The SAT solver **Zchaff** picked the assignment given by the line

```
1 2 3 4 -5 -6 -7 -8 9 10 -11 -12 13 -14 15
```

which stands for the Boolean assignment

```
x=1,  y=1,  z=1,  w=1,  p=0, q=0, r=0, s=0, t1=1, t2=1,
t3=0, t4=0, t5=1, t6=0, t7=1
```

Suppose we force the assignment of x=0 by adding the clause $\neg x$: the assignment then becomes the one shown below, where p,q,r,s are set to 1:

```
-1 -2 -3 -4 5 6 7 8 -9 -10 11 12 -13 14 15
```

18.3.6 Unsatisfiable CNF instances

Consider the unsatisfiable CNF formula represented by the DIMACS file in Figure 18.4: An important theorem pertaining to unsatisfiable

```
p cnf 4 16
1 2 3 4 0
1 2 3 -4 0
1 2 -3 4 0
1 2 -3 -4 0
1 -2 3 4 0
1 -2 3 -4 0
1 -2 -3 4 0
1 -2 -3 -4 0
-1 2 3 4 0
-1 2 3 -4 0
-1 2 -3 4 0
-1 2 -3 -4 0
-1 -2 3 4 0
-1 -2 3 -4 0
-1 -2 -3 4 0
-1 -2 -3 -4 0
```

Fig. 18.4. An unsat CNF instance

CNF formulas is the following:

Theorem 18.1. In any unsatisfiable CNF formula, for any assignment, there will be *one clause with all its variables false and another with all its variables true.*

Proof: We can prove this fact by contradiction, as follows. Suppose *Fmla* is an unsatisfiable CNF formula and there is an assignment σ under which one of the following cases arise (both violating the 'one clause with all its variables false and another with all its variables true' requirement):

- There are no clauses with all its literals assuming value true. In this case, we can complement the assignment σ, thus achieving a situation where no clause will have all *false* values.
- There are no clauses with all its literals assuming value false. In this case, *Fmla* is satisfiable.

If we feed the file in Figure 18.4 to a SAT solver, it will conclude that the formula is indeed unsatisfiable. In modern SAT solvers, it is possible to find out the *root cause* for unsatisfiability. In particular, if the clauses of Figure 18.4 were embedded amidst many other clauses, upon finding the system to be unsatisfiable, a modern SAT solving framework is capable of extracting the clauses of Figure 18.4 as the root cause. This capability is being used during formal verification to provide explanations for failed proofs (e.g., [91, 109]).

18.3.7 3-CNF, $\neq$-satisfiability, and general CNF

A 3-CNF formula is a CNF formula in which every clause has three literals. For example,

$$(a \vee a \vee \neg b) \wedge (c \vee \neg d \vee e)$$

is a 3-CNF formula. As can be seen, there is no requirement on what variables the clauses might involve. Many proofs are rendered easier by standardizing on the 3-CNF form.

In many proofs, it will be handy to restrict the satisfying assignments allowed. One useful form of this restriction is the $\neq$-satisfiability restriction. Given a 3-CNF formula, a $\neq$-assignment is one under which every clause has two literals with unequal truth values. Given a set of N clauses where the ith clause is $c_i = y_{i1} \vee y_{i2} \vee y_{i3}$, suppose we transform each such clause into two clauses by introducing a new variable z_i per original clause, and a single variable b for the whole translation:

$$c_{i1} = (y_{i1} \vee y_{i2} \vee z_i) \text{ and } c_{i2} = (\neg z_i \vee y_{i3} \vee b).$$

We can show that for any given formula $Fmla = \wedge_{i \in N} c_i$, the above described transformation is a polynomial-time mapping reduction $\leq_P$ from 3-SAT to $\neq$-sat.

Finally, a general CNF formula is one where the number of literals in each clause can vary (a special case where this number is ≤ 2 is discussed in Section 18.3.8). A conversion from general CNF to 3-CNF that does not increase the length of the formula beyond a polynomial factor and takes no more than polynomial-time is now described (for each clause with less than 3 literals, replicating some literal can increase its size to 3; *e.g.*, $(a \vee b) = (a \vee b \vee b)$). Consider a clause with l literals:

$$a_1 \vee a_2 \vee \ldots \vee a_l.$$

This clause can be rewritten into one with $l - 2$ clauses:

$$(a_1 \vee a_2 \vee z_1) \wedge (\neg z_1 \vee a_3 \vee z_2) \wedge (\neg z_2 \vee a_4 \vee z_3) \ldots (\neg z_{l-3} \vee a_{l-1} \vee a_l).$$

Applying these transformations to a general CNF formula results in one that is equivalent *as far as satisfiability goes*. This is known as *equi-satisfiable*.

18.3.8 2-CNF satisfiability

A 2-CNF formula is one where each clause has two literals. One can show that the satisfiability of 2-CNF is polynomial-time, as follows:

- Consider one of the clauses $l_1 \vee l_2$. This can be written as two implications, $(\neg l_1 \Rightarrow l_2)$ and $(\neg l_2 \Rightarrow l_1)$.
- The complete algorithm is to process each clause of a 2-CNF formula into such a pair of implications. Now, viewing each such implication as a graph edge, we can connect its source to its destination; for example, in $(\neg l_1 \Rightarrow l_2)$, connect node $\neg l_1$ to node l_2. In any situation, if a double negation is introduced, as in $\neg\neg a$, we must simplify it to a before proceeding.
- Finally, we connect pairs of edges such that the destination vertex of one edge is the source vertex of the other. For example, if $(\neg l_1 \Rightarrow l_2)$ is an edge, and $(l_2 \Rightarrow l_3)$ is another edge, connect these edges at the common point l_2.

Now, the following results can be shown:

- If the graph resulting from the above construction is cyclic and a cycle includes p and $\neg p$ for some literal p, the formula is unsatisfiable.
- If this situation does not arise, we can perform a value assignment to the variables (find a satisfying assignment) as follows:
  - If a literal x (for variable x) is followed by literal $\neg x$ in the graph, assign x false.
  - If a literal $\neg x$ (for variable x) is followed by x in the graph, assign x true.
  - Else assign x arbitrarily.

Chapter Summary

This chapter considered many topics in propositional and first order logic. The highlights were: (i) the undecidability of the validity of first order logic sentences, by reduction from PCP, and (ii) Boolean sat-isfiability from a theoretical and practical point of view. Armed with this background, we will next study the very important problem of NP-completeness.

| Exercises |

18.1. Using the operators used in the Hilbert style axiomatization of Section 18.1, describe how to implement the familiar operators $\wedge$, $\vee$, and $\oplus$ (exclusive-OR). Also implement the constants *true* and *false*.

18.2.
1. Describe what the following terms mean:
   a) axiom
   b) rule of inference
   c) theorem
   d) satisfiable
   e) proof
   f) tautology
2. Show that the following is a tautology:

$$(\neg x \Rightarrow \neg y) \Leftrightarrow (y \Rightarrow x)$$

Recall that $x \Leftrightarrow y$ means $(x \Rightarrow y) \wedge (y \Rightarrow x)$.

18.3. Prove the following using a similar approach as illustrated above (Sperschneider's book, p. 107): *Every barber shaves everyone who does not shave himself. No barber shaves someone who shaves himself. Prove that there exists no barber.*

18.4. Show that *Fmla2* is true under Interpretation 2 of page 328, describing the witness function ("$\exists F$"). Also show that it is false under Interpretation 3.

18.5. One often falls into the following trap of claiming that FOL for-mulas are decidable: the negation of every FOL formula is an FOL formula; thus one can enumerate proofs, and see whether F or $\neg F$ gets enumerated in a proof, and hence decide the validity of F. What is the fallacy in this argument? (Hint: Are the cases listed, namely F being valid and $\neg F$ being unsatisfiable, exhaustive?)

18.6. Argue that the set of all non-valid but satisfiable FOL formulas is not recursively enumerable.

18.7.

1. Minimize the Boolean expression $(\neg a + ab)c + ac$ using a Karnaugh map.
2. How many Boolean functions can you build with a 4-to-1 multiplexer that has two select inputs?

18.8. Modify the program in Figures 18.2 and 18.3 to obtain a CNF to DNF converter. Test your resulting program.

18.9. Argue that DNF satisfiability and CNF validity have linear-time algorithms.

18.10. Solve Exercise 5.29 through proof by contradiction. Encode the problem as a CNF formula, and use a SAT tool.

18.11. Solve Exercise 5.30 similar to Exercise 18.10.

18.12. Write down the CNF formula represented by the DIMACS file given in Figure 18.4.

18.13. Prove that this transformation is indeed a polynomial-time mapping reduction $\leq_P$ from 3-SAT to $\neq$-sat. In other words, prove that (i) it is a mapping reduction, and (ii) the runtime of the function involved is a polynomial with respect to the input size.

18.14. Why does the conversion from general CNF to 3-CNF described on page 341 preserve only satisfiability? (i.e., why do the formulas not emerge to be logically equivalent)

18.15. Prove that the 2-CNF satisfiability algorithm sketched in Section 18.3.8 is correct (meaning that it finds a satisfying assignment whenever one exists).

18.16. Apply the 2-CNF satisfiability algorithm of Section 18.3.8 to the following 2-CNF formulas, showing the steps as well as your final conclusion as to the satisfiability of these formulas:

- $(a \vee b) \wedge (a \vee \neg b) \wedge (\neg a \vee b)$
- $(a \vee b) \wedge (a \vee \neg b) \wedge (\neg a \vee b) \wedge (\neg a \vee \neg b)$

19

Complexity Theory and NP-Completeness

The theory of NP-completeness is about a class of problems that have defied efficient (polynomial-time) algorithms, despite decades of intense research. Any problem for which the *most efficient known algorithm* requires exponential time[1] is called *intractable*. Whether NP-complete problems will *remain* intractable forever, or whether one day someone will solve *one* of the NP-complete problems using a polynomial-time algorithm remains one of the most important open problems in computing. The Clay Mathematics Institute has identified seven Millennium Problems, each carrying a reward of $1 million (US) for the first person or group who solves it; the 'P =NP' problem is on this list. Stephen Cook provides an official description of this problem, and associated (extremely well-written) set of notes, also at the Clay web site [23].

The theory of *NP-completeness* offers a way to "bridge" these problems through efficient simulations (polynomial-time mapping reductions $\leq_P$ (Definition 16.4) going both ways between any two of these problems) such that *if* an efficient algorithm is found even for *one* of these problems, then an efficient algorithm is immediately obtained for *all* the problems in this class (recall our discussions in Chapter 1, page 11). Section 19.1 presents background material. Section 19.3 presents several theorems and their proofs. Section 19.4 provides still more illustrations that help avoid possible pitfalls. Section 19.5 discusses notions such as CoNP. Section 19.5 concludes.

---

[1] The best known algorithm for an intractable problem has complexity $\mathcal{O}(k^n)$ for an input of length n, and $k > 1$.

19.1 Examples and Overview

19.1.1 The traveling salesperson problem

The *traveling salesperson problem* is a famous example of an NPC problem. Suppose a map of several cities as well as the cost of a direct journey between any pair of cities is given.[2] Suppose a salesperson is required to start a tour from a certain city c, visit the other $n-1$ cities in some order, but visiting each city *exactly once*, and return to c while minimizing the overall travel costs. What would be the most efficient algorithm to calculate an optimal tour (optimal sequence of cities starting at c and listing every other city exactly once)?

- This problem is intractable.
- This problem has also been shown to be NPC.

The NPC class includes thousands of problems of fundamental importance in day-to-day life - such as the efficient scheduling of airplanes, most compact layout of a set of circuit blocks on a VLSI chip, etc. They are all intractable.

19.1.2 P-time deciders, robustness, and 2 vs. 3

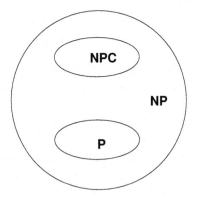

Fig. 19.1. Venn diagram of the language families P, NP, and NPC; these set inclusions are *proper* if $P \neq NP$ — which is an open question

---

[2] Assume that these costs, viewed as a binary relation, constitute a symmetric relation.

The space of problems we are studying is illustrated in Figure 19.1. *It is not know* whether the inclusions in Figure 19.1 are proper.[3] The oval drawn as P stands for problems for which polynomial time deciders exist.

Let

$$\text{TIME}(t(n)) = \{ \ L \ | \ L \text{ is a language decided by an } \mathcal{O}(t(n)) \text{ time TM } \}$$

Then,

$$P = \cup_{k \geq 0} \text{TIME}(n^k).$$

As an example, the following language $L_{0^n 1^n}$ is in P

$$L_{0^n 1^n} = \{0^n 1^n \ | \ n \geq 0\}$$

because

- it is in $\text{TIME}(n^2)$, as per the following algorithm, A1:
  - Consider a DTM that, given $x \in \Sigma^*$, zigzags on the input tape, crossing off one 0 and then a (supposedly matching) 1. If the tape is left with no excess 0s over 1s or vice versa, the DTM accepts; else it rejects.
- It is even in $\mathcal{O}(n \, log n)$ as per the following algorithm, A2:
  - In each sweep, the DTM scores off every other 0 and then every other 1. This means that in each sweep, the number of surviving 0s and 1s is half of what was there previously. Therefore, $log(n)$ sweeps are made, each sweep spanning n tape cells. The stopping criterion is the same as with algorithm A1.

Another member of P is *TwoColor* (see Exercise 19.1); here two-colorable means one can color the nodes using two colors, with no two adjacent nodes having the same color:

$$TwoColor = \{\langle G \rangle \ | \ G \text{ is an undirected graph that is two\_colorable } \}$$

We shall define NP and NPC later in this chapter.

19.1.3 A note on complexity measurement

We will not bother to distinguish between N and $N \, log(N)$, lumping them both into the polynomial class. The same is true of $N^k \, log^m(N)$ for all k and m. Our complexity classification only has two levels: "polynomial" or "exponential." The latter will include the *factorial* function, and any such function that is harder than exponential (e.g., Ackermann's function).

---

[3] When the $1M Clay Prize winner is found, they would either assert this diagram to be exact, or simply draw one big circle, writing "P" within it – with a footnote saying "NP and NPC have been dispensed with."

19.1.4 The robustness of the Turing machine model

There are a mind-boggling variety of *deterministic* Turing machines: those that have a doubly-infinite tape, those that have multiple tapes, those that employ larger (but finite) alphabets, and even conventional deterministic random-access machines such as desktop and laptop computers (given an unlimited amount of memory, of course). All this variety does not skew our two-scale complexity measurement:

 - P is invariant for all models that are polynomially equivalent to the deterministic single-tape Turing machine (which includes all these unusual Turing machine varieties)
 - P roughly corresponds to the class of problems that are realistically solvable on modern-day random-access computers.

Hence, studying complexity theory based on deterministic single-tape Turing machines allows us to predict the complexity of solving problems on real computers.

19.1.5 Going from "2 to 3" changes complexity

It is a curious fact that in many problems, going from "2 to 3" changes the complexity from polynomial to seemingly exponential. For instance, K-colorability is the notion of coloring the nodes of a graph with K colors such that no two adjacent nodes have the same color. Two-colorability is in P, while three-colorability is NPC. This is similar to the fact that the satisfiability of 2-CNF formulas is polynomial (Section 18.3.8), while that of 3-CNF formulas is NPC (Theorem 19.8). The reasons are, not surprisingly, somewhat similar.

19.2 Formal Definitions

We now proceed to define the remaining ovals in Figure 19.1.

19.2.1 NP viewed in terms of verifiers

We now present the *verifier* view of NP, with the *decider* view presented in Definition 19.5.

Definition 19.1. *(Verifier view of* NP*)* A language L is in NP if there exists a deterministic polynomial-time Turing machine V_L called the *verifier*, such that given any $w \in \Sigma^*$, there exists a string $c \in \Sigma^*$ such that $x \in L$ exactly when $V_L(w, c)$ accepts.

Here, c is called a *certificate*. It also corresponds to a "guess," as introduced in Section 15.5.1 of Chapter 15 (some other equivalent terms are *witness, certificate, evidence,* and *solution*). According to Definition 19.1, the language

$Clique = \{\langle G, k \rangle \mid G$ is an undirected graph having a k–clique$\}$

is in NP because there exists a verifier V_{Clique} such that

$Clique = \{\langle G, k \rangle \mid$ There exists c such that V_{Clique} accepts $(G, k, c)\}$.

Here, c is a sequence of k nodes. V_{Clique} is a polynomial-time algorithm that is captured by (as well as carried out by) a *deterministic* Turing machine. This DTM does the following: (i) checks that G is an undirected graph, and c is a list of k nodes in G, and (ii) verifies that the nodes in c indeed form a clique. Note that the ability to *verify* a guess in polynomial-time means that the length of c must be bounded by a polynomial in terms of input size.

   Given G and k, all known practical algorithms take exponential time to *find out* which k nodes form a clique. On the other hand, given an arbitrary list of k nodes, *verifying* whether these form a clique is easy (takes a linear amount of time). Problems in NP share this property by definition. Recall our discussions in Section 15.5.1 about Mersenne primes that also shares this property of easy verifiability.[4]

19.2.2 Some problems are outside NP

It is indeed remarkable that there are problems where even *verifying a solution* is hard, taking an exponential amount of time with respect to all known algorithms for these problems! Clearly these problems *do not* belong to NP. For example, for the $\overline{Clique}$ problem defined in Chapter 17, efficient verifiers have, so far, remained impossible to determine. Intuitively, this seems to be because languages (problems) such as $\overline{Clique}$ seem to call for an enumeration of *all candidate list of k nodes* and an assertion that each such list, in turn, does *not* form a clique.[5] It seems that the certificates for these problems must be

---

[4] Students believe that every problem assigned to them is NP-complete in difficulty level, as they have to *find* the solutions. Teaching Assistants, on the other hand, find that their job is only as hard as P, as they only have to *verify* the student solutions. When some students confound the TAs, even verification becomes hard - something discussed in Section 19.2.2.

[5] Continuing with the analogy introduced in Chapter 17, we are being asked to prove that no group of k people know each other, as opposed to proving that *some k* people know each other.

exponentially long, because they are a concatenation of an exponential number of these candidate list of nodes mentioned above. The mere act of reading such certificates consumes an exponential amount of time! Of course, these are simply conjectures: it is simply not know at present how to prove that problems such as $\overline{Clique}$ cannot have succinct (polynomially long) certificates.

To sum up, problems whose solutions are easy to verify (NP) are, in some sense, *easier* than problems whose solutions are *not* easy to verify — even though finding the solution is hard in both cases. As Pratt puts it so eloquently in his paper where he proves that *primes* are in NP [98],

> "The cost of *testing* primes and composites is very high. In contrast, the cost of *selling* composites (persuading a potential customer that you have one) is very low—in every case, one multiplication suffices. The only catch is that the salesman may need to work overtime to prepare his short sales pitch."

19.2.3 NP-complete and NP-hard

Definition 19.2. *(NP-complete)* If a language L is shown to be in NP, *and furthermore*, if it can be shown that for every language $X \in$ NP, $X \leq_P L$, then L is NPC.

Therefore, showing a problem to be in NP is a prerequisite to showing that it is NPC.

Definition 19.3. *(NP-hard)* If for all $X \in$ NP we have $X \leq_P L$, then L is said to be *NP-hard*.

From all this, "NPC" means "NPH" *and* "belongs to NP."

Note: If L is NP-hard, it means that L is *at least as hard as NP*. It is possible that L is so hard as to be *undecidable*, as is shown in Section 19.4.

19.2.4 NP viewed in terms of deciders

In Definition 19.1, NP was defined with the help of deterministic Turing machines V_L which are verifiers. There is an alternative definition of NP in terms of nondeterministic Turing machines, which is now presented, after presenting the notions of a *nondeterministic decider* and a *nondeterministic polynomial-time decider*.

Definition 19.4. *(NP decider)* An NDTM N_L with starting state q_0 is a nondeterministic decider for a language L exactly when for all $x \in \Sigma^*$, $x \in L$ if and only if N_L has an accepting computation history starting at q_0x. It is an NP (nondeterministic polynomial-time) decider for L if, for *all strings* $x \in L$, the length of the *shortest* accepting computation history starting at q_0x is $\mathcal{O}(n^k)$.

Note that we are not requiring this NDTM to terminate along all paths. Using the notion of '*shortest*,' we are able to ignore all other paths.

Definition 19.5. *(Decider view of NP)* NP is the class of decidable languages such that associated with each $L \in$ NP is a nondeterministic polynomial-time decider.

Definition 19.4 is adopted in [44], [39], and [67]. In [111], different definitions (that consider the *longest* computation) are employed. The advantages of definitions that go by the shortest accepting computation history are the following:

- It allows NDTMs to be designed without special precautions that are irrelevant in the end. In particular, we do not need to define the NDTMs to avoid paths that are unbounded in length (see Section 19.2.5 for an example).
- It helps focus one's attention on positive outcomes (the $x \in L$ case), as well as the "guess and check" principle of Section 15.5.1, and, last but not least,
- It helps present and prove Theorem 19.6 very clearly.

Also, please note the following:

- A nondeterministic polynomial-time decider Turing machine N_L has *nondeterministic polynomial* runtime. 'Nondeterministic polynomial' is a different way of measuring runtime, different from how it is done with respect to DTMs, where run times are measured in terms of the number of steps taken by a DTM from start to finish over all inputs, where each input induces exactly one computation history. In case of NP, for each input, there could be *multiple* computation histories, and we simply focus on the shortest ones.

19.2.5 An example of an NP decider

These notions are best explained using an example; we choose Figure 15.3 for this purpose. Given any string of the form ww, this machine generates one guess (refer to Section 15.5.1) that correctly identifies and checks around the midpoint, ultimately leading to acceptance.

However, this machine also has many useless guesses that lead to rejection. In fact, *there is no a priori bound on the number of times this NDTM loops back to state* q2 *before exiting the loop!* It could, therefore, be guessing any point that is arbitrarily away from the left-end of the tape to be the midpoint! Of course we could easily have defined a "better" NDTM that rejects as soon as we are off the far right-end of the input. However, such "optimizations" are not helpful in any way; keep in mind that

- NDTMs are mathematical devices that only serve one purpose: to measure complexity in a manner that implements the "guess and check" idea discussed in Section 15.5.1 (and further discussed in this chapter).
- In its role as a theoretical device, it is perfectly normal for an NDTM to have a computation tree that has *an infinite number of branches out of its nondeterministic selection state(s)*. However, since we measure the time complexity in terms of the shortest accepting computation history, these longer paths automatically end up getting ignored.

Theorem 19.6. *The verifier view of* NP *(Definition 19.1) and the decider view of* NP *(Definition 19.4) are equivalent.*

Proof outline:
    With respect to the first part of the proof, we observe that an NDTM can always be designed to have a loop similar to the self-loop at state q2 of Figure 15.3. In this loop, it can write out *any* string from the tape alphabet and then call it "the certificate c" and then verify whether it is, indeed, a certificate. Now, if there *exists* any certificate at all, then one would be found in this manner. Furthermore, if there exists a polynomially bounded certificate, then again, it would be found since we heed only the shortest accepting computation history. For the second part of the proof, we let the certificate be tree paths, as described in Section 15.6.2.

Proof: Given V_L, we can build the NDTM N_L as follows:

```
N_L =
  Accept input w;
    Employ a nondeterministic loop, and write out
      a certificate string c on the tape; c is
      an arbitrary nondeterministically chosen finite string;
      c is written after the end of w;
    Run V_L on (w,c), and accept if V_L accepts.
```

Going by our definition of the *shortest accepting computation history*, there will be one certificate that works, since V_L has a certificate that works. Therefore, N_L will consist of the certificate generation phase followed by feeding V_L with w and c. The N_L thus constructed will have a nondeterministic polynomial runtime and decides L.   □

- Given N_L, we can build the NDTM V_L as follows.

```
V_L =
  On input w,c,
    Use c to guide the selection among the
       nondeterministic transitions in N_L;
    Accept when N_L accepts.
```

In essence, c would be a sequence of natural numbers specifying which of the nondeterministic selections to make ("take the first turn; then the third turn; then the second turn; ...[6]). Now, if N_L has an accepting computation history, there is such a certificate c that leads to acceptance. Hence, the DTM V_L would be a deterministic polynomial-time verifier for w, c.   □

19.2.6 Minimal input encodings

In measuring complexity, one must have a convention with regard to n, the length of the input. The conventions most widely used for this purpose are now explained through an example.. Consider the *Clique* problem again. To measure how much time it takes to answer this membership question, one must encode $\langle G, k \rangle$ "reasonably"—in a minimality sense. In particular, we should avoid encoding $\langle G, k \rangle$ in a unary fashion. Doing so can skew the complexity. Details are in Section 19.5.

19.3 NPC Theorems and proofs

Definition 19.2 defined $L \in$ NPC in terms of a reduction from *all* $X \in$ NP. This may seem to be an infeasible recipe to follow, as there are $\aleph_0$ languages that are in NP. Historically, only *one* problem was solved using this tedious approach (detailed in Section 19.3.1). All other problems shown to be NPC were proved using Definition 19.7, which offers

---

[6] Imagine a boat in a lake being turned and pushed around by the hands of a giant, and its rudder limply rotating, following the motions of the boat. The motions of the rudder are analogous to c.

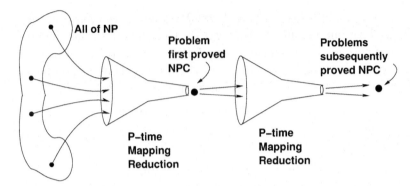

Fig. 19.2. Diagram illustrating how NPC proofs are accomplished. Definition 19.2 is illustrated by the reduction shown on the left while Definition 19.7 is illustrated on the right.

a much more practical recipe, assuming there exists at least one NPC language.[7]

Definition 19.7. *(NP-complete)* A language L is NPC if $L \in$ NP and furthermore, for some other language $L' \in$ NPC, we have $L' \leq_P L$.

This definition is equivalent to Definition 19.2 because if $L' \in$ NPC, we have $\forall X \in$ NPC $: X \leq_P L'$, and $\leq_P$ is transitive. The "funnel diagram" in Figure 19.2 illustrates this approach.

In order to identify the very first NPC problem, we do need to go by Definition 19.2. The first problem that had this 'honor' was 3-CNF satisfiability ("3-SAT"), as Cook and Levin's celebrated proof shows. Recall from Section 18.3.7 general discussions about 3-CNF.

19.3.1 NP-Completeness of 3-SAT

Theorem 19.8. 3-CNF satisfiability is NP-complete.

3-SAT is in NP

We go by Definition 19.1. Consider a satisfying assignment σ for a given 3-CNF formula φ. Clearly, σ is of polynomial length, and verifying that it satisfies φ takes polynomial-time through a verification algorithm that substitutes into φ as per σ and simplifies the formula to *true* or *false*.

---

[7] Like the proverbial 'chicken and the egg,' we assume the first egg, - er, first chicken

For any $L \in$ NP, $L \leq_P$ 3-SAT

We sketch the proof emphasizing the overall structure of the proof as well as some crucial details. For example, (i) we show how the *existence* of a deterministic polynomial time algorithm for 3-SAT implies the *existence* of such an algorithm for *any* problem in NP, and (ii) we show how, given a specific NP problem such as *Clique* and given a deterministic polynomial algorithm for 3-SAT, we can obtain a deterministic polynomial algorithm for *Clique*.

Consider some $L \in$ NP. Then there exists an NDTM decider N_L for L. What we have to show is that there *exists* a polynomial-time mapping reduction f from L to 3-CNF such that given N_L and an arbitrary $w \in \Sigma^*$, there *exists* a 3-CNF formula $\varphi_{L,w}$ that can be obtained from N_L and w using f, such that $\varphi_{L,w}$ is satisfiable if and only if $w \in L$.

> *Punchline:* Therefore, if one were to find a polynomial-time algorithm for 3-CNF satisfiability, there would now be a polynomial-time algorithm for *every* L in NP.

To prove the existence of the mapping reduction alluded to above, refer to Figure 19.3. Consider the computation of N_L on some $w \in \Sigma^*$ starting from the instantaneous description ID0 $= q_0 w$. If $w \in L$, there is an accepting computation history that starts with $q_0 w$ and is of polynomial length (we do not know this length exactly; all we know is that it is a polynomial with respect to $|w| = n$). Let this polynomial be

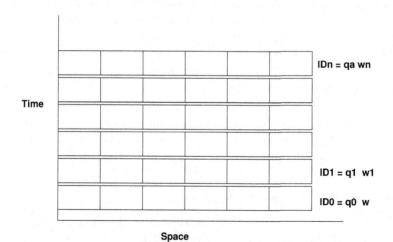

Fig. 19.3. Proof of the Cook-Levin Theorem

$p(n)$. If $w \notin L$, then no accepting computation history (of any length!) exists.

Imagine a computation history (sequence of IDs) of length $p(n)$. In it, (i) only a polynomial amount of new tape cells get written into (the "space" axis) and (ii) the accepting computation history is of polynomial length (the "time" axis; see Figure 19.3 for an illustration). This figure illustrates the computation starting from $q_0 w$ as a sequence of IDs starting with ID0 at the bottom. Now, observe that if $w \in L$, there must exist a $p(n) \times p(n)$ matrix as shown in Figure 19.3 that (i) starts with $q_0 w$ at the bottom, (ii) ends with an accepting ID at the top, and (iii) has any two adjacent rows related by a Turing machine transition rule. Now, it is clear that there are basically two rules, as mentioned in Section 17.3.6: (i) one rule corresponds to a right-hand move ($\delta(q, a) = (r, b, R)$), and (ii) another corresponding to a left-hand move ($\delta(q, a) = (r, b, L)$ for every $c \in \Gamma$). Furthermore, the effect of these moves on adjacent IDs can be captured through "windows" that change across two IDs. For instance, for ID_{i+1} obtained from ID_i through a right move, a 2×2 window comes into play, and for a left move, a 3×2 window comes into play, as mentioned in Section 17.3.6. It is also clear that the contents of each cell in this $p(n) \times p(n)$ matrix can be represented using a finite number of cell level Boolean variables.

Given all this, the crux of our proof is that given this $p(n) \times p(n)$ matrix, we can build our 3-CNF formula $\varphi_{L,w}$ involving the cell level Boolean variables such that this formula is true exactly when the matrix represents an accepting computation history (we also refer to this 3-CNF formula as a "circuit," connoting a digital combinational circuit that can serve as a decoder for an accepting computation history matrix). The actual construction of this formula is tedious and skipped here (but may be found in scores of other web references or books). On page 357, we illustrate what this formula achieves with respect to the *Clique* example.

What we have sketched thus far is the *existence* of a mapping reduction that yields $\varphi_{L,w}$ such that this formula is satisfiable exactly when $w \in L$. If $w \notin L$, there is no matrix that will represent an accepting computation history, and hence $\varphi_{L,w}$ will *not* be satisfiable.   □

A detail about not knowing $p(n)$

A point that may vex the reader is this: how do we know how big a circuit (3-CNF formula) to build, given that the circuit has to sit on top of the $p(n) \times p(n)$ matrix, hoping to decode its contents, when we don't even know $p(n)$ concretely? Fortunately, this step is not necessary—all we care for is the *existence* of a family of circuits, one circuit for each value of $p(n)$. There would, in this family, exist a circuit that "works" for every language L in NP, and correspondingly there would be a family of f functions that produced this family of circuits. Therefore, for each possible value of $p(n)$, there *exists* an f function that produces a matrix of size $p(n) \times p(n)$ and a 3-CNF formula that acts on this matrix; and hence, the desired mapping reduction $L \leq_P$ 3-CNF *exists* for each $L \in$ NP. The *existence* of this mapping reduction for every $L \in$ NP allows us to claim that 3-SAT is in NPC.

What if 3-SAT is in *P*?

It is good to be sure that formal proofs mean something concrete; to this end, we subject our discussions above to an acid test. Suppose 3-SAT is in P (which is an open question; but we entertain this thought to see what happens) and let the decider be $DP_{3-\text{CNF}}$. Because of what NP-completeness means, we should now be in a position to argue that there exists a deterministic polynomial-time algorithm for any $L \in$ NP. How do we achieve that?

Fortunately, this result is immediate. Since the polynomial $p(n)$ exists, a mapping reduction to yield $\varphi_{L,w}$ exists. We can obtain this formula using the mapping reduction, feed it to $DP_{3-\text{CNF}}$, and return the accept/reject decision of $DP_{3-\text{CNF}}$. Therefore, a deterministic polynomial-time algorithm for an arbitrary $L \in$ NP exists, which would then mean $P = NP$.

Illustration on *Clique*

We would like to take our acid test even further: suppose 3-SAT is in P; let us find a polynomial algorithm for *Clique*. Since *Clique* $\in$ NP, it has a nondeterministic polynomial time decider, say N_{Clique}. We can design a specific NDTM decider for *Clique*. One of the most straightforward designs for N_{Clique} would be to have an NDTM nondeterministically write out k nodes on the tape and check whether these nodes indeed form a k-clique. Now, there are *many* (exponentially many!) choices of k nodes to write out on the tape. However, after writing out one of those guesses on the tape, N_{Clique} would engage in a polynomially bounded checking phase. The 3-CNF formula $\varphi_{L,w}$ that would be synthesized for this example will have the following properties:

- It will be *falsified* if the first ID (bottom row of the matrix) is not the starting ID which, in our example, would be $q_0 \langle G, k \rangle$.

- It will be falsified if any two adjacent rows of the matrix are not bridged by a legitimate transition rule of the NDTM.
- It will be falsified if the final row (topmost) is not an accepting ID.
- The formula will straddle a matrix of size $p(n) \times p(n)$, where the value of $p(n)$ will be determined by the nature of the algorithm used by N_{Clique} for $Clique$, and the value of k. In particular, $p(n)$ will equal the number of steps taken to write out some sequence of k nodes nondeterministically, followed by the number of steps taken to check whether these nodes form a clique.

In short, an accepting computation history will deposit a bit-pattern inside this matrix to make the Boolean formula emerge true. Said another way, the *satisfiability* of the formula will indicate the *existence* of an accepting computation history (the existence of a selection of k nodes that form a clique). Therefore, if DP_{3-CNF} exists, $Clique$ can be solved in polynomial-time.

> To reflect a bit, the existence of DP_{3-CNF} is a *tall order* because it gives us a mechanism to encode exponential searches as polynomially compact formulas (such as $\varphi_{L,w}$ did) and conduct this exponential search in polynomial-time! This is one reason why researchers strongly believe that deciders such as DP_{3-CNF} do not exist.

19.3.2 Practical approaches to show NPC

The most common approach to show a language L to be NPC is to use Definition 19.7 and reduce 3-SAT to L, and then to show that $L \in$ NP. This has led many to observe that 'NP-complete problems are 3-SAT in disguise.' Many other source languages for reduction (besides 3-SAT) are, of course, possible.

Illustration 19.3.1 Let $Hampath =$

$\{\langle G, s, t \rangle \mid G$ is a directed graph with a Hamiltonian path from s to $t\}$.

It can be shown that $Hampath \in$ NP, and further 3-SAT$\leq_P Hampath$. This establishes that $Hampath$ is NPC.

Illustration 19.3.2 Let us prove that $Clique$ is NPC. First of all, $Clique \in$ NP as captured by the verifier V_{Clique} on page 349. To show that $Clique$ is NPH, we propose the mapping reduction from 3-SAT into

phi = (X1 ∨ X1 ∨ X2) ∧ (X1 ∨ X1 ∨ ~X2) ∧ (~X1 ∨ ~X1 ∨ X2) ∧ (~X1 ∨ ~X1 ∨ ~X2)

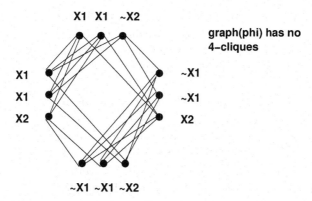

Fig. 19.4. The Proof that *Clique* is NPH using an example formula $\varphi = (x_1 \vee x_1 \vee x_2) \wedge (x_1 \vee x_1 \vee \neg x_2) \wedge (\neg x_1 \vee \neg x_1 \vee x_2) \wedge (\neg x_1 \vee \neg x_1 \vee \neg x_2)$

Clique as captured in Figure 19.4. Basically, for every clause, we introduce an "island" of nodes with each node in the island labeled with the same literal as in the clause. There are no edges among the nodes of an island. Between two islands, we introduce an edge between every pair of literals that can be simultaneously satisfied (are not complementary literals).

Suppose $\varphi \in$ 3-SAT. This means that there is an assignment that satisfies every clause. Let l_i be the literal that is set to true by the assignment in clause c_i, for every $i \in C$, where C is the number of clauses (for uniqueness, we may select the literal with the lowest index that is set to true in each clause). In this case, by construction, $graph(\varphi)$ will have a clique connecting the nodes $l_0, l_1, \ldots, l_{C-1}, l_0$.

Suppose $\varphi \notin$ 3-SAT. Now suppose we assume that $graph(\varphi)$ has a k-clique. The existence of a k-clique means that by following the edges of the clique, it should be possible to pick one literal per clause such that all these literals (and hence these clauses) can be simultaneously satisfied. However, from Theorem 18.1, we know that given any unsatisfiable CNF formula φ, one can pick an arbitrary assignment σ, and be assured that $\sigma(\varphi)$ (the formula under the assignment) has one clause all of whose literals are true, and another clause all of whose literals are false. The clause that has all its literals false will prevent there being a k-clique. To confirm all this, in Figure 19.4 we observe that there are no 4-cliques.

19.4 NP-Hard Problems can be Undecidable (Pitfall)

What happens if someone shows L to be NPH but neglects to show
$L \in$ NP, and yet claims that $L \in$ NPC? To show the consequences of this
mistake rather dramatically, we will show that the language of Diophantine equations, *Diophantine*, is NP-hard (NPH). Briefly, *Diophantine*
is the set of Diophantine equations that have integer roots. An example
of such an equation is $6x^3z^2 + 3xy^2 - x^3 - 10 = 0$. This language was
shown to be undecidable by Yuri Matijasević in a very celebrated theorem. Hence, if someone forgets to show that a language L is in NP, and
yet claims that L is NPC, he/she may be claiming that something undecidable is decidable! (Recall that all NPC problems are decidable.) In
short, NP-completeness proofs *cannot* be deemed to be correct unless
the language in question is shown to belong to NP.

19.4.1 Proof that Diophantine Equations are NPH

We follow the proof in [27] to show that the language *Diophantine*
below is NPH:

$Diophantine = \{p \mid p$ is a polynomial with an integral root$\}$

The mapping reduction 3-SAT $\leq_P$ *Diophantine* is achieved as follows.
Consider a 3-CNF formula φ:

- Each literal of φ, x, maps to integer variable x.
- Each literal $\overline{x}$ maps to expression $(1 - x)$.
- Each $\vee$ in a clause maps to . (times).
- Each clause is mapped as above, and then *squared*.
- Each $\wedge$ maps to +.
- The resulting equation is set to 0.
- Example: map

$$\varphi = (x \vee y) \wedge (x \vee \overline{y}) \wedge (\overline{x} \vee \overline{y})$$

to

$$E = (x.y)^2 + (x.(1-y))^2 + ((1-x).(1-y))^2 = 0.$$

- To argue that this is a mapping reduction, we must show that φ is
  satisfied iff E has integral roots. Here is that proof:
  - For the forward direction, for any assignment of a variable v to
    true, assign v in the integer domain to the integer 0; if v is *false*,
    use integer 1. In our example, $x = true, y = false$ satisfies
    φ, and so choose $x = 0, y = 1$ in the integer domain. This
    ensures that $(x.y)^2$ is zero. Proceeding this way, every satisfying
    assignment has the property of leaving the entire summation of
    expressions 0, thus satisfying the integer equation.

– For the reverse direction, note that $E = 0$ means that each product term in the integer domain is 0 (since squares can't be negative). For example, if xy is a product term in the summation of E, we may have $x = 45$ and $y = 0$. The Boolean assignment for this case is found as follows: for every integer variable x that is zero, assign the corresponding Boolean variable x to $true$; for integer variable x that is non-zero, assign the Boolean variable x to $false$. For example, if we have $x = 0$ in a product term $x.y$, we assign Boolean x to $true$. This ensures that $(x \vee y)$ is true. Also, in $x.(1-y)$ if $x = 45$ and $y = 1$, we assign y to $false$ and x to false. This ensures that $(x \vee \neg y)$ is true. We can easily check that this construction ensures that $E = 0$ exactly when the corresponding φ has a satisfying assignment.

19.4.2 "Certificates" of Diophantine Equations

In order to visualize the transition from being NPC, to being outside NP but still decidable, and finally to being undecidable, let us discuss Diophantine equations in the context of certificates. Consider the language $Hampath$; clearly, every member of this language has a polynomial certificate. This certificate is a simple path (path without repeating nodes) connecting s and t that visits every other node. Languages such as $\overline{Hampath}$ do not have, as best as is known, polynomial certificates. However, exponentially long certificates do exist; these certificates list every simple path connecting s and t, thus providing cumulative evidence that there *does not* exist a Hamiltonian path.

It is evident that $Diophantine$ is recursively enumerable (it is TR) but not recursive (it is not decidable). One may attempt to build a nondeterministic machine M_{Dio}, of as yet unclear status, to process membership of a given Diophantine equation in $Diophantine$: M_{Dio} guesses a certificate in the guess-generation phase consisting of guessed values for the variables of the equation. M_{Dio} then plugs in these values and checks whether the given equation is satisfied (equals 0). For an undecidable languages such as $Diophantine$, certificates exist, but are unbounded for the case where the equation has a solution. When the equation has no solution, the certificates are infinite (one has to list every possible value for the variables and show that they do not solve the equation).

In summary, polynomial, exponential, and unbounded certificates correspond to three classes of hardness.

19.4.3 What other complexity measures exist?

There are many other complexity metrics such as space complexity and circuit (parallel) complexity. It must be relatively obvious what the term 'space complexity' means: how much *space* (memory) does an algorithm require? In this context, please note that *space is reusable while time is not.* This means that the term space complexity refers to the *peak* space requirement for an algorithm. There is also the fundamentally important result, known as *Savich's Theorem*, that says that nondeterministic Turing machines can be simulated on top of deterministic Turing machines with only a polynomial added cost. This is in contrast with time complexity where we do not know whether nondeterministic Turing machines can be simulated on a deterministic Turing machine with only a polynomial added cost.

The term 'circuit complexity' may be far from obvious to many. What it pertains to is, roughly, *how easy a problem is to parallelize.* In circuit complexity, the intended computation is modeled as a Boolean function, and the *depth* of a combinational circuit that computes this function is measured. Problems such as *depth-first search* are, for instance, not easy to parallelize (log-depth circuits cannot be found), whereas *breadth-first search* easy to parallelize under this complexity measure. Log-depth circuits, in a sense, help assess how easy it is to divide and conquer a problem.

19.5 NP, CoNP, etc.

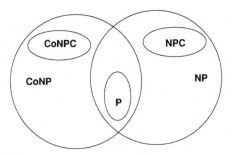

Fig. 19.5. The language families P, NP, and NPC. All these set inclusions are likely to be proper

A language L is said to be CoNP exactly when $\overline{L}$ is in NP. Similarly, L is said to be CoNPC exactly when $\overline{L}$ is in NPC. Figure 19.5 depicts

these additional language classes and their likely containments. To illustrate these ideas, consider the following languages which are both subsets of positive natural numbers $\{1, 2, 3, \ldots\}$:
The language

$$Primes = \{n \mid (n > 1) \wedge (\exists p, q > 1 : n = p \times q \Rightarrow p = 1 \vee q = 1\}.$$

The language
$$Composites = \overline{Primes},$$

where the complementation is with respect to positive naturals.
Composites is clearly in NP because there exists a P-time verifier for this language, given a certificate which is a pair of natural numbers suggested to be factors. In [98], Pratt proves that *Primes* are also in NP; he shows this result by demonstrating that there are polynomially long proofs for primes (given a prime p, a polynomially long sequence of proof steps can serve to demonstrate that p is such). Furthermore, he showed that such a proof for *Primes* can be checked in polynomial-time. Now, *Composites* is in CoNP because *Primes* is in NP, and *Primes* is in CoNP because *Composites* is in NP. The question now is: could either of these languages be NPC? Theorem 19.9 below, shows that even if there exists one such language, then NP and CoNP would become equal—a result thought to be highly unlikely. Surely enough, in 2002, Agrawal et al. [1] proved that *Primes* are in P (and hence *Composites* are also in P). Theorem 19.9 has helped anticipate the direction in which some of the open problems in this area would resolve.

Illustration 19.5.1 *(A Caveat)* Please bear in mind that recognizing a number to be composite in polynomial-time *does not*, by itself, give us the ability to *find* its prime factors in polynomial-time. Therefore, all public key crypto systems are still safe, despite Agrawal et al.'s result. Factoring a composite number into its prime factors can be expressed as a language

$$PrimeFactors = \{(x, i, b) \mid i^{th} \text{ bit of prime factorization of } x \text{ is } b\}.$$

Here, it is assumed that the prime factors of x are arranged in a sequence. Clearly, we do not want the *PrimeFactors* language to be in P. It can be easily shown that this language is in NP, however. □

Theorem 19.9. L is in NPC and L is in CoNP if and only if NP = CoNP.

Proof:

- To show that if L is NPC and CoNP then NP=CoNP.

  - Assume L is NPC; therefore,
      - L is in NP
      - For all L_1 in NP, we have $L_1 \leq_P L$. Now, assuming $\overline{L}$ is in NP (because L is in CoNP), we have $\overline{L} \leq_P L$.
  - Now, we are about to embark on showing the NP =CoNP part. For that, consider an arbitrary L' in NP. Then $L' \leq_P L$.
  - Now, using the result of Exercise 16.9, $\overline{L'} \leq_P \overline{L}$. Also $\overline{L'} \leq_P \overline{L} \leq_P L$.
  - Now, since there is an NP decider for L, there is an NP decider for $\overline{L'}$ also, using the above mapping reduction chain; in other words L' is in CoNP.
  - Now, consider an arbitrary L' in CoNP. This means that $\overline{L'}$ in NP. Since L is NPC, we have $\overline{L'} \leq_P L$. From this we have $L' \leq_P \overline{L}$.
  - Using the fact that $\overline{L} \leq_P L$, we have $L' \leq_P \overline{L} \leq_P L$, or that there is an NP decider for L'.
  - Hence, NP = CoNP.

- To show that if NP =CoNP, then there exists an L that is NPC and CoNP. This is straightforward: consider any NPC language L; it would be CoNP because L is in NP and NP =CoNP.

Chapter Summary

We discussed the theory of NP-completeness, going through practical techniques to show that a problem is NP-complete. We now discuss the question of input encodings postponed in Section 19.2.6. In the setting of input encodings, there are basically two classes of problems:

- Strongly NPC: Those problems where the problem remains NPC even if the input is encoded in unary. Almost every NPC problem we have studied (e.g., *Clique*, 3-SAT, etc.), is strongly NPC. In addition, the 3-partition problem (discussed momentarily), several problems in the context of the game of Tetris [34], and several scheduling problems are strongly NPC.
- Not Strongly NPC, or *pseudo polynomial*: There are problems where encoding the input in unary can give a polynomial algorithm. The 2-partition problem is an example which has a pseudo polynomial algorithm.

The 2-partition problem is: Given a finite set $A = \{a_1, a_2, \ldots, a_n\}$ of positive integers having an overall sum of $2b$, is there a subset A' of

A that sums exactly to b (in other words, $A \setminus A'$ and A' sum to b)? Note that we only determine the existence of such a subset—not *which* subset it is. Analogously, the 3-partition problem seeks three disjoint subsets that contain all the elements of A and sum to equal values.

2-partition is known to be NP-complete. However, there is a straightforward dynamic programming algorithm to solve the 2-partition problem, which is now briefly discussed. Stating things in genera, let g be the 'goal' in terms of a subset of the a_i's adding up to g. Let $T(j,g)$ denote the assertion that the sum of $\{a_1, \ldots, a_j\}$ is exactly g. We now write a recursive recipe to compute the truth of $T(i+1,g)$. This falls into two cases:

1. We do not include a_{i+1}, and T(i,g); or
2. We include a_{i+1}, and the remaining elements add up to $g - a_{i+1}$.

We build a dynamic programming table following the above recurrence, as follows:

- $T(i+1,g) = T(i,g) \lor ((a_{i+1} \le g) \land T(i, g - a_{i+1}))$
- $T(1,g) = ((g = 0) \lor g = a_1)$

Now, the answer we seek—whether there exists a subset of a_1 through a_n that adds up to b—is the value of $T(n,b)$ when the above algorithm finishes.

This algorithm has complexity $\mathcal{O}(n.b)$, as there are that many entries to be filled in the memoization table T. Note that by encoding the problem in $O(n.b)$ bits, we can achieve polynomial-time solution. However, a *reasonable encoding* of this problem takes only $n \, log(b)$ bits. Therefore, by "bloating" the input representation, 2-partition can be solved in polynomial-time. No such luck awaits strongly NP-complete problems—they cannot be solved in polynomial-time even with a unary input representation (the most bloated of input representations). Further work on this topic may be easily found on the internet.

Exercises

19.1. Show that $TwoColor \in P$.

19.2. Suppose we write a program that traverses a "tape" of n cells, numbered 1 through n. The program performs n traversals of the tape, with the ith traversal sequentially examining elements i through n. What is the runtime of such a program in the Big-O notation?

19.3. 1. Let $k = 2$. Estimate the magnitudes of x^k (polynomial) k^x (exponential) complexity growth for $x = 1, 2, 5, 10, 50$ and 100.

2. Estimate $2^{2^{2^{2^{\cdots^{2^2}}}}}$ (*i* times) for various *i* (note how to read this tower: begin with 2, and keep taking '$2^{previous}$').

19.4.

1. Draw an undirected graph of five nodes named a, b, c, d, e such that every pair of nodes has an edge between them (such graphs are called "cliques" - the above being a 5-clique).
2. What is the number of edges in an *n*-clique?
3. Which *n*-cliques are planar (for what values of *n*)?
4. What is a *Hamiltonian cycle* in an undirected graph?
5. Draw a directed graph G with nodes being the subsets of set $\{1, 2\}$, and with an edge from node n_i to node n_j if either $n_i \subseteq n_j$ or $|n_i| = |n_j|$. $|S|$ stands for the *size* or *cardinality* of set S.
6. How many strong components are there in the above graph? A strong component of a graph is a subset of nodes that are connected pairwise (reachable from one another).
7. What is the asymptotic time complexity of telling whether a directed graph has a cycle?

19.5. A Hamiltonian cycle in a graph with respect to a given node *n* is a tour that begins at *n*, visits all other nodes exactly once, returning to *n*. In a 5-clique, how many distinct Hamiltonian cycles exist? How about in a an *n*-clique?

19.6.

1. Suppose you are sent into a classroom where *n* (honest) students are seated, and are patiently awaiting your arrival. You are charged by your boss with determining whether *some k* of these students know each other – any such subset of *k* students will do. Suppose you pick exactly one random subset of *k* of these students and each pair within this subset tells you that they know each other. Can you now report back your answer to the boss?
2. The above is a *nondeterministic* algorithm to check whether a graph has a *k*-clique. What would a *deterministic* algorithm be?
3. Suppose what you are charged with is to assure that *no k-subset* is such that all its members know each other pairwise. Suppose you pick exactly one random subset of *k* of these students and listen to them pairwise as to whether they know each other or not. Can you now report back any answer to your boss? How many more queries would you need as a function of *n* and *k*?

19.7. Define the language *HalfClique* to be the set of input encodings $\langle G \rangle$ such that G is an undirected graph having a clique with at least

$n/2$ nodes, where n is the number of nodes in G. Show that $HalfClique$ is NPC.

19.8. Show that $\neq$-sat, as defined in Section 18.3.7, is NPC.

19.9. *Problems pertaining to NPC abound in various places; rather than repeat them, we leave it to the student / teacher to find and assign them suitably. We close off by assigning a rather interesting proof to read and understand.*

In his MS thesis, Jason Cantin (Wisconsin) proves that the problem of verifying memory coherence is NPC [15, 16]. Read and understand his proof.

20

DFA for Presburger Arithmetic

To motivate the subject matter of this and the subsequent chapters, let us briefly reflect on the topics we have studied thus far. We have largely studied computation from the point of view of standard machines such as finite automata, push-down automata, and Turing machines. In Chapter 11, we took our first step towards presenting another perspective of computation, namely one based on mathematical logic. Specifically, in Chapter 11, we demonstrated that

- Boolean propositions (expressed through Boolean formulas) can be represented using binary decision diagrams (BDDs).
- Each BDD b representing a Boolean formula f is a minimized DFA that recognizes the finite language of bit-strings that satisfy f.
- We introduced the notion of state transition systems, which can be used to represent the behavior of many real-world systems such as synchronous hardware and even recreational games.
- We showed that the allowed moves of transition systems can be encoded using Boolean formulas, based on the convention of using two classes of Boolean variables capturing the *current* and the *next* state values. Such formulas can, then, be encoded using BDDs.
- We showed how to compute the set of all states reached by a transition system through logical manipulations performed on two BDDs, one representing the "present state" and another representing the "machine" (or transition system). Essentially, given these BDDs, one conjoins them and projects out the present state variables through existential quantification to obtain the next state. The present and next states are pooled, thus forming the set of all visited states. When this pool stops growing, we essentially reach a fixed-point amounting to the set of all reachable states.

We connected these ideas to the real world by demonstrating how to encode the familiar game of tic-tac-toe using BDDs, as well as computing all possible draw configurations in one fell swoop. In Chapter 18, we examined the power of machines to decide the validity of first-order logic sentences, and proved that Turing machines cannot be programmed to algorithmically carry out this decision. In Chapter 19, we showed that Turing machines can decide Boolean logic, but the apparent complexity is exponential (the NPC class). Chapters 21 through 23 will again pursue the automaton/logic connection, but in the context of verifying reactive systems.

This chapter illustrates another instance of automaton-logic connection. In contrast to the undecidability of full first-order logic discussed in Section 18.2.3, there is a fragment of first-order logic called *Presburger arithmetic* that is *decidable*. A widely used decision procedure for Presburger arithmetic consists of building a DFA for the satisfying instances of Presburger formulas. In this connection, an interesting contrast with Chapter 11 is the following. BDDs are essentially DFA whose languages are a *finite set* of *finite strings*,[1] with each string specifying assignments to the Boolean variables of the BDD. In contrast, the languages of the Presburger DFA built in this chapter are *finite or infinite* sets of *finite* but *unbounded strings*. Each string encodes the assignments to the quantified *individual* variables of a Presburger formula. As another contrast, in Chapter 23, Section 23.3, we introduce Büchi automata which are machines that accept only *infinitely long* strings.

In this chapter we present the following results:

- We define the syntax of Presburger arithmetic formulas.
- We introduce, largely through examples,[2] a technique by which DFA can be made to accept bit-serial presentations of tuples of natural numbers.
- Using this method, we present a conversion algorithm from a given Presburger formula f to a DFA d such that the satisfying assignments for the free variables of f correspond to the language of d.
- We further demonstrate that logical operations and automaton operations have a natural strong correspondence. In particular, we sketch the following results:

---

[1] In fact, the length of these strings is bounded by the number of BDD variables, which is usually a small number - no more than a few hundred in practice.

[2] We believe that our choice of extremely simple examples will help the reader study this fascinating topic with ease. We are indebted to Comon et al.'s [25] very readable presentation of this topic from which we borrow heavily.

- The DFA for the conjunction of two Presburger formulas f_1 and f_2 may be obtained by first obtaining the DFA d_1 and d_2 for the individual formulas and then performing a DFA product construction for intersection. Similar results are obtained for disjunction and negation.
- Existential quantification can be modeled by hiding the quantified symbol from the interface of the corresponding automaton. The result may be an NFA that can then be determinized.

20.1 Presburger Formulas and DFAs

20.1.1 Presburger formulas

The basic terms in Presburger arithmetic consist of first-order variables x, y, x_1, x', ..., the constants 0, and 1, and sums of basic terms. For instance, $x + x + 1 + 1 + 1$ is a basic term, which we also write as $2x + 3$. The *atomic formulas* are equalities and inequalities between basic terms. For instance, $x + 2y = 3z + 1$ is an atomic formula. The *formulas* of Presburger logic are first-order formulas built on the atomic formulas. For instance,

$$\forall x.\exists y.(x = 2y \lor x = 2y + 1)$$

is a formula. The *free variables* of formula φ are defined as usual: for instance,

$$FV(\varphi_1 \lor \varphi_2) = FV(\varphi_1) \cup FV(\varphi_2), \text{ and } FV(\exists x.\varphi) = FV(\varphi) \setminus \{x\}.$$

A formula without free variables is called a *sentence*. Sentences are true (valid) or false (non-valid).

Any first-order formula with nested quantifiers, for example

$$\forall x.(p(x) \Rightarrow [\exists y.q(x, y)])$$

may be rewritten as a *logically equivalent* formula, in this example as

$$\forall x.\exists y.[p(x) \Rightarrow q(x, y)]$$

using a transformation process called *prenexing* or *arriving at the prenex normal form*. As another example, the prenex normal form for the formula

$$\forall x.([\exists y.q(x, y)] \Rightarrow p(x))$$

```
(*----------------------------------------------------------------------*)
(* Prenexing code in Ocaml                                              *)
(* ''_'' stands for wild-card, and ''rec'' helps define a recursive function. *)
(*----------------------------------------------------------------------*)
let rec prenex(fmla) =
  match fmla with
    | FORALL(t1, f1) -> FORALL(t1, prenex(f1))
    | EXISTS(t1, f1) -> EXISTS(t1, prenex(f1))
    | AND(f1, f2)    ->
        let pf1 = prenex(f1) in
          (match pf1 with
            | FORALL(t1', f1') -> prenex(FORALL(t1', AND(f1', f2)))
            | EXISTS(t1', f1') -> prenex(EXISTS(t1', AND(f1', f2)))
            | _                ->
              (let pf2 = prenex(f2) in
                (match pf2 with
                  | FORALL(t2', f2') -> prenex(FORALL(t2', AND(pf1, f2')))
                  | EXISTS(t2', f2') -> prenex(EXISTS(t2', AND(pf1, f2')))
                  | _                -> AND(pf1,pf2))
              )
          )
    | OR(f1, f2)     -> (* ... similar to AND ... *)

    | NOT(f1) ->
        let pf1 = prenex(f1) in
          (match pf1 with
            | FORALL(t1', f1') -> prenex(EXISTS(t1', NOT(f1')))
            | EXISTS(t1', f1') -> prenex(FORALL(t1', NOT(f1')))
            | _                -> NOT(pf1)
          )
    | IMPLIES(f1,f2) ->
        let pf1 = prenex(f1) in
          (match pf1 with
            | FORALL(t1', f1') -> prenex(EXISTS(t1', IMPLIES(f1', f2)))
            | EXISTS(t1', f1') -> prenex(FORALL(t1', IMPLIES(f1', f2)))
            | _                ->
              (let pf2 = prenex(f2) in
                (match pf2 with
                  | FORALL(t2', f2') -> prenex(FORALL(t2', IMPLIES(pf1, f2')))
                  | EXISTS(t2', f2') -> prenex(EXISTS(t2', IMPLIES(pf1, f2')))
                  | _                -> IMPLIES(pf1,pf2))
              )
          )

    | _ -> fmla (* The final default match is a wild-card match that returns ''fmla'' *)
(*----------------------------------------------------------------------*)
```

Fig. 20.1. Prenexing code in Ocaml

is

$$\forall x.\forall y.[q(x,y) \Rightarrow p(x)].$$

Notice how the quantifier changes from $\exists$ to $\forall$ when lifted out of the antecedent of an implication, since the antecedent of an implication has an implicit negation ($a \Rightarrow b$ is equivalent to $\neg a \lor b$). After prenexing, we are left with a list of quantified variables called the *prefix* and a quantifier-free inner formula called the *matrix*. While prenexing is not essential in the Presburger formula to DFA transformation to be described,[3] it is quite helpful for the purposes of exposition. Another reason why we present prenexing is that it is an important algorithm to know about. The full set of rewrite rules for prenexing are captured in the Ocaml program in Figure 20.1.

The interpretation domain of formulas is the set of natural numbers Nat in which 0, 1, +, =, and $\leq$ have their usual meanings. A *solution* of a formula is an assignment of the free variables of the formula in Nat which satisfies the formula.

20.1.2 Encoding conventions

We now define the encoding conventions for natural numbers employed in our automaton construction. We define natural numbers in base 2 using a bit-serial format, with the least significant bit (LSB) appearing leftmost. This means that when viewed as inputs of DFA, numbers will be consumed LSB-first. For example, 13 is written as 1011 (or as 10110, or in general, with as many zeros to the right). Pairs of natural numbers are represented as a sequence of tuples, and in general k-tuples of natural numbers are represented as a sequence of k-tuples. k-tuples of zeros may be added to the right without changing the meaning of these representations. For example, $\langle 13, 6 \rangle$ is represented as $\langle 1, 0 \rangle \langle 0, 1 \rangle \langle 1, 1 \rangle \langle 1, 0 \rangle$.

20.1.3 Example 1 — representing $x \leq 2$

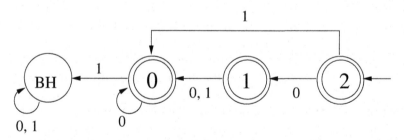

Fig. 20.2. Presburger Formula $x \leq 2$ and its DFA

Figure 20.2 shows how a DFA for $x \leq 2$ is represented. Before seeing any of the bits of x (that will, as explained above, arrive LSB-first), the value of "x seen so far" is 0. At this stage, the imbalance between x and

---

[3] Prenexing may even hurt the performance of the Presburger decision procedure to be presented in this chapter, as it postpones quantification steps, and quantification steps can help get rid of variables.

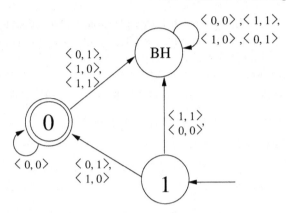

Initial state: 1, Final state : 0

```
T1: 1 -- <0,1> --> 0      T6: 0 -- <0,0> --> 0
T2: 1 -- <1,0> --> 0      T7: 0 -- <0,1> --> BH
T3: 1 -- <1,1> --> BH     T8: 0 -- <1,0> --> BH
T4: 1 -- <0,0> --> BH     T9: BH -- any   --> BH
T5: 0 -- <1,1> --> BH     Ta: BH -- any   --> BH
```

Fig. 20.3. DFA for $x + y = 1$. The transition names are referred to in Section 20.1.5.

2 is 2—which is what the initial state of the automaton is labeled with. When a 0 bit arrives, the magnitude of "x seen so far" is still 0. As far as the next bit of x yet to arrive is concerned, it possesses a positional weight that is two times the weight of the LSB. However, we want to restore an 'inductive' situation in our DFA diagram; therefore, our convention will be that we will '*div*' (divide with truncation) both sides of the equation by 2. This results in the weight imbalance reducing to 1—exactly equal to the state label of the state to which state 2 transitions. On the other hand, if the LSB were to be 1, we advance to state 0 which is $(2-1)$ *div* 2. It is also apparent that a sequence such as 011 represents, in the new format, the number 6. Since $6 \leq 2$ does not hold, the machine moves to a "black hole" state labeled with BH, and stays there forever. The language of the DFA in Figure 20.2 includes all bit-serial sequences representing natural numbers that satisfy this formula; for instance, $01 = 2_{10}$ as well as $10 = 1_{10}$ are both in the language of this DFA.

NFA for $\exists y.(x + y) = 1$:

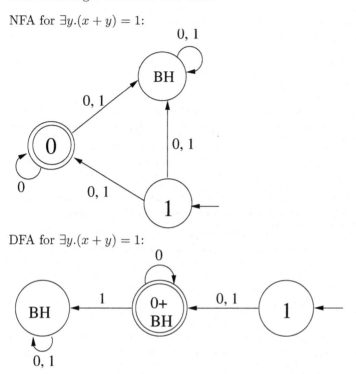

DFA for $\exists y.(x + y) = 1$:

Fig. 20.4. NFA for $\exists y.x + y = 1$ and its equivalent DFA

20.1.4 Example 2 — $\forall x.\exists y.(x + y) = 1$

We continue example-driven, now demonstrating how to show the non-validity of the sentence $\forall x.\exists y.(x + y) = 1$. To that end, we start with the matrix sub-formula $(x + y) = 1$, and build up the whole formula, as will now be described.

Subformula $(x + y) = 1$:

As illustrated in Figure 20.3, we build a DFA that can handle the bit-serial left-to-right arrival of the bits comprising $x + y = 1$. When $\langle 1, 1 \rangle$ arrives, the machine is thrown into the black hole (BH) state because the left-hand side of the equation adds up to at least 2 now, and will from now on never equal the right-hand side. The reason we mention *at least* is because the LSB of x and the LSB of y are both 1s, as captured by the pair $\langle 1, 1 \rangle$ that has just arrived; and hence, even if the entire sequence of tuples yet to arrive were to be $\langle 0, 0 \rangle$, the left-hand side would be 2. The accepting runs are either $\langle 0, 1 \rangle$ or $\langle 1, 0 \rangle$ followed by an arbitrary number of $\langle 0, 0 \rangle$s.

Subformula $\exists y.(x + y) = 1$:

The meaning of $\exists y.x + y = 1$ is that one should take the disjunction of $x + y = 1$ with y set to 0 and 1, in turn (y is a Boolean variable owing to our following a bit-serial format). In the automaton world, this effect would be attained if we project away the second component of the pairs from the alphabet (e.g., $\langle x, y \rangle$ would become just x). The resulting machine—an NFA now—is shown in Figure 20.4. Determinizing, we obtain the DFA also shown in the same figure. This DFA shows that if the LSB of x is a 0 or a 1, the machine can be in a final state (y can be internally chosen to be 1 or 0, respectively, thus still satisfying the equation).

Handling the complete formula

The full formula $\forall x.\exists y.(x + y) = 1$ is handled by treating $\forall x.P$ as $\neg(\exists x.\neg P)$. The corresponding automaton operations are automaton complementation, hiding, determinization, and finally complementation once again, as Figure 20.5 shows.

Throughout these conversions, we have maintained a correspondence between formulas and automata. Because of this, when converting sentences to DFA, there can only be two outcomes: if the sentence is true, the final DFA will have a single state that is initial and final. This is a DFA that, "upon power-up," instantaneously accepts. If the sentence is false, the final DFA will have a single state that is only initial and not final. This is a DFA that, "upon power-up," instantaneously rejects. The proof that these are the only two outcomes possible for Presburger *sentences* is left to the reader (Exercise 20.1). Section 20.2 discusses a pitfall to be avoided while using this algorithm.

20.1.5 Conversion algorithm: Presburger formulas to automata

Let us review the conversion algorithm presented thus far, again keeping the example $(x + y) = 1$ in mind, referring to Figure 20.3. The machines we build employ bit-serial conventions, meaning: (i) they begin in a start state where initially it has not seen any inputs (by convention, we assume that $x = y = 0$). If the formula is satisfied for these

Starting DFA for $\exists y.(x + y) = 1$:

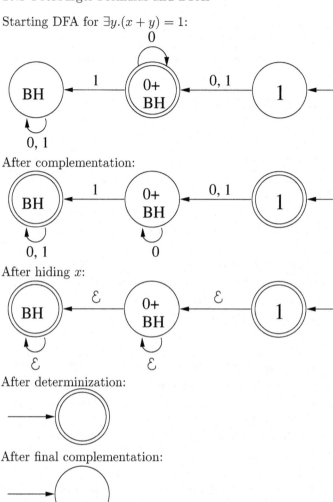

Fig. 20.5. Obtaining the DFA for $\forall x.\exists y.(x + y) = 1$ from the DFA in Figure 20.4 through complementation, hiding, determinization, and complementation

values of x and y, the start state is also a final state. In this case, since $0 + 0 \neq 1$, the start state is not a final state. Furthermore, the start state gets labeled as "1." This is because each state will be labeled by the "imbalance" between the right-hand side and the left-hand side of the equation $x + y = 1$ divided by 2. This quantity is signed negative if the left-hand side is "heavier," and positive if the left-hand side is "lighter." The division by 2 happens because as far as the *next* bit to arrive is concerned, it will treat the imbalance as occurring in a position of half the weight. The general algorithm for handling quantifier-free formulas (matrices) is now described. Quantifications are handled as already illustrated.

General equation format: The general equation we are dealing with is

$$a_1 x_1 + a_2 x_2 + \ldots + a_n x_n = b (\text{where } a_1, \ldots, a_n, b \in Z).$$

Here Z stands for integers (positive and negative numbers). Remember that we allow $x_1, \ldots, x_n$ to range over Nat only.

Example: Our equation is $x + y = 1$.

Initial state rule: The initial state is always q_0. In our example, the initial state is 0 (we simply express the subscript, omitting the "q" part).

State Transition Equation: Let the set of states be Q. State $q_c \in Q$ will evolve to state $q_d \in Q$ upon arrival of bit-tuple θ, written $(q_c, \theta, q_d) \in \delta$, provided the following side conditions are true:

- The state transition exists *only if* the LSB of the equation is satisfied. In other words, $a_1 \theta_1 + a_2 \theta_2 + \ldots + a_n \theta_n = c$ must be true. If this condition is not satisfied, the transition upon θ goes to the black hole state "BH." Here, $\theta = \langle \theta_1, \theta_2, \ldots, \theta_n \rangle$.

  In our example, transition T3 goes to BH because the LSB of the addition $1 + 1$ does not equal the LSB of 1. The same is the reason for T4 going to BH.

- If the above condition holds, then $d = (c - a_1\theta_1 - a_2\theta_2 \ldots - a_n\theta_n)$ *div* 2.

  Consider T1 for the sake of illustration. The above equation yields 0 which is where this transition goes. Consider T5. The above equation yields -1, and so the transition goes to the BH state, because the arrival of further bits of x and y only makes the left-hand side of the equation heavier (more unbalanced).

In summary, these two equations describe: (i) how to determine whether a state transition exists, and (ii) how to decide which state the transition leads to.

20.2 Pitfalls to Avoid

Let us do another example that points out a real pitfall. Consider the formulas $F_f = \forall x. \exists y. (x - y) = 1$ and $F_t = \forall y. \exists x. (x - y) = 1$. Over natural numbers, F_f is false (consider $x = 0$) while F_t is true. Following the method for DFA construction all the way through, however,

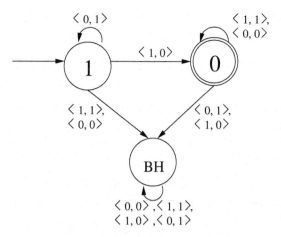

Fig. 20.6. Study of $\forall x.\exists y.\ x - y = 1$ versus $\forall y.\exists x.\ x - y = 1$

Figure 20.6 (which only represents the common matrix of these formulas) will reduce to the *same* machine with empty language as shown in Figure 20.5. In other words, our reasoning shows that F_f and F_t are false. Why?

20.2.1 The restriction of equal bit-vector lengths

Let us think a bit more carefully about how the DFA construction method treats the matrix $x - y = 1$. This DFA is based on the following conventions:

- Bit-sequences are to be interpreted in unsigned binary. For example, 1 represents the value 1 in decimal while 11 represents 3 in decimal.
- We will accept x and y bit serially with the *number of bits of x and y seen at any stage being equal.* Suppose we have seen n 1s corresponding to y. What should be the n-long bit sequence corresponding to x such that $x - y = 1$? We immediately realize that there is no such x!

Therefore, we have to read $\forall y.\exists x.(x - y) = 1$ as

$$\forall y.\exists x.[(x - y) = 1\ \wedge\ eqlen(x, y)],$$

where *eqlen* is a predicate that asserts that the bit-serial vectors modeling x and y have the same length. With this interpretation, both formulas emerge false. In other words, for any run of y's of all 1s (equal in magnitude to $2^n - 1$ for various n), there is no run of x of the same length that exceeds this magnitude.

Chapter Summary

This chapter provided another example of how exceedingly powerful and versatile *finite* automata are! In particular, we demonstrated two facts:

- That the validity problem of Presburger arithmetic can be modeled and solved using DFAs, and
- The procedure we employed shows the nice duality that exists between logical operators and automata-theoretic operators.

An efficient computer tool named Mona [70] employs many of these ideas, and employs an augmented form of BDDs to reason about Presburger logic sentences. Despite the non-elementary complexity (defined by the "towers of 2" formula given in Exercise 19.3), this tool has been able to handle many real-world problems in formal verification.

| Exercises |

20.1. Prove the assertion in Page 376 that only two outcomes are possible when converting any Presburger sentence into a DFA.

20.2. Apply the DFA construction method of Section 20.1.4 on the Presburger sentence

$$\exists x. \exists y. \exists z.\ (x + 2y = 3z + 1)$$

and determine whether this sentence is true or false.

20.3. First inspect the sentence

$$\exists x. \exists y. \forall z.\ (x + 2y = 3z + 1)$$

and determine whether it is true or false. Next, apply the DFA construction method and check your answer.

20.4. Carry the DFA construction method presented in Section 20.1.4 all the way through for the formulas F_f and F_t of Section 20.2, and verify the assertion that their languages emerge to be empty. In the light of this result, explain the need for the *eqlen* assertion introduced in Section 20.2.1.

20.5. Consider the addition of the multiplication ('$*$') operator into the "Presburger" arithmetic language. Can one still employ DFAs to reason about sentences in this language? What problems do you anticipate? Consider two cases: (i) only multiplications with constants are allowed (e.g., $2 * x$), (ii) arbitrary multiplications are allowed (e.g., $x * y$).

21

Model Checking: Basics

21.1 An Introduction to Model Checking

The development of model checking methods is one of the *towering* achievements of the late 20th century in terms of debugging concurrent systems. This development came about in the face of the pressing need faced by the computing community in the late 1970's for effective program debugging methods. The correctness of programs is *the* central problem in computing, because there are often very designs that are correct, and vastly more that are incorrect. In some sense, defining what 'correct' means is half the problem - proving correctness being the other half of the problem.

The quest for high performance and flexibility in terms of usage (e.g., in mobile computing applications) require systems (software or hardware) to be designed using multiple computers, processes, threads, and/or function units, thus quickly making system behavior highly concurrent and non-intuitive. With the rapid progress in computing, especially with the availability of inexpensive microprocessors, the computer user community found itself – in the late 1970's – in a position where it had plenty of inexpensive hardware but close to no *practical* debugging methods for concurrent systems! We will now examine the chain of events that led to the introduction of model checking in this setting.

The enterprise of *sequential program verification* pioneered, among others, by Floyd [40], Hoare [55], and Dijkstra [37] was soon followed by the quest for *parallel program correctness*, pioneered, among others, by Owicki and Gries [93], and Lamport [73]. The difficulty of these methods when applied to real-world programs led to alternate proposals, such as relying on social processes [35]. Unfortunately, the arbitrariness hidden in such proposals makes it difficult to situate computation engineering in the same plane as other engineering disciplines where design verification against rigorous specifications is the *rule* rather than the *exception*. Contrast the following guarantees: where the former is likely to be offered by a

> "If I press the eject button, I am guaranteed to be safely ejected from a burning airplane in less than 5 seconds."

versus

> "If I am lucky to be in a plane that was debugged by an expert reader of a program who happened to spot a bug, then I might get ejected in a reasonable amount of time."

In one thread of work that was evolving in the late 1970's, some scientists, notably Pnueli [97], had the vision of focusing on *concurrency*. In a nutshell, by focusing on *control* and not *data*, it becomes possible to model a system in terms of finite-state machines, and then employ decision procedures to check for its reactive properties. Even after such simplifications, system control tends to be highly non-intuitive, and *hence simply not amenable to any reasonable social processes*. Automated analysis of finite-state models can, on the other hand, automatically hunt bugs and report them back. Pnueli's vision lead to Manna, Pnueli, and many others developing temporal logic *proof* systems [79, 80].

We must admit that in this historical "sampler" that we are presenting, it is entirely possible that we have overlooked some key references, despite our best efforts to prevent any such omissions. One such unfortunately omitted citation from many recent works on model checking pertains to Carl Adam Petri's [96] seminal work. Petri not only proposed many basic ideas in concurrency and work-flow as early as 1963, but also saw the importance of focusing on control flow and synchronization. Similarly, Hoare [56] and Milner [85] pioneered much of the understanding of concurrency in terms of *process algebras*. However, none of the early works had emphasized algorithmic approaches similar to model checking, which is our focus in this chapter.

The breakthrough towards algorithmic methods for reasoning about concurrent systems (as opposed to the initial proof theoretic methods) was introduced in the work of Clarke and Emerson [18], Queille and Sifakis [99], and Clarke, Emerson, and Sistla [19]. This line of work also received multiple fundamental contributions, notably from Vardi and Wolper who introduced an automata theoretic approach to automatic program verification [120], and a team of researchers at AT&T

Bell Laboratories, notably by Holzmann, Peled, Yannakakis, and Kurshan [58, 51, 72, 59], who developed various ways to build finite-state
machine models and formally analyze them. Known as *model checking*, these methods relied on (i) creating a finite state model of the
concurrent system being verified, and (ii) showing that this model possesses desired temporal properties (expressed in temporal logic). Graph
traversal algorithms were employed in lieu of deductive methods, thus
turning the whole exercise of verification largely into one of building
system models as graphs, and performing traversals on these graphs
without encountering state explosion.

State explosion—having to deal with an exponential number of
states—is an unfortunate reality of model checking methods because
finite-state models of concurrent systems tend to *interleave* in an exponential number of ways with respect to the number of components
in the system. Effective methods to combat state explosion became the
hot topic of research – but meanwhile model checking methods were
being applied to a number of real systems, with success, finding deep-
seated bugs in them! In [14, 13], Bryant published many seminal results
pertaining to binary decision diagrams (BDD), and following his popu-
larization of BDDs in the area of hardware verification, McMillan [83]
wrote his very influential dissertation on *symbolic model checking*. This
is one line of work that truly made model checking feasible for certain
"well structured," very large state spaces, found in hardware modeling.
The industry now employs BDDs in symbolic trajectory evaluation
methods (e.g., [2]).

Model checking has truly caught on in the area of hardware verification, and promises to make inroads into software verification—the
area of "software model checking" being very actively researched at the
time of writing this very sentence. In particular, Boolean satisfiability
(SAT) methods are being widely researched, as already discussed in
Section 18.3. In modern reasoning systems, SAT and BDD methods
are being used in conjunction with first-order (e.g., [92, 110]) reasoning
systems, for example in tools such as BLAST [53]. In addition, higher-
order logic (e.g., [3, 47, 94]) based reasoning systems also employ BDD,
SAT, and even model checking methods as automated proof assistants
within them. As examples of concrete outcomes, we can mention two
success stories:

Model checking in the design of modern microprocessors: All
modern microprocessors are designed to be able to communicate with
other microprocessors through shared memory.[1] Unfortunately since

---

[1] Often these other microprocessors are situated on the same silicon chip

only one processor can be writing at any memory location at a given time, and since "shared memory" exists in the form of multiple levels of caches, with further levels of caches being far slower to access than nearly levels of caches, extremely complex protocols are employed to allow processors to share memory. Even one bug in one of these protocols can render a microprocessor useless, requiring a redesign that can cost several 100s of millions of dollars. No modern microprocessor is sold today without its cache coherence protocol being debugged through model checking.

Model checking in the design of device drivers: Drivers for computer input/output devices such as Floppy Disk Controllers, USB Drivers, and Blue-tooth Drivers are extremely complex. Traditional debugging is unable to weed out hidden bugs unless massive amounts of debugging time are expended. Latent bugs can crash computers and/or become security holes. Projects such as the Microsoft Research SLAM project [9] have technology transitioned model checking into the real world by making the Static Driver Verifier [8] part of the Windows Driver Foundation [122]. With this, and other similar developments, device-driver writers now have the opportunity to model-check their protocols and find deep-seated bugs that have often escaped, and/or have taken huge amounts of time to locate using traditional debugging cycles.

Has the enterprise of model checking succeeded? What about social processes? We offer two quotes:

Model checking has recently rescued the reputation of formal methods [64].

Don't rely on social processes for verification [38].

In this chapter and Chapter 22, we offer a tiny glimpse of model checking basics as well as tools. The rest of this chapter is organized as follows. Section 21.2 examines reactive systems. Section 21.3 discusses the verification of safety and liveness properties. Section 21.4 illustrates these ideas on one example—namely the *Dining Philosophers* problem of Dijkstra [37]—using the Promela language and SPIN model checker.

21.2 What Are Reactive Computing Systems?

Reactive computing systems are hardware/software ensembles that maintain an ongoing interaction with their environment, coordinating as well as facilitating these interactions. They are widely used in all walks of life—often in areas that directly affect life. Examples of reactive systems include device drivers that control the operation of computer peripherals such as disks and network cards, embedded control systems used in spacecraft or automobiles, cache coherency controllers that maintain memory consistency in multiprocessor machines, and even certain cardiac pacemakers that measure body functions (such as body electric fields and exercise/sleep patterns) to keep a defective heart beating properly. Model checking has already been employed in most of these areas, with its use imminent in critical areas such as medical systems. Clearly, knowing a little bit about the inner workings of a model checker can go a long way towards their proper usage.

21.2.1 Why model checking?

The design of most reactive systems is an involved as well as exacting task. Hundreds of engineers are involved in planning, analyzing, as well as building and testing the various hardware and software components that go into these systems. Despite all this exacting work, at least two vexing problems remain:

- Reactive systems often exhibit nasty bugs only when field-tested. Unfortunately, at such late stages of product development, identifying the root cause of bugs as well as finding solutions or workarounds takes valuable product engineering time. A manufacturer caught in this situation can very easily lose their competitive advantage, as these late life-cycle bug fixes can cost them dearly—especially if they miss critical market windows.
- The risk of undetected bugs in products is very high,[2] in the form of law-suits and recalls. Since software testing methods are seldom exhaustive, product managers have a very difficult time deciding when to begin selling products.

Formal methods based on *model checking* are geared towards eliminating the above difficulties associated with late cycle debugging. While model checking is not a panacea, it has established a track record of finding many deep bugs missed by weeks or months of testing. Specifically,

- model checking is best used when a reactive protocol is in its early conceptual design stages. This is also the most cost-effective point at which to eliminate deep conceptual flaws.

---

[2] Software is often like a bridge that does not fail when subject to 100 tons or 101 tons of weight, but suddenly collapse when 101.1 tons of weight are applied.

- model checking can return answers — either successful verification outcomes or high level counterexamples — often in a matter of a few minutes to a few hours. In contrast, testing can wastefully explore vast expanses of the state-space over weeks or months of testing. Error location can also become nightmarishly hard during testing, as the state-space sizes are large, and because an astronomically large number of computation steps may be executed from when the actual erroneous steps were carried out until when the system crashes or other symptoms of "ill health" are manifested.

In reality, the success of model checking can be attributed to several facts. Many of these facts are just pure "common sense," not exclusive to model checking in any way. Yet, it has been observed that model checking facilitates the use of such common sense! We now list the "virtues of model checking," starting from the most pragmatic and going towards the more mathematical reasons.

Successive refinement:

Human thought seldom advances[3] without the benefit of symbolic thought or successive refinement. Once one erects symbols and defines "rules of the game," one can begin playing. In the same sense, rather than remain frozen with indecision in the face of full design complexity, designers adopting model checking methods have at their disposal a vehicle for testing early prototypes and evaluating design alternatives. The models that a designer builds for reactive systems are largely finite-state machines. These finite-state machines can either be standard ones or embellished with extra information pertaining to communication or computation.

Exhaustively verify simplified models:

Even though a manually created design model can be defective, creating one actually allows the designer to unload a piece of their mounting mental burden, and test the integrity of their thoughts through model checking experiments. The modern tendency in this area is to supplant manual model construction with abstraction refinement techniques that can help one gradually discover, through tool assistance, what a suitable abstract model is. This approach promises to considerably reduce the level of effort needed in creating abstract models.

---

[3] Children are known to have difficulties with symbolic thought, and have been observed to try inserting their feet into pictures of shoes.

State-space analysis tools are the 'heart' of any model checker. These tools help exhaustively analyze finite-state machine models through a combination of techniques that help reduce memory requirements as well as overall computation time. As Rusbhy points out [104], experience has shown that the exhaustive analysis of finite-state models of reactive systems can often lead to the discovery of unexpected interactions ("corner cases") and bugs much more readily than testing methods can, for the same amount of resources (human and computer time) spent.

21.2.2 Model checking vs. testing

Model checking is a technique for verifying *finite-state models* of reactive computing systems for *proper behavior*. The intuitive notion of 'proper behavior' is, in turn, formalized through the use of either temporal logic (e.g., linear-time temporal logic) or automata (e.g., Büchi automata) which are used to express the *desired* properties of these systems. In contrast, testing is a well-established and expansive area central to product integrity, and employs both formal and informal criteria for coverage. Neither approach excludes the other; in fact, some of the most promising recent results are in combining model checking and testing ideas, for instance as in [10].

Traditional testing-based debugging methods are known to miss bugs due to many reasons. Typical systems being tested contain an astronomical number of states: 2^{10^9} states, for instance, in a memory of capacity 1MB! While engineers are known to hand-simplify designs before testing them, in model checking, such simplifications are often done much more aggressively, to the point of turning control branches into nondeterministic selections. Such simplifications help abstract (or "smoothen") the state-space modeled. BDDs (Chapter 12), and many other symbolic representation methods for state spaces, have the (curious) property that by adding *more* information (which helps overapproximate state spaces, perhaps by adding some infeasible states) the actual BDD sizes are dramatically reduced. Hence, even though the number of states modeled may increase, the memory for representing the states diminishes.

Two additional important benefits due to the use of nondeterminism are: (i) failure possibilities are introduced without increasing model or state-space sizes, and (ii) the effect of testing for all values of certain critical system parameters is obtained through nondeterminism. Of course these benefits come at a cost, and a designer who understands this cost/benefit ratio can often swing the overall balance in favor of benefit. To understand these discussions, consider the process of debugging of a concurrent system that uses a '*send*' communication primitive in a library. Assume that *send* uses a FIFO buffer, and would block if the buffer were to be full. Further, suppose that such blocked sends are considered errors. If one were to allocate a certain number of buffer locations and test in the traditional sense, one would have to wait for the right combinations of global states to arise in which the send buffer in question becomes full, and then only be able to observe the error. Furthermore, the input/state combinations under which this error can be observed during testing may differ from those in the real system because, clearly, the designer would have downscaled all system parameters before testing begins. To make matters worse, it is easy to eliminate errors while downscaling systems in an *ad hoc* manner.

In contrast to the testing approach described above, a model checking based approach to solving the same problem would consist of: (i) not modeling buffer capacities, and (ii) nondeterministically triggering a *buffer full* condition at every possible point in the state-space (this is called *over-approximation* of the reachable state-space). The advantage of this approach is that all possible buffer capacities are being simulated in one fell swoop. The obvious disadvantage of this approach is that a *buffer full* may be simulated by the nondeterministic selection mechanism when not warranted (e.g., at the very first *send*), and the test engineer is forced to study and overrule many false alarms reported by the model checker. This is often a small price to be paid for the thoroughness of coverage offered by a model checker, especially given that testing may not be feasible at all. We are really talking about modeling several thousands of lines of actual code (which may be impossible to test in any real sense) by a less than 100-line model checker input description. In this setting, repeated running of model checking and overruling errors is actually feasible.

It is possible that, despite the best precautions, over-approximation based model checking can inundate the engineer with false alarms. There are many recently emerging solutions to this problem, the most important of which is an approach based on *abstraction refinement*. This approach promises to enhance the power of model checking, while helping us handle larger models and at the same time avoiding inaccuracies arising due to hand abstractions created by the designer. In the counterexample guided abstraction refinement (CEGAR, [21]) approach for abstraction/refinement, the initial abstraction is deliberately chosen to be too coarse. This abstraction is successively refined by heuristically discovering aspects of the system which need to be captured in detail. The end result is often that systems are verified while keeping much of their behavior highly abstract, thus reducing the overall complexity of verification.

21.3 Büchi automata, and Verifying Safety and Liveness

Büchi automata are automata whose languages contain only infinite strings. The ability to model infinite strings is important because of the fact that all bugs can be described in the context of infinite executions. We now elaborate on these potentially unusual sounding, but rather simple, ideas. Broadly speaking, all errors (bugs) in reactive systems can be classified into two classes: *safety* (property) violations and *liveness* (property) violations.

- Safety violations are bugs that can be presented and explained to someone in the form of *finite* executions (finite sequence of states) ending in erroneous states. Some examples of systems that exhibit safety violations are the following:
  - two people who, following a faulty protocol, walk opposite in a narrow dark corridor and collide;
  - an elevator which, when requested to go to the 13th floor, proceeds to do so with its doors wide open;
  - a process P which acquires a lock L and dies, thus permanently blocking another process, say Q, from acquiring L.

  All finite executions of the form $s_1 \ldots s_k$ can be modeled as infinite executions that infinitely repeat the last state, namely $s_1 s_2 \ldots (s_k)^\omega$. Modeling finite executions as infinite executions allows one to employ Büchi automata.
- Liveness violations are bugs that can be presented and explained to someone only in the form of an infinite execution in which a desired state never occurs. In practice, liveness violations are those that end in a bad "lasso" shaped cyclic execution path which does not contain the desired state. Examples of liveness violations are:
  - two people who, following a faulty protocol, engage in a perpetual 'dance,' trying to pass each other in a narrow well-lit corridor;
  - an elevator that permanently oscillates between the 12th and 14th floors when requested to go to the 13th floor;
  - A process P which acquires a lock L precisely before another process Q tries to acquire it, and releases the lock precisely after Q has decided to back off and retry; this sequence repeats infinitely.

All infinite executions in finite state systems can be cast into the form $s_1 s_2 \dots (s_j \dots s_k)^\omega$ where the *reachable bad cycle* $(s_j \dots s_k)^\omega$ is called the "lasso." Liveness verification of finite state systems reduces to finding one of these reachable lassos.

Model checking methods based on the use of Büchi automata help detect safety and liveness violations through language containment of Büchi automata as is described in Section 21.4.1.

Outline of the rest of the chapter

In Section 21.4, we present the example of three dining philosophers—a slight variant of the example presented by Dijkstra in [37] to illustrate the principles of concurrent programming and synchronization. We do not study the whole gamut of solutions that have been discussed for this problem for nearly four decades. Instead we simply use this example for the sake of illustration, since it is such a well-known example. In fact, *we deliberately present a buggy solution* and demonstrate how model checking can be employed to detect this bug. Section 21.4.1 informally presents how to express models as well as properties as automata in the syntax of Promela and find a liveness bug in the philosophers example.

21.4 Example: Dining Philosophers

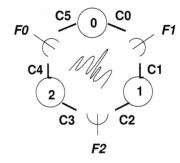

Fig. 21.1. Three dining philosophers

Imagine three philosophers numbered P0, P1, and P2 seated around a table with a bowl of spaghetti in the middle (Figures 21.1 and 21.2). They decide to eat out of the same spaghetti bowl as well as share forks as per the following rules:[4] P0 would eat with F0 and F1 in his right

---

[4] Foregoing basic hygiene...

```
mtype = {are_you_free, yes, no, release}
byte progress; /* SPIN initializes all variables to 0 */
proctype phil(chan lf, rf; int philno)
{ do
  :: do
     :: lf!are_you_free ->
        if
        :: lf?yes -> break
        :: lf?no
        fi
     od;
     do
     :: rf!are_you_free ->
        if
        :: rf?yes -> progress = 1 -> progress = 0
                      -> lf!release    -> rf!release -> break
        :: rf?no  -> lf!release   ->   break
        fi
     od
  od
}
proctype fork(chan lp, rp)
{ do
  :: rp?are_you_free -> rp!yes ->
     do
     :: lp?are_you_free -> lp!no
     :: rp?release       -> break
     od
  :: lp?are_you_free -> lp!yes ->
     do
     :: rp?are_you_free -> rp!no
     :: lp?release       -> break
     od
  od
}
init {
   chan c0 = [0] of { mtype }; chan c1 = [0] of { mtype };
   chan c2 = [0] of { mtype }; chan c3 = [0] of { mtype };
   chan c4 = [0] of { mtype }; chan c5 = [0] of { mtype };
   atomic {
     run phil(c5, c0, 0); run fork(c0, c1);
     run phil(c1, c2, 1); run fork(c2, c3);
     run phil(c3, c4, 2); run fork(c4, c5); }
}
never { /* Negation of []<> progress */
 do
 :: skip
 :: (!progress) -> goto accept;
 od;
 accept: (!progress) -> goto accept;
}
```

Fig. 21.2. Promela model for three dining philosophers

and left hands (respectively), while P1 would eat with F1 and F2, and P2 with F2 and F0. Clearly, when one philosopher 'chows,' the other two have to lick their lips, patiently waiting their turn. It is assumed that despite their advanced state of learning, they are disallowed from trying to eat with one fork or reach across the table for the disallowed fork. In order to pick up a fork, a philosopher sends a message to the fork through the appropriate channel. For example, to acquire F0 or F1, respectively, P0 would send an `are_you_free` request message through channel C5 or C0, respectively. Such a request obtains a "yes" or a "no" reply. If the reply is "yes," the philosopher acquires the fork; otherwise he retries (if for the left fork) or gives up the already acquired left fork and starts all over (if for the right fork). Consequently, the deadlock due to all the philosophers picking up their left forks and waiting for their right forks does not arise in our implementation. After finishing eating, a philosopher puts down the two forks he holds by sequentially issuing two `release` messages, each directed at the respective forks through the appropriate channels.

Figure 21.2 describes how the above protocol is described in the Promela modeling language. Consider the `init` section where we create channels c0 through c5. Each channel is of zero size (indicated by [0]) and carries `mtype` messages. Communication through a channel of size 0 occurs through *rendezvous* wherein the sender and the receiver both have to be ready in order for the communication to take place. As one example, consider when `proctype phil` executes its statement `lf!are_you_free`. We trace channel `lf` and find that it connects to a philosopher's left-hand side fork. This channel happens to be the same as the `rp` channel as far as fork processes are concerned (rp stands for the 'right-hand philosopher' for a fork). These connections are specified in the `init` section. Hence, `proctype fork` must reach `rp?are_you_free` statement at which time both `lf!are_you_free` and `rp?are_you_free` execute jointly. As a result of this rendezvous, proctypes `phil` as well as `fork` advance past these statements in their respective codes.

Continuing with the `init` section, after "wiring up" the phil and fork processes, we run three copies of `phil` and three copies of `fork` in parallel (the `atomic` construct ensures that all the `proctype` instances start their execution at the same time). The `never` section specifies the property of interest, and will be described in Sections 21.4.1 and 22.1. Now we turn our attention to the proctypes `phil` and `fork`.

Proctype `phil` consists of one endless outer `do/od` loop, inside which are two sequential `do/od` loops. The first of these loops has as its first statement `lf!are_you_free`. This statement acts as a *guard*, in the sense that the execution of this statement allows the subsequent statement to be engaged in — in our example, this happens to be a nested `if/fi` statement (sequencing is indicated by `->`; in Promela, one may use `->` and `;` interchangeably to denote sequencing). An `if/fi` offers a selection of guards, and nondeterministically selects one of the arms that can execute (a `do/od` is like an `if/fi` with a `goto` leading back to the beginning). Hence, if `lf?yes` rendezvous with `rp!yes`, the first inner `do/od` loop breaks, and the second `do/od` loop is engaged. In this, the guard is `rf!are_you_free`. If the reply obtained is `rf?yes`, the philosopher in question can begin eating. To indicate that a philosopher has successfully started eating, we set a `progress` flag to 1 and then to 0. The attainment of `progress=1` will be monitored by a `never` automaton. The forks are released and the execution continues with another iteration of the outer `do/od` loop. Notice that if the second arm of the `if/fi` statement is chosen (through `lf?no`), the `lf!are_you_free` is retried. Therefore, `phil` repeatedly requests its left fork, and if/when successful, requests its right fork. If a `rf?no` is obtained when requesting the right fork, notice that the already acquired left fork is released via `lf!release` and the whole process is restarted.

The process of acquiring both forks in the manner described above is based on *two-phase locking*—an algorithm known to be able to avoid deadlocks. The idea is (i) to acquire all the required resources in some sequential order, and (ii) when some resource is found unavailable, instead of holding on to all the resources acquired thus far and waiting for the unavailable resource (which can cause deadlocks), we release all the resources acquired thus far and restart the acquisition process. It is easy to see that while this approach avoids deadlocks, it can introduce livelocks. To detect the livelock in our example, we employ a *property automaton* expressed in the `never` section of a Promela model. Our property automaton (or "`never`" automaton) is designed to check whether the `progress` bit is set to true *infinitely often*. Even if one `phil` proctype violates this assertion—meaning that it only sets `progress` true a *finite* number of times—we want the bug to be reported. We describe this property automaton as well as how it finds the liveness violation in the next section. In Chapter 22, we explain the relationship between property automata as well as linear-time temporal logic assertions.

21.4.1 Model (proctype) and property (never) automata

A typical Promela model consists of multiple proctypes and one never
automaton. The proctypes are concurrent processes whose atomic
statements interleave in all possible ways. The collection of proctypes
represents the "system" or "model." Formally, one can say that the
asynchronous or *interleaving* product of the model automata represents
the set of all behaviors of interest. A never or *property* automaton observes the infinite runs of this interleaving product (all model runs).
Conceptually, after every move of the model (a move of one of the constituent proctypes), one move of the property automaton is performed
to check whether the model automaton move left the system in a good
state. Formally, one can say that the *synchronous* product of the model
and property automata are performed. In Figure 21.2, the property automaton represents the *complement*[5] of the desired property—hence,
the keyword never. One of the desired properties in our example is
that progress is infinitely often set true. This means that infinitely
often, every philosopher gets to eat. If there is even one philosopher
who, after a finite amount of eatings, is never allowed to eat again, we
certainly would like to know that. The never automaton accepts a run
(or a computation) if the run visits one of the *final* states infinitely. The
never automaton defined in Figure 21.2 can nondeterministically loop
in its initial do/od loop. In fact, it can even stay in its initial state if
progress is false. However, it may also choose to go to the final state
labeled accept when progress is false. Here, it can continue to visit
accept so long as progress is false. Hence, this never automaton accepts only an infinite sequence of progress being false—precisely when
our desired liveness property is violated.

When we ran the description in Figure 21.2 through the SPIN model
checker (which is a model checker for descriptions written in Promela),
it quickly found a liveness violation which indicates that, indeed, there
exists at least one philosopher who will starve forever. SPIN produces
a *message sequence chart* (MSC) corresponding to this error; this is
shown in Figure 21.3.

An MSC displays the "lifelines" (time lines of behaviors) of all the
processes participating in the protocol. Ignoring the never and init
lifelines that are at the left extremity, the main lifelines shown are verti-
cal lines tracing the behaviors of various process instances, specifically
phil:1, fork:2, phil:3, fork:4, phil:5, and fork:6. Reading this
MSC, we find out how the bug occurs:

---

[5] The complement of the property automaton is expected to be given by the user
as complemented automata are, in the worst case, exponentially bigger.

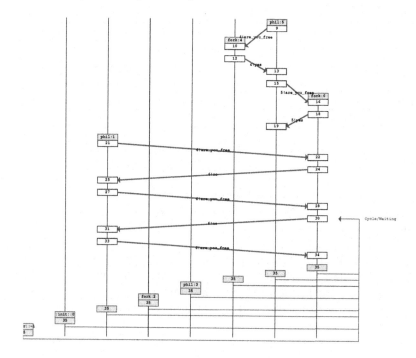

Fig. 21.3. Message sequence chart showing liveness violation bug

- `phil:5` acquires `fork:4` by sending it an **are_you_free** message and obtaining a **yes** reply (the channel names have been converted into internal channel numbers in this MSC). It also acquires `fork:6` in the same manner.
- `phil:1` attempts to acquire `fork:6` by sending an **are_you_free** and obtaining a **no** *repeatedly*. This is a liveness violation because `phil:1` is starving.

Chapter Summary

In this chapter, we examined the basic differences between informal and formal approaches to debugging concurrent reactive systems. We presented an overview of the Promela modeling language and the SPIN verifier through a simple example. In the next chapter, we delve slightly deeper into how temporal specifications are written, and some of the algorithms used in enumerative model checking approaches that explicitly enumerate states.

Exercises

21.1. Write a Promela model for a two-process mutual exclusion algorithm that is based on Dekker's solution (many operating systems books contain a description of this algorithm). State one safety and one liveness property, express these using **never** automata, and verify these properties in turn.

21.2. Repeat Exercise 21.1, but following Peterson's mutual exclusion algorithm, again described in most books on operating systems.

21.3. Modify the dining philosophers protocol to eliminate the liveness bug described in Section 21.4.1. Also make sure that your solution is deadlock free.

21.4. Even sequential programs are easily gotten wrong. This exercise shows that even though transformational programs are most often required to be analyzed through Floyd-Hoare-Dijkstra's logic [40, 55, 37], they can sometimes be easily verified through *finite-state* methods as well.

In an old textbook on Algorithms (name withheld), the following Bubble sort program is offered, accompanied by the assertion, "It takes a moment's reflection to convince oneself first that this works at all, second that the running time is quadratic."

```
procedure bubblesort;
  var j, t: integer;
  begin
  repeat
    t:=a[1];
    for j:=2 to N do
      if a[j-1]>a[j] then
      begin t:=a[j-1]; a[j-1]:=a[j]; a[j]:=t end
  until t=a[1];
  end;
```

- Examine the above program and find a bug such that the program can exit with an *unsorted* array(!)[6]
- Next, run the following Promela model encoding this Bubblesort program and find the bug in it.

```
#define Size 5
#define aMinIndx 1
#define aMaxIndx (Size-1)
```

---

[6] It may be that the author meant to write a few side conditions, but going by exactly what he wrote, I assert that there is a bug in the program.

```
/* Gonna "waste" a[0] because Sedgewick uses 1-based arrays */
active proctype bubsort()
{ byte j, t;    /* Init to 0 by SPIN */
  bit a[Size]; /* Use 1-bit abstraction */

  /* Nondeterministic array initialization */
  do ::break ::a[1]=1 ::a[2]=1 ::a[3]=1 ::a[4]=1 od;

  t=a[aMinIndx];
  j=aMinIndx+1;

  do /* First ''repeat'' iteration */
  :: (j >(aMaxIndx)) -> break /*-- For-loop exits --*/
  :: (j<=(aMaxIndx)) ->
     if
     :: (a[j-1] > a[j]) -> t=a[j-1]; a[j-1]=a[j]; a[j]=t
     :: (a[j-1] <= a[j])
     fi;
     j++
  od;

  do /* Subsequent ''repeat'' iterations */
  :: t!=a[1] ->
     t=a[aMinIndx];
     j=aMinIndx+1;
     do
     :: (j >(aMaxIndx)) -> break /*-- For-loop exits --*/
     :: (j<=(aMaxIndx)) ->
        if
        :: (a[j-1] > a[j]) -> t=a[j-1]; a[j-1]=a[j]; a[j]=t
        :: (a[j-1] <= a[j])
        fi;
        j++ /*-- for-index increments --*/
     od            /*-- end of for-loop --*/

  :: t==a[1] -> break
  od;

  t=1; /*-- Comb from location-1 to look for sortedness --*/
  do
  :: t < aMaxIndx-1 -> t++
  :: t > aMinIndx    -> t--
  :: a[t] > a[t+1]   -> assert(0) /*- announce there is a bug! -*/
  od
}
```

21.5. Modify the Bubblesort algorithm of Exercise 21.4, recode in Promela, and prove that it is now correct.

Model Checking: Temporal Logics

This chapter presents two temporal logics, namely computational tree logic (CTL) and linear-time temporal logic (LTL). Section 22.1 presents Kripke structure, 22.1.2 presents computations versus computation trees, and 22.1.3 show how LTL and CTL assertions serve as *Kripke structure classifiers*. Section 22.1.4 presents key differences between CTL and LTL through some examples. Chapter 23 will present Büchi automata as well as model checking algorithms.

22.1 Temporal Logics

22.1.1 Kripke structures

Kripke structures[1] are finite-state machines used to model concurrent systems. A Kripke structure is a four-tuple $\langle S, s_0, R, L \rangle$ where S is a *finite* set of states, s_0 is *an* initial state, $R \subseteq S \times S$ is a *total* relation known as the *reachability relation*, and $L : S \to 2^P$ is a labeling function, with P being a set of *atomic propositions*. It is standard practice to require that R be total, as Kripke structure are used to model infinite computations (in modeling finite computations, terminal states are equipped with transitions to themselves, thus in effect making R total). The fact that we choose one initial state is a matter of convenience.

Figure 22.1 (bottom) presents two Kripke structures. In the Kripke structure on the left, $S = \{s0, s1, s2\}$, $s_0 = s0$, R is shown by the directed edges, and L by the subsets of $\{a, b\}$ that label the states. In its behavior over time, this Kripke structure asserts a and b true in

---

[1] Named after Saul Kripke, who is famous for his contributions to philosophy and modal logic. Temporal logic can be viewed as an offshoot of modal logic where the possible "worlds" are various points in time.

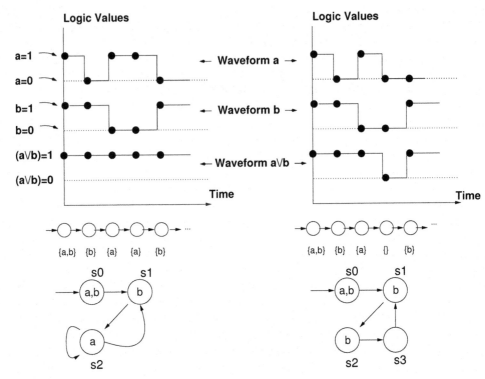

Fig. 22.1. Two Kripke structures and some of their computations. In the Kripke structure on the left, the assertion 'Henceforth $(a \vee b)$' is true.

the first time step, b true in the second time step, and a in the third time step. In the fourth time step, it may assert *either* a or b. One of these sequences is shown in the middle – namely, choosing to assert a in the fourth time step, and b in the fifth. When a label is omitted (e.g., there is no b in state s2), we are effectively asserting $\neg b$. Therefore, the assertion in state s3 is $\neg a \wedge \neg b$. The corresponding "waveforms[2]" (as if viewed on an oscilloscope) are shown at the top of this figure.

From these discussions it is clear that we model the *behavior* of systems in terms of the truth values of a set of Boolean variables. This is the most common approach taken when modeling systems as finite-state machines. We therefore let P be the set of these Boolean variables. There are of course other ways to model systems. For instance, if one

---

[2] Notice that we 'connect the dots' only for visual clarity. The values of the signals shown are actually undefined in between various discrete time points. Depending on the actual context, intermediate values may or may not be defined. For example, if a signal is the output of a clocked flip-flop, it may be valid even between the points. Even in this case, it is only sampled at discrete points.

were to employ a set of Boolean variables B and a set of integer variables I to model a system, P would consist of B as well as propositions over integers, such as $i > 2$ or $i = j + 1$. In any case, knowing the truth of the atomic propositions P, it is possible to calculate the truth of any Boolean proposition built over P.

Each state $s \in S$ is an *equivalence class* of time points, while R models how time advances. Whenever the system is at a state $s \in S$, a fixed set of Boolean propositions determined by L are true. Since S is finite, we can immediately see that the temporal behavior of systems modeled using Kripke structures recur over time, thus basically generating a *regular language* of truth values of the propositions in P. For example, Figure 22.1 shows how the truth of the Boolean proposition $a \vee b$ varies with time. Since either a or b is always true, $a \vee b$ is always true for the Kripke structure on the left; that is not the case for the Kripke structure on the right, as we can enter state s3 where both a and b are false.

This chapter is concerned with how we can study such Kripke structures through formal analysis. We will conduct this analysis by evaluating *temporal logic* properties with respect to Kripke structures. We will study both *linear* sequences of runs through Kripke structures, such as captured by waveforms, as well as *branching* behaviors (to be discussed under the heading of *computation trees*).

22.1.2 Computations vs. computation trees

In general, it is not sufficient to view the behavior of a system in terms of computational sequences, such as captured by waveforms. To see why, consider Figure 22.2, where two scenarios of a user interacting with a machine (say, a vending machine) are shown in the top and bottom half. In each scenario, the behavior shown to the left of the $\|$ sign is that of the vending machine, while the behavior shown to the right of the $\|$ sign is that of the human customer. The upper vending machine *accepts* a coin-drop (shown as a), and decides to allow the user to *bang open* its trap door (b). Since there are two (nondeterministic) ways to open the trap door, the user may be faced with a candy (c) or a donut (d) – *never* both at the same time! Since the user is always insistent on collecting a candy (c) after a b event, the system will deadlock if the machine chooses d instead of a c. Unfortunately, these "internal choices" made by the machine cannot[3] often be controlled from outside, and so such machines do exist in real life. The machine in the bottom half is not

---

[3] Perhaps depending on the "mood" of the machine or weather conditions...

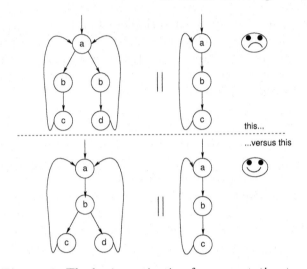

Fig. 22.2. The basic motivation for computation trees

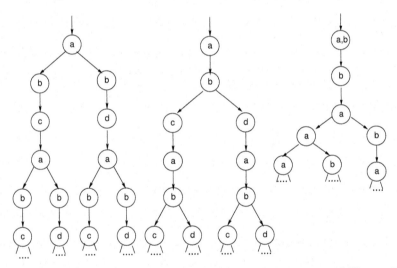

Fig. 22.3. Computation trees for the two Kripke structures of Figure 22.2, as well as the left-hand side Kripke structure of Figure 22.1

prone to this deadlock, as it always leaves c and d enabled after a b. How does one model this subtle, but important difference between the two machines? Modeling the finite behaviors of the machines through regular expressions, one obtains $(ab(c + d))*$ as the regular expression for *both* machines — and so the use of regular expressions does not help distinguish the machines.[4]

Let us therefore take another approach to see how we can distinguish these machines. In this approach, we unwind the machines into their infinite computation trees, as shown in Figure 22.3 for the Kripke structures of Figure 22.2 and the left-hand side Kripke structure of Figure 22.1. With these computation trees at hand, one can then ask questions such as:

> "At all times, after seeing a *b* event being offered, does there exist a path leading into a future where the *c* event is *guaranteed* to be offered?"

It is clear that the Kripke structures of Figure 22.2 are distinguished by this query: the bottom machine provides this guarantee by always allowing the user to pull out *c*, while the top machine might sometimes precommit to offering only a *d*. Computational tree logic (CTL) considers not only the truths of atomic propositions along paths as linear-time temporal logic (LTL) does, but also has the ability to consider: (i) whether there *exists* (E) paths at a state that satisfy a given criterion, or whether for *all* (A) paths at a state, some criterion holds. However, this is *not* to say that CTL is superior to LTL. There are many path-centered properties such as *fairness* that are expressible only in LTL and not in CTL. We will compare these logics in Section 22.1.4 as well as in Section 23.3.1. We now come to the main idea behind LTL and CTL that allows them to be studied on an equitable basis.

22.1.3 Temporal formulas are Kripke structure classifiers!

CTL formulas can be viewed as *classifiers of Kripke structures* or as *classifiers of computation trees*. Both these views are equivalent (knowing the Kripke structure, we can obtain the computation tree, and vice versa). LTL formulas can also be viewed in the same manner. We now explain these ideas, beginning with CTL first.

---

[4] Since *infinite* behaviors of reactive systems are of interest, LTL would view the vending machines as having the set of behaviors described by $(ab(c + d))^\omega$ (see Section 2.8 for an explanation of ω) while CTL would view them in terms of infinite computation trees.

CTL formulas are Kripke structure classifiers

Given a CTL formula φ, all possible computation trees fall into two bins—*models* and *non-models*.[5] The computation trees in the *model* ('good') bin are those that satisfy φ while those in the *non-model* ('bad') bin obviously falsify φ.

Consider the CTL formula AG (EF (EG a)) as an example. Here,

- 'A' is a *path quantifier* and stands for *all paths* at a state
- 'G' is a *state quantifier* and stands for *everywhere along the path*
- 'E' is a *path quantifier* and stands for *exists a path*
- 'F' is a *state quantifier* and stands for *find* (or *future*) along a path
- 'X' is a *state quantifier* and stands for *next* along a path

The truth of the formula AG (EF (EG a)) can be calculated as follows:

- In all paths, everywhere along those paths, EF (EG a) is true
- The truth of EF (EG a) can be calculated as follows:
  - There exists a path where we will find that EG a is true.
  - The truth of EG a can be calculated as follows:
    * There exists a path where a is globally true.

In CTL, one is required to use path quantifiers (A and E) and state quantifiers (G, F, X, and U) in combinations such as AG, AF, AX, AU, EG, EF, EX, and EU. More details are provided in Section 22.1.7. In other temporal logics such as CTL*, these operators may be used separately; see references such as [20] for details.

Coming back to the examples in Figure 22.1, AG (EF (EG a)) — which means,

> *Starting from any state s of the system ($s0$, $s1$, or $s2$ in our case), one can find a future reachable state t such that starting from t, there is an infinite sequence of states along which a is true*

is true of the Kripke structure on the left, but not the one on the right. This is because wherever we are in the "machine" on the left, we can be permanently stuck in state s2 that emits a. In the machine on the right, a can never be permanent. As another example, with respect to the Kripke structures of Figure 22.2, the assertion AG($b \Rightarrow$ EX c) ("wherever we are in the computation, if b is true now, that means

---

[5] It is understood that for any formula φ involving variables $vars(\varphi)$, the Kripke structures that are considered to be models or non-models include all of the variables $vars(\varphi)$ in their set of variables. In other words, these Kripke structures must assign all these variables; they may assign more variables, but not less.

that there exists at least one next state where c is true") is true of the
bottom Kripke structure but not the top Kripke structure.

LTL formulas are Kripke structure classifiers

Turning back to LTL, at its core it is a logic of *infinite computations*
or *truth-value sequences* ("waveforms"). For example, the LTL formula
$\Box(a \lor b)$ (also written as G $(a \lor b)$ or "henceforth $a \lor b$") is true with
respect to the computation (waveform) shown on the left-hand side of
Figure 22.1 against $(a \lor b) = 1$, while it is false with respect to the
waveform shown on the right.

It is customary to view LTL also as a Kripke structure (or computation tree) classifier. This is really a simple extension of the basic idea
behind LTL. Under this view, an LTL formula φ is true of a computation tree if and only if *it is true of every infinite path in the tree*. As
an example, no LTL formula can distinguish the two Kripke structures
given in Figure 22.2, as they both have the same set of infinite paths.

22.1.4 LTL vs. CTL through an example

The differences between LTL and CTL are actually quite subtle.
Consider Figure 22.4, and the CTL formula AG (EFx). In order for

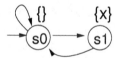

Fig. 22.4. AG (EF x) is true, yet there is a computation where x is permanently false

this formula to be true over this Kripke structure, in all paths starting from s0, everywhere along those paths, EFx must be true. This, in
turn, means that there must *exist* a path where x is eventually *found*.
Starting either from s0 or s1, we see that there exists a path on which
x is found true eventually. Hence, AG (EFx) is true of this Kripke
structure.

Let us try to pretend that the CTL formula AG (EF x) is equivalent to the LTL formula G (F x). An LTL formula is true of a Kripke
structure if and only if it is true of *every* infinite path in the Kripke
structure considered separately! Now consider the infinite path of spin-
ning in state s0. For this infinite path, $\neg x$ is permanently true. Therefore, we will violate Fx, and hence G (Fx). In short, this *one path*

serves as the "death knell" for G (F x) with respect to this Kripke structure—whatever other paths might exist in this Kripke structure. Section 23.3.1 in Chapter 23 will have more to say about LTL versus CTL. We now introduce these logics formally.

22.1.5 LTL syntax

LTL formulas φ are inductively defined as follows, through a context-free grammar:

$$
\begin{array}{ll}
\varphi \rightarrow x, & \text{a propositional variable} \\
\quad | \; \neg\varphi & \text{negation of an LTL formula} \\
\quad | \; (\varphi) & \text{parenthesization} \\
\quad | \; \varphi_1 \vee \varphi_2 & \text{disjunction} \\
\quad | \; G\varphi & \text{henceforth } \varphi \\
\quad | \; F\varphi & \text{eventually } \varphi \; (\text{"future"}) \\
\quad | \; X\varphi & \text{next } \varphi \\
\quad | \; (\varphi_1 U \; \varphi_2) & \varphi_1 \text{ until } \varphi_2 \\
\quad | \; (\varphi_1 W \; \varphi_2) & \varphi_1 \text{ weak-until } \varphi_2
\end{array}
$$

Note that G is sometimes denoted by $\square$ while F is denoted by $\Diamond$. We also introduce the X operator into the syntax above to capture the notion of *next* in a time sense. It is clear that in many real systems— for example, in globally clocked digital sequential circuits—the notion of next time makes perfect sense. However, in reasoning about the executions of one sequential process P_i among a collection of parallel processes $P_1, \ldots, P_n$, the notion of next time does not have a unique meaning. As far as P_i is concerned, it has a list of candidate "next" statements to be considered; however, these candidate statements may be selected only after an arbitrary amount of interleavings from other processes. Hence, what is "next" in a local sense (from the point of view of P_i alone) becomes "eventually" in a global sense. While conducting verification, we will most likely not be proving properties involving X. However, X as a temporal operator can help expand other operators such as G through recursion. With this overview, we now proceed to examine the semantics of LTL.

22.1.6 LTL semantics

Recall that the semantics of LTL are defined over (infinite) computational sequences. LTL semantics can be defined over computation trees by conjoining the truth of an LTL formula over every computational sequence in the computational tree. Let $\sigma = \sigma^0 = s_0, s_1, \ldots$, where the superscript 0 in σ^0 emphasizes that the computational sequence begins at state s_0. By the same token, let $\sigma^i = s_i, s_{i+1}, \ldots$, namely the infinite sequence beginning at s_i. By $\sigma \models \varphi$ we mean φ is true with respect to computation σ; $\sigma \not\models \varphi$ means φ is false with respect to computation σ. Here is the inductive definition for the semantics of LTL:

$$\begin{aligned}
\sigma &\models x & &\text{iff } x \text{ is true at } s_0 \text{ (written } s_0(x)) \\
\sigma &\models \neg\varphi & &\text{iff } \sigma \not\models \varphi \\
\sigma &\models (\varphi) & &\text{iff } \sigma \models \varphi \\
\sigma &\models \varphi_1 \vee \varphi_2 & &\text{iff } \sigma \models \varphi_1 \vee \sigma \models \varphi_2 \\
\sigma &\models G\varphi & &\text{iff } \sigma^i \models \varphi \text{ for every } i \geq 0 \\
\sigma &\models F\varphi & &\text{iff } \sigma^i \models \varphi \text{ for some } i \geq 0 \\
\sigma &\models X\varphi & &\text{iff } \sigma^1 \models \varphi \\
\sigma &\models (\varphi_1 U \varphi_2) & &\text{iff } \sigma^k \models \varphi_2 \text{ for some } k \geq 0 \text{ and } \sigma^j \models \varphi_1 \text{ for all } j < k \\
\sigma &\models (\varphi_1 W \varphi_2) & &\text{iff } \sigma \models G\varphi_1 \vee \sigma \models (\varphi_1 U \varphi_2)
\end{aligned}$$

LTL example

Consider formula GF x (a common abbreviation for G(F x)). Its semantics are calculated as follows:

$$\sigma \models GFx \text{ iff } \sigma^i \models Fx, \text{ for all } i \geq 0$$
$$\sigma^i \models Fx \text{ iff } \sigma^j \models x, \text{ for some } j \geq i$$

Putting it all together, we obtain the meaning as:

> x is true infinitely often—meaning, *beginning at no point in time is it permanently false.*

22.1.7 CTL syntax

CTL formulas γ are inductively defined as follows:

$\gamma \to x$		a propositional variable	
	$\neg\gamma$	negation of γ	
	(γ)	parenthesization of γ	
	$\gamma_1 \vee \gamma_2$	disjunction	
	AG γ	on all paths,	everywhere along each path
	AF γ	on all paths,	somewhere on each path
	AX γ	on all paths,	next time on each path
	EG γ	on some path,	everywhere on that path
	EF γ	on some path,	somewhere on that path
	EX γ	on some path,	next time on that path
	A$[\gamma_1 U \gamma_2]$	on all paths,	γ_1 until γ_2

$E[\gamma_1 \ U \ \gamma_2]$	on some path,	γ_1 until γ_2
$A[\gamma_1 \ W \ \gamma_2]$	on all paths,	γ_1 weak-until γ_2
$E[\gamma_1 \ W \ \gamma_2]$	on some path,	γ_1 weak-until γ_2

22.1.8 CTL semantics

The semantics of CTL formulas are defined over computation trees. Our approach to defining this semantics recurses over the structure of computation trees. We will also employ fixed-points to capture this semantics precisely. Other approaches may be found in references such as [20] and [83]. Some of the reasons for taking a different approach[6] are the following:

- Some approaches (e.g., [20]) introduce a more general temporal logic (namely CTL*) and then introduce CTL and LTL as special cases. We wanted to keep our discussions at a more elementary level.
- Our definitions make an *explicit* connection with the standard fixed-point semantics of CTL (discussed in Chapter 23).
- The view of LTL and CTL being Kripke structure classifiers (see Section 22.1.3) is more readily apparent from our definitions.

Let a computation tree be denoted by τ. Here are the notations we shall use (see Figure 22.5):

- $\tau = \tau^\varepsilon$. We are going to "exponentiate" τ with tree paths as described in Section 15.6.2, with ε denoting the empty sequence.
- The state at τ^ε is $s(\tau^\varepsilon)$.
- τ^ε has $\beta(\tau^\varepsilon) + 1$ branches numbered 0 through $\beta(\tau^\varepsilon)$. In effect, β specifies the arity at every node of the computation tree. Since each Kripke structure has a total state transition relation R, $\beta(\tau) \geq 0$ for any computation tree τ.
- Branch $0 \leq j \leq \beta(\tau^\varepsilon)$ leads to computation tree τ^j (for example, τ^0 is the computation tree rooted at the first child of τ^ε, τ^1 is the computation tree rooted at the second child of τ^ε, etc).
- Generalizing this notation, for a sequence of natural numbers π, τ^π denotes the computation tree arrived at by traversing the branches identified in π. We call these the (computation) trees *yielded* by π (for example, $\tau^{1,0,1,1,2}$ is a computation tree obtained by traversing the tree path described by $1, 0, 1, 1, 2$, starting from τ^ε).
- For an arbitrary sequence of natural numbers π, τ^π is not defined if any of the natural numbers in the sequence π exceeds the corresponding β value, *i.e.*, the arity of the tree at a certain level. We

---

[6] Our definitions appear to resemble *co-inductive* definitions [86].

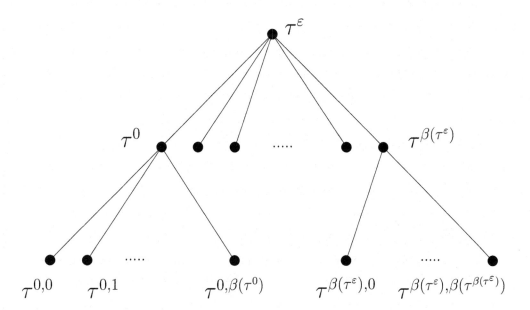

Fig. 22.5. Illustration of basic notions relating to computation trees

will make sure not to exceed the arity. Hence, $s(\tau^{\pi})$ is the state that the computation tree τ^{π} is rooted at, and $\beta(\tau^{\pi}) + 1$ is the number of children it has.

Basically, there are as many computation trees as states. Each state uniquely defines a tree. Each tree is represented by a state at its root. Given that Kripke structures only have a finite number of states, it is easy to see that with respect to any computation tree τ, only finitely many *distinct* computation trees may be yielded. Therefore, along every path, the computation trees tend to recur. In the Kripke structure on the left-hand side of Figure 22.1, τ^{ε} is the computation tree rooted at s0. Taking the self-loop at state s2 as its 0-th child, $\tau^{0,0,0}$ is the computation tree rooted at s2. This is the same computation tree as $\tau^{0,0,0,0}$, or as $\tau^{0,0,0,1,0}$. Given all this, we will end up defining the semantics of CTL through mutual recursion over a finite number of computation trees. This recursion can then be solved by appealing to the least fixed-point or the greatest fixed-point.

More notation: For a computation tree $\tau = \tau^{\varepsilon}$ associated with a Kripke structure (S, s_0, R, L), define $\Pi(\tau)$ to be a set of tree paths such that for every state $s \in S$, there is exactly one path in $\Pi(\tau)$ that corresponds to the path taken by standard depth first search (DFS) applied to the Kripke structure to reach s. For example, for the Kripke structure of Figure 22.1, $\Pi(\tau) = \{\varepsilon, 0, 00\}$.

Note: We really do not care about the exact search method; we merely want $\Pi(\tau)$ to be a unique set. We use this definition of Π towards the end of this page.

We first define the semantics of the propositional logic subset of CTL formulas:

$$\tau^\pi \models x \qquad \text{iff } x \text{ is true at } s(\tau^\pi) \text{ (written } s(\tau^\pi)(x))$$
$$\tau^\pi \models \neg\gamma \qquad \text{iff } \tau^\pi \not\models \gamma$$
$$\tau^\pi \models (\gamma) \qquad \text{iff } \tau^\pi \models \gamma$$
$$\tau^\pi \models \gamma_1 \vee \gamma_2 \qquad \text{iff } \tau^\pi \models \gamma_1 \text{ or } \tau^\pi \models \gamma_2$$

We now define the CTL semantics for the remaining CTL formulas, beginning with the AG operator.

Notation: For sequences of natural numbers α and β, we write $\alpha.\beta$ to denote their concatenation. For example, if $\alpha = 1, 2$ and $\beta = 3$, then $\alpha \, . \, \beta = 1, 2, 3$. Similarly, $1, 2 \, . \, 3, 4 = 1, 2, 3, 4$.

- AG :

$$\tau^\pi \models \text{AG}\gamma \quad \text{iff } \tau^\pi \models \gamma \text{ and } \tau^{\pi.j} \models \text{AG } \gamma \text{ for all } 0 \leq j \leq \beta(\tau^\pi).$$

The first question that probably pops into one's mind is: *does this recursion define a unique semantics?* To answer this question, notice that the quantity being defined, generically written as the configuration "$\tau^\pi \models Fmla$", will recur since there are only a finite number of computation trees. In effect, we have a standard finitary mutual recursion for which we can provide a fixed-point semantics. Illustration 22.1.1 will clarify these ideas through an example.

The *intended semantics* for 'AG' is that of the *greatest fixed-point* (the greatest fixed-point defines a fixed-point that is higher than any other fixed-point in the implication order defined in Illustration 4.5.3). To cut a long story short, when a configuration "$\tau \models Fmla$" recurs, one must substitute *true* for the second occurrence, unravel the formula up to and including the second occurrence, and thus obtain a *closed-form solution.*[7] One can see that AG indeed computes the finitary conjunction computed by the following equivalent formulation:

---

[7] We refer the reader to [20] and/or [83] for an explanation of why greatest fixed-point computations begin with the "bottom" or "seed" element being *true*. See also the discussion on this topic in Chapter 6. Likewise, for the least fixed-point computation, we use the seed of *false*.

$\tau^\pi \models \mathrm{AG}\gamma$ iff $\tau^{\pi.\sigma} \models \gamma$ for all $\sigma \in \Pi(\tau^\pi)$.

Therefore, $\mathrm{AG}\gamma$ evaluates γ at every state of the computation tree and takes the conjunction of the results.

We now proceed to define the remaining CTL operators through recursion, pointing out in each case which fixed-point is to be taken.

- AF :

$\tau^\pi \models \mathrm{AF}\gamma$ iff $\tau^\pi \models \gamma$ or $\tau^{\pi.j} \models \mathrm{AF}\gamma$ for all $0 \le j \le \beta(\tau^\pi)$.

Here, the *least fixed-point* is what is intended. This means that when a configuration "$\tau^\pi \models Fmla$" recurs, we must substitute *false* for all such repeated occurrences, thus obtaining a closed-form solution. Again, Illustration 22.1.1 provides a simple example. Notice that the AF obligation is required of *every path* going forwards. In other words, in all paths, γ holds *somewhere*.

- EG :

$\tau^\pi \models \mathrm{EG}\gamma$ iff $\tau^\pi \models \gamma$ and $\tau^{\pi.j} \models \mathrm{EG}\gamma$ for some $0 \le j \le \beta(\tau^\pi)$.

Here, the *greatest fixed-point* is what is intended. In other words, when a configuration "$\tau \models Fmla$" recurs, we must substitute *true* for all such repeated occurrences, thus obtaining a finitary conjunction along *some* path.

- EF :

$\tau^\pi \models \mathrm{EF}\gamma$ iff $\tau^\pi \models \gamma$ or $\tau^{\pi.j} \models \mathrm{EF}\gamma$ for some $0 \le j \le \beta(\tau^\pi)$.

Here, the *least fixed-point* is what is intended, thus obtaining a finitary disjunction along *some path*.

- AU :

$\tau^\pi \models \mathrm{A}[\gamma_1 \ \mathrm{U} \ \gamma_2]$ iff $\tau^\pi \models \gamma_2$ or $\tau^\pi \models \gamma_1$ and $\tau^{\pi.j} \models \mathrm{A}[\gamma_1 \mathrm{U} \gamma_2]$
for all $0 \le j \le \beta(\tau^\pi)$.

Here, the *least fixed-point* is what is intended. The least fixed-point ensures that the recursion will "bottom out" with γ_2 holding somewhere along all paths.

- EU :

$\tau^\pi \models \mathrm{E}[\gamma_1 \ \mathrm{U} \ \gamma_2]$ iff $\tau^\pi \models \gamma_2$ or $\tau^\pi \models \gamma_1$ and $\tau^{\pi.j} \models \mathrm{E}[\gamma_1 \mathrm{U} \gamma_2]$
for some $0 \le j \le \beta(\tau^\pi)$.

Here, the *least fixed-point* is what is intended. The least fixed-point ensures that the recursion will "bottom out" with γ_2 holding somewhere along some path.

- AW :

$$\tau^\pi \models A[\gamma_1 \ W \ \gamma_2] \quad \text{iff } \tau^\pi \models AG\gamma_1 \text{ or } \tau^\pi \models A[\gamma_1 \ U \ \gamma_2].$$

We define AW by permitting the possibility of γ_2 not happening. The idea behind EW is also similar. In effect, these are weakenings of AU and EU. Section 23.2.4 shows that this subtle variation is caused merely by changing which fixed-point we choose: least fixed-point for AU, and greatest fixed-point for AW. This is the reason why 'U' is termed the *strong* until while 'W' is termed the *weak* until.

- EW :

$$\tau^\pi \models E[\gamma_1 \ W \ \gamma_2] \quad \text{iff } \tau^\pi \models EG\gamma_1 \text{ or } \tau^\pi \models E[\gamma_1 \ U \ \gamma_2].$$

Illustration 22.1.1 Consider a Kripke structure with states $\{s_0, s_1, s_2\}$ and reachability relation $s0 \rightarrow s1$, $s0 \rightarrow s2$, $s1 \rightarrow s0$, and $s2 \rightarrow s0$. The labeling over one variable x is $L(s2) = \{x\}$, $L(s1) = L(s0) = \{\}$. For example, $s2(x) = true, s1(x) = false, s0(x) = false$.

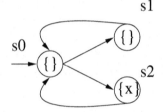

We show how to evaluate AG (EF x):

$$
\begin{aligned}
s0 \models AG \ (EF \ x) = \ & \wedge \ \ s0 \models (EFx) \\
& \wedge \ \ s1 \models AG(EFx) \\
& \wedge \ \ s2 \models AG(EFx)
\end{aligned}
$$

$$
\begin{aligned}
s1 \models AG \ (EF \ x) = \ & \wedge \ \ s1 \models (EFx) \\
& \wedge \ \ s0 \models AG(EFx)
\end{aligned}
$$

$$
\begin{aligned}
s2 \models AG \ (EF \ x) = \ & \wedge \ \ s2 \models (EFx) \\
& \wedge \ \ s0 \models AG(EFx)
\end{aligned}
$$

Let

$$R = \langle \ s0 \models AG(EFx), \ s1 \models AG(EFx), \ s2 \models AG(EFx) \ \rangle$$

Also let $\#_1$, $\#_2$, and $\#_3$ represent tuple component selectors. Then we have the following recursion in terms of R which represents the above "tuple of solutions:"

$$R = \langle\ s0 \models (\mathrm{EF}x)\ \wedge\ \#_1(R)\ \wedge\ \#_2(R),$$
$$\quad s1 \models (\mathrm{EF}x)\ \wedge\ \#_2(R),$$
$$\quad s2 \models (\mathrm{EF}x)\ \wedge\ \#_0(R)$$
$$\rangle.$$

The conjunctions can be solved using the greatest fixed-point (GFP) iteration, starting from "seed" $\langle T, T, T\rangle$, standing for a tuple of three trues. The GFP iteration proceeds as follows, with the R approximants named R_0, R_1, R_2, etc:

$$R_0 = \langle\ T,\ T,\ T\ \rangle.$$

$$R_1 = \langle\ s0 \models (\mathrm{EF}x),\ s1 \models (\mathrm{EF}x),\ s2 \models (\mathrm{EF}x)\ \rangle.$$

$$R_2 = \langle\ s0 \models (\mathrm{EF}x)\ \wedge\ s1 \models (\mathrm{EF}x)\ \wedge\ s2 \models (\mathrm{EF}x),$$
$$\quad s1 \models (\mathrm{EF}x)\ \wedge\ s0 \models (\mathrm{EF}x),$$
$$\quad s2 \models (\mathrm{EF}x)\ \wedge\ s0 \models (\mathrm{EF}x)$$
$$\rangle.$$

$$R_3 = \langle\ s0 \models (\mathrm{EF}x)\ \wedge\ s1 \models (\mathrm{EF}x)\ \wedge\ s2 \models (\mathrm{EF}x),$$
$$\quad s0 \models (\mathrm{EF}x)\ \wedge\ s1 \models (\mathrm{EF}x)\ \wedge\ s2 \models (\mathrm{EF}x),$$
$$\quad s0 \models (\mathrm{EF}x)\ \wedge\ s1 \models (\mathrm{EF}x)\ \wedge\ s2 \models (\mathrm{EF}x)$$
$$\rangle.$$

The GFP has been attained.

Now, we need to solve for $\mathrm{EF}x$. Proceeding as before, define

$$S = \langle\ s0 \models (\mathrm{EF}x),\ s1 \models (\mathrm{EF}x),\ s2 \models (\mathrm{EF}x)\ \rangle.$$

The recursive equation for S is

$$S = \langle\ s0 \models x\ \vee\ \#_1(S)\ \vee\ \#_2(S),$$
$$\quad s1 \models x\ \vee\ \#_0(S),$$
$$\quad s2 \models x\ \#_0(S)$$
$$\rangle.$$

The LFP iteration proceeds as follows:

$$S_0 = \langle\ F,\ F,\ F\ \rangle$$

$$S_1 = \langle\ s0 \models x \vee F \vee F,\ \ s1 \models x \vee F,\ \ s1 \models x \vee F\ \rangle.$$
i.e.,

$$S_1 = \langle\ s0 \models x,\ s1 \models x,\ s2 \models x\ \rangle.$$

$$S_2 = \langle\ s0 \models x \ \lor\ s1 \models x \ \lor\ s2 \models x,$$
$$s1 \models x \ \lor\ s0 \models x,$$
$$s2 \models x \ \lor\ s0 \models x$$
$$\rangle.$$

$$S_3 = \langle\ s0 \models x \ \lor\ s1 \models x \ \lor\ s2 \models x,$$
$$s0 \models x \ \lor\ s1 \models x \ \lor\ s2 \models x,$$
$$s0 \models x \ \lor\ s1 \models x \ \lor\ s2 \models x$$
$$\rangle.$$

The LFP has been attained. So now, coming back to R, the item of interest, and substituting for S in it, and simplifying, we get:

$$R = \langle\ s0 \models x \ \lor\ s1 \models x \ \lor\ s2 \models x,$$
$$s0 \models x \ \lor\ s1 \models x \ \lor\ s2 \models x,$$
$$s0 \models x \ \lor\ s1 \models x \ \lor\ s2 \models x$$
$$\rangle.$$

and so, $s0 \models AG(EFx)$ is $(s0 \models x \ \lor\ s1 \models x \ \lor\ s2 \models x)$. □

CTL example

Consider formula AG (EFx). Its semantics is calculated as follows:

$$\tau^\varepsilon \models AG\ EFx \text{ iff } s(\tau^\varepsilon)(EFx) \text{ and}$$
$$\tau^{\varepsilon.j} \models AG\ EFx \text{ for all } 0 \le j \le \beta(\tau^\varepsilon).$$

This simply means that EFx is true at every state of the computation tree. This assertion is *satisfied* by the computation tree generated by the Kripke structure shown in Figure 22.4, as was discussed in Section 22.1.4.

Chapter Summary

This chapter presented the syntax and semantics of LTL and CTL, both formally and informally. The next chapter will further concretize this body of knowledge by presenting three model checking algorithms: an enumerative algorithm for CTL, a symbolic algorithm for CTL, and an enumerative algorithm for LTL.

Exercises

22.1. Based on the definitions in Section 22.1.1, argue that all computational trees are infinite. Also argue that the number of children of any node of a computation tree is finite ("finitary branching").

22.2. In both LTL and CTL, show that the "until" operator 'U' can help realize the 'G' and 'F' operators.

22.3. Explain the semantics of all sixteen combinations of $[\gamma_1 \ U \ \gamma_2]$ where γ_1 and γ_2 range over $\{true, false, \gamma, \neg\gamma\}$. Explain which familiar temporal logic operator (if any) is modeled by each of the cases considered. Explain the overall semantics in all cases.

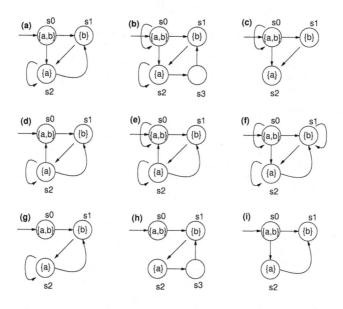

Fig. 22.6. Nine diagrams, some of which are Kripke Structures

22.4. Consider the nine state machines (a) through (i) in Figure 22.6. They are all drawn with respect to two variables, a and b. Which of them are Kripke structures and which are not?

22.5. We now discuss how to construct an LTL formula such that it is true only of structure (b) of Figure 22.6, and not of structure (a). We include an encoding of these Kripke structures in Promela as well as an LTL query that demonstrates this assertion.

```
/* Declare states that'll be used in Kripke structures */
mtype = {s0, s1, s2, s3}

/* Declare sa,... standing for kripke structure a, ... */
mtype = {sa, sb, sd, se, sf, sg, sh}
```

```
byte state=s0;  /* Init state to s0 */
bit a=1;        /* Initial values of a and b */
bit b=1;

proctype generic(mtype structure)
{ if
  :: structure==sa ->
        do
        :: d_step{state==s0;a=1;b=1} -> d_step{state=s1;a=0;b=1}
        :: d_step{state==s0;a=1;b=1} -> d_step{state=s2;a=1;b=0}

        :: d_step{state==s1;a=0;b=1} -> d_step{state=s2;a=1;b=0}

        :: d_step{state==s2;a=1;b=0} -> d_step{state=s2;a=1;b=0}
        :: d_step{state==s2;a=1;b=0} -> d_step{state=s1;a=0;b=1}
        od
  :: structure==sb ->
        do
        :: d_step{state==s0;a=1;b=1} -> d_step{state=s0;a=1;b=1}
        :: d_step{state==s0;a=1;b=1} -> d_step{state=s1;a=0;b=1}
        :: d_step{state==s0;a=1;b=1} -> d_step{state=s2;a=1;b=0}

        :: d_step{state==s1;a=0;b=1} -> d_step{state=s2;a=1;b=0}

        :: d_step{state==s2;a=1;b=0} -> d_step{state=s2;a=1;b=0}
        :: d_step{state==s2;a=1;b=0} -> d_step{state=s3;a=0;b=0}

        :: d_step{state==s3;a=0;b=0} -> d_step{state=s1;a=0;b=1}
        od
  fi
 /*
  :: structure==sd -> ... similar ... */
}


init
{ run generic(sb) }

/* sb satisfies this, and not sa */
/*Type 'spin -f "formula"' and cut&paste the resulting never aut. */
/*-----------------------------------------------------------------*/
never {/* !( !(<>([](a && b))) -> ((<>([]a)) || (<> (!a && !b)))) */
T0_init:
        if
        :: (! ((a)) && (b)) -> goto accept_S373
        :: ((((a)) || ((b)))) -> goto T0_init
        fi;
accept_S373:
        if
        :: ((((a)) || ((b)))) -> goto T0_init
```

```
        fi;
}
/*--- in contrast, both sa and sb satisfy this --->
never {  /* !( !(<>([](a && b))) -> ( (<>([]a)) || (<> (b)) ) ) * /}
<---*/
```

- Understand the Promela code for the Kripke structures. In particular note that the **never** automaton specifies a property true of **sb** and not **sa**. This **never** automaton was synthesized using SPIN's LTL translator.
- There is also another property given in the comments such that the property is true both of **sa** and **sb**. Translate this property into a Büchi automaton using SPIN's LTL translator and check the results on **sa** and **sb**.
- Come up with assertions that pairwise distinguish the remaining Kripke structures and test your understanding using Promela and SPIN.

22.6. Consider the Kripke structures (d) and (f) of Figure 22.6.

- Obtain two LTL formulas P_d and P_f that are true of (d) and false of (f); these formulas must capture as many of the features of structure (d) as possible. Demonstrate using SPIN that P_d and P_f are both true of (d) and are both false of (f).
- Now obtain Q_d and Q_f that are both false of (d) and both true of (f), and demonstrate as above using SPIN.
- Now write down R_d and R_f which are two CTL formulas that are both true of (d) and both false of (f) and check using NuSMV.
- Now write down S_d and S_f which are two CTL formulas that are both false of (d) and both true of (f) and check using NuSMV.

Here is a hint of how one might code things in NuSMV (please find out which Kripke structure is modeled by this):

```
MODULE main
VAR state: {s0, s1, s2};

ASSIGN init(state) := s0;

next(state) := case state = s0 : {s1, s2};
                    state = s1 : s2;
                    state = s2 : {s2, s1};
               esac;

DEFINE a := (state = s0) | (state = s2);
       b := (state = s0) | (state = s1);

SPEC AG (EF (a & b))
```

Model Checking: Algorithms

This chapter closes off our presentation of model checking by presenting three different model checking algorithms at an intuitive level. Section 23.1 presents an enumerative algorithm for CTL model checking. Section 23.2 presents a symbolic algorithm for CTL model checking. Section 23.3 introduces Büchi automata, discusses how Boolean operations on Büchi automata are performed, and that nondeterministic Büchi automata are not equivalent to deterministic Büchi automata. It also presents an enumerative algorithm for LTL model checking. All algorithms are presented through intuitive examples; however, we do present salient details such as the exact fixed-point computation process (for symbolic CTL model checking) or the nested depth-first search process (for enumerative LTL model checking). For further details, please see references such as [20, 83].

23.1 Enumerative CTL Model Checking

We now present the basic ideas behind an *enumerative* (or explicit state) method for CTL model checking through one example, reinforcing ideas presented in Chapter 22. In the enumerative approach, explicit graph-based algorithms are employed, with states typically stored in hash tables. This is as opposed to representing and manipulating states as well as state transition systems using BDDs. Our presentation of enumerative CTL model checking contains excerpts from the paper [19] by Clarke, Emerson, and Sistla.

Let us consider the problem of verifying the formula below at the starting state of the Kripke structure presented in Figure 23.1:

```
(AG (OR (NOT T1) (AF C1)))
```

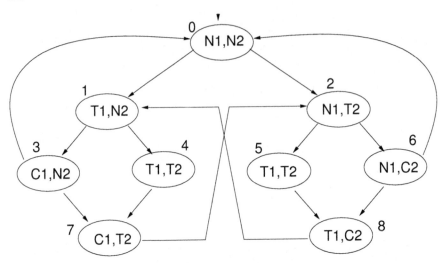

Fig. 23.1. A Kripke structure for critical sections

The algorithm proposed by [19] accomplishes this task, and in the process actually labels *all* states of this Kripke structure with *all* formulas, as well as subformulas of this formula, thanks to a recursive depth-first traversal over the formula structure. The algorithm runs in polynomial time with respect to the size of the Kripke structure and the CTL formula to be checked. However, please bear in mind that the size of the Kripke structure may be *exponential* in the number of constituent processes, and so our polynomial is actually over an exponential.

The Kripke structure of Figure 23.1 represents the execution of two parallel processes with alternating priorities, taking N to stand for "in non-critical region," T for "trying to enter," C for "in critical region," and the numbers adjoining N, T, and C representing process IDs.

The first step of the model checking algorithm is to number the formula and its subformulas as shown in Figure 23.2. Basically, each subformula acquires a higher number in this scheme than all its including formulas:

The next step of the model checking algorithm consists of executing the **for** loop shown in Figure 23.3 that considers the subformulas of the given formula bottom-up, and labels each state of the FSM with those subformulas that are true in that state (a '*' in the table indicates that the formula in the column heading is true in the state shown on the left).

In our example, we would invoke **label_graph** on formulas C1, (AF C1), T1, (NOT T1), (OR ...), and (AG ...), in that order. While doing

Formula number	Formula	Subformula List
1	(AG (OR (NOT T1)(AF C1)))	(2)
2	(OR (NOT T1)(AF C1))	(3 5)
3	(NOT T1)	(4)
4	T1	nil
5	(AF C1)	(6)
6	C1	nil

Fig. 23.2. Formula and its Subformulas

so, we generate and fill the table given in Figure 23.4. Each column under a subformula indicates whether the subformula holds in various states. Because of the order in which the subformulas are considered, the table can be filled using a recursive depth-first traversal. We present the pseudocode executed by `label_graph(AF C1)` in Figure 23.3. For further details, please see [19]. Figure 23.4 summarizes the results of model checking on this example. We see that the property of interest is true at state 0, and so is true of the given Kripke structure (in fact, this property is true of every state).

23.2 Symbolic Model Checking for CTL

We will now present symbolic model checking algorithms for CTL through examples. Our first example is to evaluate EG p on the Kripke structure of Figure 23.5(a), where p will be $a \oplus b$.

23.2.1 EG p through fixed-point iteration

The steps below show how to use the BED tool as a "calculator" in this process.

• First we obtain a fixed-point formulation for EG p.

$$\text{EG } p = p \wedge (\text{EX (EG p)})$$

• We then realize that this can be solved through greatest fixed-point iteration (iterating with "seed" equal to true). Iterating with a seed of "false" yields a solution for EG p equal to *false*—clearly not its intended semantics!

• We realize that we will need the TREL of this Kripke structure. We obtain it in the same manner as already explained in Section 11.3.1 of Page 192. The BED tool session is below:

```
for i := length(f) step -1 until 1
  do label_graph(fi)

procedure label_graph(f)
var b : boolean; % dummy var
{ %-- Do a case analysis based on the principal operator

CASE: principal operator of "f" is AF :
begin
  ST := empty_stack;
  for all s in S do marked(s) := false; -- S' are the set of states
  for all s in S do                      --   of the Kripke structure.
    if not marked(s) then AF(f, s, b);
%-- Other cases are not shown...
}

procedure AF(f, s, var b) -- var parameters can return values
{if marked(s) then
  {if labeled(s,f) then
     { b := true; return }
    b := false; return
  }
 marked(s) := true;
 if labeled(s, f) then
   { add_label(s, f); b := true; return }
 push(s, ST);
 for all s1 in successors(s) do
  { AF(f, s1, b1);
     if not(b1) then
       { pop(ST); b := false; return }
  }
 pop(ST); b := true; add_labels(s, f); return
}%-- procedure AF
```

Fig. 23.3. Algorithm used in CTL model checking illustrated on the "AF" operator

```
bed> var a a1 b b1
var a a1 b b1
bed> let TREL =
    (not(a) and b and a1 and not(b1)) or (a and not(a1) and b1) or
    (a and not(b) and b1)                or (a and not(b) and a1)
bed> upall TREL
Upall( TREL ) -> 53
bed> view TREL ... (displays the BDD)
```

State number	Atomic Propositions	C1	(AF C1)	T1	(NOT T1)	(OR (NOT T1) (AF C1))	(AG (OR (NOT T1) (AF C1)))
0	N1,N2		*		*	*	*
1	T1,N2		*	*		*	*
2	N1,T2		*		*	*	*
3	C1,N2	*	*		*	*	*
4	T1,T2		*	*		*	*
5	T1,T2		*	*		*	*
6	N1,C2		*		*	*	*
7	C1,T2	*	*		*	*	*
8	T1,C2		*	*		*	*

Fig. 23.4. Table for our CTL model checking example

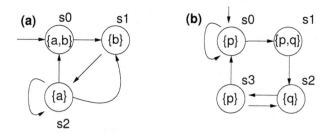

Fig. 23.5. Kripke structures that help understand CTL

• In the BED syntax, $a \oplus b$ is written a != b. Now we perform the fixed-point iteration assisted by BED. We construct variable names that mnemonically capture what we are achieving at each step:

```
EG_a_xor_b_0 = true -- first approximant
```

```
EG_a_xor_b_1 = (a != b) and (EX true) -- second approximant
```

This simplifies to (a != b), as (EX true) is true.

Now, in order to determine EG_a_xor_b_2, we continue the fixed-point iteration process, and write

```
EG_a_xor_b_2 = (a != b) and EX (a != b)
```

At this juncture, we realize that we need to calculate EX (a != b). This can be calculated using BED as follows:

```
bed> let EX_a_xor_b = exists a1. exists b1. (TREL and (a1 != b1))
bed> upall EX_a_xor_b
bed> view EX_a_xor_b
```

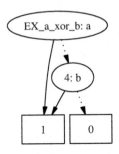

Fig. 23.6. BDD for EX (a != b)

Section 23.2.2 explains the details of why this represents `EX_a_xor_b`. The BDD in Figure 23.6 confirms that `EX (a != b)` is true at those states where a or b are true — i.e., the states {s0, s1, s2}. Now we carry on with the fixed-point iteration:

```
bed> let EG_a_xor_b_2 = (a != b) and EX_a_xor_b
bed> upall EG_a_xor_b_2
bed> view EG_a_xor_b_2
```

We see that `EG_a_xor_b_2` becomes equal to `(a != b)`, and hence we stabilize at this fixed-point.

The set of states in which $EG\,a \oplus b$ is true is $a \oplus b$, which includes states s1 and s2. *Notice that state s0 is avoided because $a \oplus b$ is false there.*

23.2.2 Calculating EX and AX

Calculating EX for a simple example

Let's take a simpler example to begin with. Suppose for some machine M, whose state is described using two bits a and b, the TREL is as below, where a1 and b1 describe the next states:

```
TREL = (!a and !b and a1 and b1) or (!a and b and !a1 and b1)
```

Basically, machine M has the following moves:

```
00 -> 11 and 01 -> 01.
```

Suppose we want to find out *all* starting states from which M has *at least one move* that ends up in a state which satisfies (a != b) *i.e.*, EX (a != b). Here is a portrayal of the situation:

Present State	Next State	Next States Satisfying (a != b)
00	11	This next state does not satisfy
01	01	This next state does satisfy

So we basically perform

```
STARTSTATES = exists a1. exists b1. ( TREL and  (a1 != b1) ).
```

This simplifies to !a and b. Indeed, these are the states that have a move to a state satisfying (a1 != b1).

If we wanted to calculate those states that have at least one next state that satisfies !a1 and !b1, we would have obtained STARTSTATES = false (try it using BED). This is of course correct because there are no "next states" where !a1 and !b1.

Calculating AX

If we have to calculate AX p, we would employ duality and write it as

```
!(EX !p)
```

This approach will be used in the rest of this book.

Calculating EX for the example of Section 23.2.1

As said earlier on Page 424, the following formula represents EX_a_xor_b:

```
EX_a_xor_b = exists a1. exists b1. ( TREL and  (a1 != b1) ).
```

TREL captures constraints between a, b, a1, and b1, saying which a,b pairs can move to which other a,b pairs. Of course, the next values of a and b are modeled using a1 and b1. Now, in the resulting new states, a and b must satisfy the formula (a != b). Since it is the *next* state that must satisfy the XOR constraint, we write a1 != b1. However, it is the *present* state we must return. So, we constrain the next state and then quantify away a1 and b1.

23.2.3 LFP and GFP for 'Until'

We now will perform the least fixed-point iteration sequence for calculating A[p U q] with respect to Kripke structure (b) of Figure 23.5. This will be followed by the greatest fixed-point iteration sequence for A[p U q] with respect to the same Kripke structure. As Section 22.1.8 observed, these respectively obtain the *strong* Until and *weak* Until semantics.

The recursion over which these semantics are calculated is the following:

$$A[pUq] = q \lor (p \land AX (A[pUq]))$$

We will calculate AX using duality, as explained in Section 23.2.2.

23.2.4 LFP for 'Until'

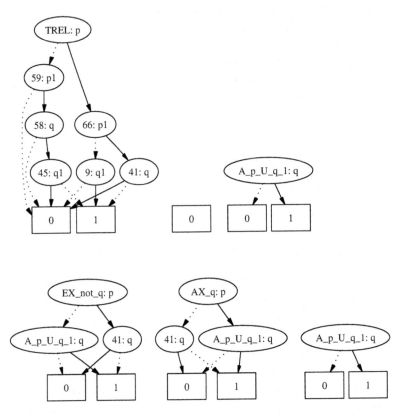

Fig. 23.7. The Strong Until (U) iteration that reaches a fixed-point

```
bed> var p p1 q q1
bed> let TREL = (p and not(q) and p1 and not(q1))
              or (p and not(q) and p1 and q1)
              or (p and q and not(p1) and q1)
              or (not(p) and q and p1 and not(q1))
              or (p and not(q) and not(p1) and q1)
              or (p and not(q) and p1 and not(q1))

bed> upall TREL
Upall( TREL ) -> 67
bed> view TREL
bed> let A_p_U_q_0 = false
bed> let AX_A_p_U_q_0 = false
bed> let A_p_U_q_1 = (q or (p and AX_A_p_U_q_0))
```

```
bed> upall A_p_U_q_1
Upall( A_p_U_q_1 ) -> 3
bed> view A_p_U_q_1
bed> let EX_not_q = exists p1. exists q1. (TREL and !q1)
bed> upall EX_not_q
Upall( EX_not_q ) -> 80
bed> view EX_not_q
bed> let AX_q = !EX_not_q
bed> upall AX_q
Upall( AX_q ) -> 82
bed> view AX_q
bed> let A_p_U_q_2 = (q or (p and AX_q))
bed> upall A_p_U_q_2
Upall( A_p_U_q_2 ) -> 3
bed> view A_p_U_q_2 --> gives ''q'', hence denotes {S1,S2} -- LFP
```

The results of all the **view** commands are pooled in Figure 23.7. The
final fixed-point reached is q, which means that states s1 and s2 are
included in it. Clearly, these are the only states from which it is the
case that *all paths* satisfy A[pUq]. Therefore, we have the following:

> Starting from either s0 or s3, there is the possibility of never
> encountering a q, and hence these states are eliminated by the
> strong until semantics of AU.

23.2.5 GFP for Until

Taking the greatest fixed-point, we obtain the 'W' (weak Until) semantics. We give the BED scripts, leaving it to the reader to try them.
We indeed reach the greatest fixed-point of $q \lor p$, which covers states
{S0,S1,S2,S3}. We also see that from these states, we either have p
holding forever, or p holding until q holds.

```
bed> let A_p_U_q_0 = true
bed> let AX_A_p_U_q_0 = true
bed> let A_p_U_q_1 = (q or (p and AX_A_p_U_q_0))
bed> upall A_p_U_q_1
Upall( A_p_U_q_1 ) -> 72
view A_p_U_q_1
bed> let EX_not_p_or_q = exists p1. exists q1. (TREL and !(p1 or q1))
bed> upall EX_not_p_or_q
Upall( EX_not_p_or_q ) -> 0
bed> let AX_p_or_q = !EX_not_p_or_q
bed> upall AX_p_or_q
Upall( AX_p_or_q ) -> 1
bed> view A_p_U_q_1
bed> let A_p_U_q_2 = (q or (p and AX_p_or_q)) --> reached
     Fixed-point (q or p) which denotes {S0,S1,S2,S3}
```

23.3 Büchi Automata and LTL Model Checking

We now revert to a more detailed look at LTL. In particular, we present
material leading up to a study of enumerative LTL model checking
using Büchi automata.

Büchi automata come in the deterministic and nondeterministic varieties; the latter are strictly more expressive than the former. Therefore, by the term "Büchi automata," we will mean nondeterministic
Büchi automata (NBA). NBA are structures $(Q, \Sigma, \delta, q_0, F)$ where Q
is a *finite non-empty* set of states, Σ is a *finite non-empty* alphabet,
$\delta : Q \times \Sigma_\varepsilon \to 2^Q$ is a *total* transition function, $q_0 \in Q$ is an initial state,
and $F \subseteq Q$ is a *finite, possibly empty* set of final states. An *infinite*
run (or computation) $\sigma = s_0, s_1, \ldots$ is in the language of a Büchi automaton if there is an infinite number of occurrences of states from F
in σ. Notice that the graphs in Figure 23.8, when viewed as an NFA,

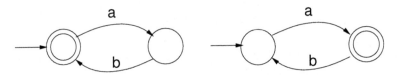

Fig. 23.8. Transition graphs read as an NFA or as an NBA

have different languages while they have the same language $(ab)^\omega$ when
viewed as an NBA. In Section 23.3.2, we define the notion of ω-regular
sets which are similar to regular sets, except, in addition to a Kleene
star operator, we also have an *infinite iterate* operator, $()^\omega$. This will
help us define the notion of infinite runs.

23.3.1 Comparing expressiveness

From the discussions in Sections 22.1.3 and 22.1.4, it is clear that when
LTL and CTL formulas are viewed as Kripke structure classifiers (each
formula partitions all Kripke structures into two bins "models" and
"non-models"), LTL tends to forget the branching structure, focusing
only on the "straight-line runs" while CTL considers the branching
structures. For example, an important fact about the example in Figure 22.4 is that in whichever state the computation is situated, there
exists a path leading to a state where x is true.

We now state some facts pertaining to the expressiveness of LTL,
CTL, and Büchi automata. Consider all possible Kripke structures

to be the set K. Let $K^V \subseteq K$ be Kripke structures that involve all the variables of V (assign all the variables in V, and possibly more). Let $k_a, k_{a_1}, k_{a_2}, \ldots, k_b, k_{b_1}, k_{b_2}, \ldots$ refer to various subsets of K^V (with $a, b, a_1, b_1, \ldots$ ranging over a suitable index set), and let $\overline{k_a}, \overline{k_{a_1}}, \overline{k_{a_2}}, \ldots, \overline{k_b}, \overline{k_{b_1}}, \overline{k_{b_2}}, \ldots$ be the complements of these sets relative to K^V. Given all this, we state a few facts pertaining to the expressiveness of various temporal logics and automata; for details, see [59] as well as [112].

- There exist sets $k_a, k_{a_1}, \ldots$ such that for any one of these sets, say k_{a_j}, there exists an LTL formula $\varphi_{a_j}^{LTL}$ that regards all of k_{a_j} as its models and all of $\overline{k_{a_j}}$ as its non-models exactly when there exists a CTL formula $\varphi_{a_j}^{CTL}$ that also regards k_{a_j} as its models and all of $\overline{k_{a_j}}$ as its non-models. In other words, partitions such as k_{a_j} capture temporal patterns that are equally expressible in both these logics.
  - An example of a temporal fact that corresponds to such a partitioning would be "henceforth a is true."
- There are partitionings, say k_{b_j}, such that there exist only LTL formulas $\varphi_{b_j}^{LTL}$ that regard all of k_{b_j} as its models and all of $\overline{k_{b_j}}$ as its non-models. In other words, the partitioning expressed by k_{b_j} captures temporal patterns expressible only in LTL.
  - An example of a temporal fact that corresponds to such a partitioning would be "infinitely often a." Any CTL formula one attempts to write down to effect such a classification would end up "lying"—it would either put a Kripke structure that is correct with respect to this behavior into the non-model bin or put a Kripke structure that is incorrect with respect to this behavior into the model bin.
- There are partitioning k_{c_j} for which only CTL formulas exist and no LTL formulas exist.
  - A temporal fact that corresponds to such a partitioning is the AG (EFx) example considered earlier.
- There exist CTL* formulas in all the above cases. In fact, using a logical connective, one can combine an LTL formula that has no equivalent CTL formula and a CTL formula that has no equivalent LTL formula into a CTL* formula that can now express something that neither LTL nor CTL can express.
- The **never** claim language of Promela allows (negated) Büchi automata to be specified. All LTL specifications have corresponding Büchi automata specifications, *but not vice versa*. An example of a

temporal pattern expressible using Büchi automata but not linear-time temporal logic is

p *can* hold after an even number of execution steps, but *never* holds after an odd number of steps.

- A logic called modal μ-calculus properly subsumes all of the above logics in terms of expressiveness.

23.3.2 Operations on Büchi automata

The language of a Büchi automaton is a ω-regular set of strings over a given set Σ. Following [115], we now inductively define ω-regular sets; this definition involves the notion of a *regular set* over Σ as defined in Chapter 10. For completeness we repeat that definition also here (albeit in terms of context-free productions); we employ two non-terminals re and ore standing for regular expressions and ω-regular.

> **Please note:** $ab^\omega c$ does not make sense as an infinite string, as there is "no sensible way" in which we can affix c after an infinite run of b's. In other words, infinite sequences will be "*tail infinite.*"

$$re \rightarrow \quad \emptyset \mid \varepsilon \mid a \in \Sigma \mid (re) \mid re_1 + re_2 \mid re_1 re_2 \mid re^*$$
$$ore \rightarrow re^\omega \mid re\ ore \mid ore_1 + ore_2$$

With respect to the above grammar, we can now define operations on Büchi automata.

Union of Büchi Automata: We can stitch the two Büchi automata in question using ε, much like we perform the corresponding operation with NFAs.

Concatenation: We may stitch an NFA and a Büchi automata in that order, as specified in the grammar for ω-regular sets.

Complementation: Büchi automata are closed under complementation, although the resulting Büchi automata are exponentially bigger, and the algorithm for complementation is extremely complex (references such as [43, 49] provide some details). For this reason, users are often required to manually come up with negated Büchi automata, as is the case with **never** automata in SPIN.

In Section 23.3.3 we prove that DBAs are not closed under complementation.

Intersection: The standard intersection algorithm no longer works, as one has to keep track of when a single infinite run is in the language of the constituent Büchi automata, visiting final states infinitely in both of them. This algorithm is given in many references (e.g., see [20]).

23.3.3 Nondeterminism in Büchi automata

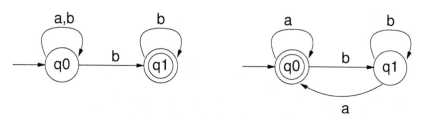

Fig. 23.9. An NBA accepting finitely many a's (left) and a DBA accepting infinitely many a's (right)

We now prove that there is at least one NBA—in particular, the one on the left-hand side of Figure 23.9—that has no corresponding DBA. The language of this NBA is $(a + b)^* b^\omega$. Suppose there is a DBA D with this language. Clearly, then, b^ω is in $L(D)$. Hence, there exists a number $n_1 \in Nat$ such that after encountering b^{n_1}, the DBA will be in some final state – say x_1 – of D for the first time. Note that if D never visits a final state, it cannot have any string in its language, let alone b^ω. Moreover, any given string takes D from a given state s to a unique state s'.

Now, consider string $b^{n_1} a b^\omega$: it is also in $L(D)$. Hence there exists n_2 such that $b^{n_1} a b^{n_2}$ takes D to the first final state after x_1 – say x_2 – in D. Continuing this way, $b^{n_1} a b^{n_2} a \ldots b^{n_k}$ first takes D to some final state x_k in D. But all these states x_i cannot be distinct, as the number of states is finite. Assume that state x_i is the first one to repeat, and the first repeated occurrence is identified by x_i'. This means that there is an accepting cycle $q_0 \ldots x_i \ldots x_i'$ where the loop $x_i \ldots x_i'$ includes an a. In turn, this means that there is some string with an infinite number of a's in $L(D)$. This is a direct contradiction with respect to what the language $L(D)$ is supposed to be. Therefore, a DBA D does not exist for this language.

Also, note that DBAs are not closed under complementation. The machines in Figure 23.9 are complements, and yet the one on the right-hand side is a DBA whereas we just now proved that the machine on the left has no equivalent DBA.

23.4 Enumerative Model Checking for LTL

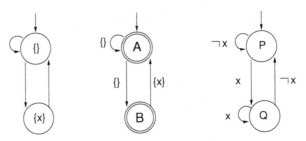

Fig. 23.10. A Kripke structure (left), its corresponding Büchi automata (middle), and a property automaton expressing GFx (right)

We now provide a glimpse of the enumerative model checking algorithm used by model checkers such as SPIN. Consider the Kripke structure given in Figure 23.10 on the left-hand side. In its initial state, x is false. In one of the computations of this Kripke structure, x would remain false forever; in all others (an infinite number of them exist), x would become infinitely often true. This Kripke structure can be viewed as a Büchi automata as shown in the middle. Notice how we go from state labels to edge labels, as is our convention when diagramming the state transition relation of automata. We are, in effect, declaring the alphabet of these Büchi automata to be powersets of atomic propositions. Also, by virtue of the fact that all states are final states, we are also asserting that *all infinite runs* of such Büchi automata are of interest to observe.

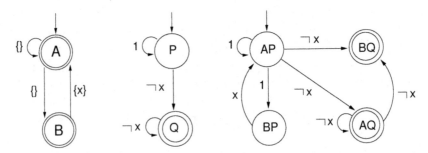

Fig. 23.11. System automaton (left), complemented property automaton (middle), and product automaton (right)

When it comes to formal verification, we would like to state properties of interest in either temporal logic or in terms of Büchi automata. From the discussions in Section 23.3.1, we can gather that anything that can be stated in LTL has a corresponding Büchi automata version. An example of this appears in the right-hand side of Figure 23.10. Here, we present a Büchi automata that expresses GFx. In other words, the language of this Büchi automata is exactly those computations that satisfy the LTL formula GFx.

23.4.1 Reducing verification to Büchi automaton emptiness

Notice that the "property automaton" (P) on the right-hand side of Figure 23.10 includes *all* runs that satisfy GFx. Therefore, to determine whether a given Kripke structure ("system" S) satisfies a property P, we can check whether

$$L(S) \subseteq L(P).$$

This check is equivalent to

$$L(S) \cap \overline{L(P)} = \emptyset.$$

Figure 23.11 shows the system automaton (left), the complemented property automaton (middle), and the product automaton (right) realizing the intersection ($\cap$) operator, above. Since the "system automaton" has all its states being final states, this intersection is obtained through the standard NFA intersection algorithm. In particular, a state of the product machine is final if and only if the constituent states of the component machines are both final states.

We notice that the intersection gives a Büchi automata whose language is not empty. In particular, it has an accepting run where the product state AP leads to product state AQ which repeats infinitely. Hence,

$$L(S) \cap \overline{L(P)} \neq \emptyset.$$

When this condition is true, a bug has been found (*i.e.*, the property has been violated). This accepting run can be displayed as an error-trace or a MSC. Consequently, the debugging of concurrent systems can be reduced to emptiness checking of Büchi automata.

Explicit enumeration algorithms, such as used in tools like SPIN, find property violations and depict them in an intuitive fashion, such as shown in the MSC of Figure 21.3. In summary, here are the steps they employ (for details, see references such as [20]):

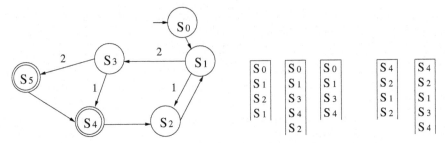

Fig. 23.12. Searching of a product graph for cycles: Example product graph (left), group of stacks for the outer DFS (middle), and group of stacks for the inner DFS (right)

- Obtain the system automaton by performing an interleaving (asynchronous) product of the system **proctypes**. In effect, we pick one of the enabled processes nondeterministically, and perform one if its enabled moves nondeterministically. This is called taking the "system step," and corresponds to the interleaving product mentioned in Section 21.4.1.
- Next, we take a "property step" by taking the **never** automaton (negated property automaton) and move it one step. In effect, the condition of the property automaton move must be enabled in the state resulting from the earlier system step, in order for the property automaton to be taking a step. This corresponds to taking the synchronous product as mentioned in Section 21.4.1.
- Generate the entire reachable graph by alternating the system- and property- steps using depth-first search (DFS). A hash table is employed to record states already visited, thus cutting off search at these re-visitations. This depth-first search can be thought of as a mechanism for enumerating the final states of the synchronous product machine in *postorder*.
- Whenever the DFS backs off a final state, a *nested* DFS is spawned. The task of this "inner" DFS is to look for Büchi automata "lassos" (accepting cycles). In [30, 20], it has been shown that if the product Büchi automata has at least one accepting cycle, then one such cycle will be found by this procedure.

In a sense, this is an on-the-fly cycle detection approach that often works much more efficiently than an offline cycle detection procedure such as Tarjan's strongly connected components (SCC) algorithm. We illustrate this nested depth-first search on a simple example in Figure 23.12. We supply all details below, including details glossed over above. Please pay attention to "outer DFS hash table"—ODHT, and "inner DFS hash table"—IDHT, below. Basically, we enter states into the same hash table, but tag it with an additional bit saying whether the state was inserted during an outer or an inner DFS:

- Suppose the synchronous product generates the graph shown on the left-hand side of Figure 23.12.
- The "outer" DFS will perform many traversals, as shown by the stack histories shown in the middle of Figure 23.12. These traversals are as follows (we assume that when at state S1, the first edge to be searched is the one indicated with a "1" on it, and the second edge has a "2"; likewise for state S3 as well):
  - Search S0, S1, take outgoing edge "1" and continue with S2, and back to S1. The stack snapshot at this point is shown in the left-hand stack of the middle group of stacks in Figure 23.12. This closes a cycle, and so DFS unwinds the stack up to state S1. ODHT now has states S0, S1, S2 in it.
  - Search forward from S1 through outgoing edge marked "2", going up to state S3, then taking "1" to go to S4, and finally to S2. The stack snapshot at this point is shown in the middle stack of the middle group of stacks in Figure 23.12. Since S2 is in the ODHT, the DFS is cut off, and the search *wants to unwind* to state S3.
  - However, we notice that we are backing off from S4, which is a final state. Hence, we spawn an *inner DFS*.
- The inner DFS proceeds in the manner shown in the right-hand group of two stacks shown in Figure 23.12. In particular,
  - The inner DFS proceeds forward from S4, producing the stack history S4, S2, S1, and taking the "1" edge, back to S2. *All these states are inserted into the* IDHT, and that is why we did not cut off at S2 after S4. This did not find an accepting cycle "lasso."
  - The inner DFS backs off to state S1 and continues via the "2" edge to S3, and then to S4.
  - At this point, the inner DFS has reached a state that is already in the stack, and so it "knows" that a cycle has been closed!
  - Also since the inner DFS was triggered from a final state, we know that we have found a "lasso" that includes a final state.
  - Last but not least, since the outer DFS had performed a postorder enumeration of the final states, a *reachable lasso* has been found.
  - Therefore, we conclude that

$$L(S) \cap \overline{L(P)} \neq \emptyset$$

and so the property in question has been violated.
- An error-trace is generated and printed, following the DFS stack history, in the form of a MSC.

Chapter Summary

We presented an enumerative model checking algorithm for CTL, and a symbolic model checking algorithm for CTL. The interactive sessions in the BED tool present intermediate BDDs created during the fixed-point iteration. Then we presented LTL and Büchi automata, and presented an enumerative algorithm for LTL model checking.

Exercises

23.1. Write the pseudo-code of an algorithm (similar to Figure 23.3) for procedure EG that handles an EG formula. Then show the details of checking
(EG (OR N1 (OR T1 C1))) by filling a table similar to Figure 23.2.

23.2. Consider the "release" (R) operator whose definition is as follows. A[p R q] means that q holds along the path up to and including the first state where p holds. However, p is *not required to hold eventually*. Express the semantics of 'R' more rigorously. Now, find a fixed-point formulation for 'R', and illustrate it on the same example as illustrated in Section 23.2.3. If there are two natural interpretations for the LFP and GFP, consider both, and then explain what their semantics are. Present all the BED sessions you employed in arriving at your answers.

23.3. Write three LTL properties that try to capture "*p can* hold after an even number of execution steps, but *never* holds after an odd number of steps," and point out where they go wrong. Now draw a Büchi automata that specifies this pattern.

23.4. Assume that you are given a digital circuit to verify and you have access to only a linear-time temporal logic model checker. You are, however, required to state and prove a CTL correctness assertion with respect to some of the input/output ports of this circuit. This CTL assertion is known to not have any equivalent LTL assertions. Can you proceed to check these assertions, somehow, using the given LTL model checker? Assume that you have access to the internal nodes of the circuit, and so you are allowed to state LTL assertions not only with respect to the input/output ports, but also with respect to the internal nodes of the circuit.

If it is theoretically possible to check the CTL assertions in terms of a (perhaps larger) set of LTL assertions, think of a simple practical example where such a situation can be faced (employing simple resettable counters and logic gates), and describe, in terms of this example, what your solution will be.

23.5. Do a literature search and find out whether the property expressed in Exercise 23.3 is expressible in (a) CTL, (b) CTL*, (c) μ-calculus.

23.6. Argue that *if* one of the Büchi automata being subject to the intersection operation has only final states (it has no non-final states), *then* the standard intersection algorithm for NFA can be used to intersect two Büchi automata. Provide an example illustrating your answer.

23.7. Change the property to be verified in Figure 23.10 from GFx to FGx, and rework all the steps of that example, including details of the nested DFS algorithm. Verify your answer by inspection (see if FGx is supposed to be true, and compare with what you got when you applied the nested DFS algorithm).

24
Conclusions

This book is about "automata" in the broad sense of the word introduced in Chapter 1. We viewed automata and logic as two sides of the same coin. Connections between automata and logic were emphasized multiple times. We often took an informal approach to drive the main point across as intuitively and in as friendly a fashion as possible. We believe that the material presented here will help someone with virtually no formal computation engineering background to pick up a considerable amount of that topic and apply it towards debugging sequential and concurrent hardware and software systems through automated finite-state methods known as model checking.

A journey of a 1000 miles begins with one step (ancient Chinese proverb); however, such a journey must somehow end! We close off, rather unceremoniously, picking four important (but random) topics to reiterate.

Beginner's mistakes

Experience has shown that a *large majority* of proofs written by beginners are flawed. For instance, it is easy to simply forget to prove one of the directions of an *iff* proof. Other common mistakes include using unproven facts, using totally circular definitions, etc. While there is no point fretting over one's past mistakes, good ways to avoid fundamental mistakes are to proofread one's proof presentations, write it out **neatly** (preferably in a *bulleted* fashion), and presenting one's proofs to others. Other ways to avoid these mistakes include a careful observation of how "common situations" are handled, and remember them by heart. For instance, while negating $A \Rightarrow B$, since A is already "under a negation" (because $A \Rightarrow B \equiv \neg A \vee B$), it does not get negated in the final result, which is $A \wedge \neg B$.

Mistakes Due to Inattention to Detail

After passing the initial hurdles, the second category of "mistakes" are simply due to inattention to detail, or to unfamiliarity with the particular domain. Here are some examples of common mistakes (the details are unimportant now; only the pattern of the mistakes needs to be noted):

1. *Misusing the unidirectional Pumping Lemma:* In Chapters 12 and 13, we studied several Pumping Lemmas of the form $A \Rightarrow B$ as well as $A \Leftrightarrow B$. An example of a Pumping Lemma of the form $A \Rightarrow B$ is $Regular(L) \Rightarrow C$ where C is a big complicated formula. The recommended usage is to assume for some language L that $Reg(L)$ holds, and to find that $\neg C$ (i.e., a contradiction), thus proving $\neg Reg(L)$. Many students abuse this argument to try and *prove* $Reg(L)$.

2. *Forgetting to do full case analysis:* Even after grasping the Pumping Lemma, many students forget to do full case analysis to derive contradictions in all cases. Similar mistakes are often repeated in doing proofs by *reducibility,* where also contradictions must be obtained in *all possible cases.*

3. *Forgotten cases of definitions:* Here, my favorite is relating to NP-complete (NPC) proofs, where the full definition of NPC is not kept in mind. Many students forget to show that the given problem *is in the class* NP, thus leaving open the question of whether the problem is even decidable (*i.e.,* it is not enough to establish polynomial reducibility).

4. *Abuse of "hard" theorems:* Theorems such as the Rice's Theorem are often not well understood, and hence abused. They often treat it as if it were a "magic" theorem or "panacea," which it isn't. For instance, many students are known to apply Rice's Theorem to context-free languages and even regular languages in a loose manner. Rice's Theorem basically states the impossibility of building an algorithmic classifier for Turing machines based on the language recognized by these Turing machines, if the classification attempted is anything but trivial. They do not remember that it is the Turing machine codes ("programs") that we need to classify, based on the language of these Turing machines, should we run these Turing machines; we are not classifying DFA or PDA.

Here, there is no quick answer except to gain experience, as well as be *formal* about definitions.

Experts' mistakes

Humans are fallible - expert or not! Lamport [74] observes that about a third of the proofs published in mathematical journals are flawed. In fact, everyone knows that, and the attitude at this level is not to "denigrate" or "get even" with the person, but to help fix the proofs. Therefore, the best advice one can pass on is

> "How do I present my proofs so that it helps others easily spot my mistakes and correct them?"

Proofs do have value even when flawed. Proofs are, after all, programs, albeit in a mathematical language, and hence prone to 'programming errors.' While mathematical notation helps you avoid nasty errors due to pointers overflowing their ranges, etc., a poorly written formal description is no better than a program written at an insidiously low level and hence fraught with many dangers.

Whether to trust machine checked proofs

Mechanical theorem proving has been developed to be able to machine check proofs. Other tools such as model checkers and finite-state reasoning tools have also been developed to verify correctness. In a well-developed mechanical reasoning system, all the critical reasoning steps are carried out inside a *logic kernel* that is very compact, and hence subject to much scrutiny. In Chapter 11, as well as Chapters 21 through 23, we obtained a glimpse of the inner workings of many mechanical correctness checking tools. While these tools are no panacea, they have often helped find *very subtle* flaws that, for all practical purposes, cannot be detected through casual inspection or repeated simulation using *ad hoc* test cases. Continued development of these tools and techniques is indeed a sign of maturation of computation engineering. For additional details, please refer to [41].

The End. Fini. Thank you for your company in this rather long journey.[1] If you spot any mistakes,[2] or have suggestions, kindly drop me a line, won't you? My email address and the book website appear on page 443. Many thanks in advance!

---

[1] Unless you are peeking at this page without reading most others!

[2] The safest place for a fly to sit is on the fly-swatter handle! – *Quoted from "The Logic of Bugs" by Gerard Holzmann, Foundations of Software Engineering, 2002.*

A

Book web site and tool information

A.1 Web site and e-mail address

The author maintains a web site for this book to post errata, updates, and copies of some of the tools and scripts. It is

`www.cs.utah.edu/ganesh-comp-engg-book`.

The author welcomes your comments and suggestions. Please send them to ganesh-comp-engg-book@cs.utah.edu.

A.2 Software tool usage per chapter

8 through 11:

- `dot` and `graphviz` tools mentioned in these chapters are downloadable from `www.graphviz.org`.
- `grail`, originally developed at the University of Western Ontario by Darrel Raymond and Derick Wood (`http://www.csd.uwo.ca/Research/grail/index.html`). A copy courtesy of Andrei Paun and Shenghua Ni will be placed on the website of this book also.
- Two scripts, `fa2grail.perl` and `grail2ps.perl`, kept at the website of this book.

11: • BED, a BDD manipulation package by Henrik Reif Andersson and Henrik Hulgaard, is available at `http://www.itu.dk/research/bed/`. It will be kept on the website of this book also.

15: • JFLAP, developed by Susan Rodger's team, is downloadable from `http://www.jflap.org/`.

18 and 19:

- Zchaff, developed by Sharad Malik's group, is downloadable from
  `http://www.princeton.edu/~chaff/`. There are many other SAT
  solvers that serve the same purpose, such as MiniSAT
  (`http://www.cs.chalmers.se/Cs/Research`
  `/FormalMethods/MiniSat/`).
- Ocaml is used in a few examples, and may be downloaded from
  `caml.inria.fr`.

21: - SPIN is used in this chapter, and may be downloaded from
  `www.spinroot.com`

22: - This chapter may also involve the use of BED.

A.3 Possible Syllabi

This book includes more material than can be comfortably taught in a
typical academic semester. Subsets of this book may be used to meet
the needs of the following types of courses:

- An undergraduate course on basic discrete mathematics, finite automata, and logic may be taught along the following lines:
  - Week 1: Mathematical Preliminaries (Chapter 2).
  - Week 2: Main topics from Binary Relations (Chapter 4), including Pre- and Partial Orders, Reflexive and Transitive Closure.
  - Week 3: Proof Methods (Chapter 5), skipping Induction Principles.
  - Week 4-5: Strings and Languages (Chapter 7), perhaps skipping Homomorphisms.
  - Week 6: Machines, Languages, and DFA (Chapter 8).
  - Week 7: NFA and Regular Expressions (Chapter 9).
  - Week 8: Operations on Regular Machinery (Chapter 10), including illustrations using Grail.
  - Week 9: The Pumping Lemma (Chapter 12).
  - Week 10: Context-free Languages (Chapter 13).
  - Week 11: Push-down Automata and Context-free Grammars (Chapter 14).
  - Week 12: Turing Machines (Chapter 15).
  - Week 13: Selected topics from Basic Notions in Logic (Chapter 18).
  - Week 14: Complexity Theory and NP-Completeness (Chapter 19).
- An undergraduate course on automata, languages, Turing machines, mathematical logic, and applied Boolean methods may be taught along the following lines:

- Week 1: Assign exercises from Mathematical Preliminaries (Chapter 2). Then convey the main ideas from Cardinalities and Diagonalization (Chapter 3), including a proof of the Schröder-Bernstein Theorem.
- Week 2: Binary Relations (Chapter 4). Also dive straight into Section 5.3.1 and work out a proof of the equivalence between the two induction principles (Arithmetic and Complete).
- Week 3: Strings and Languages (Chapter 7).
- Week 4: Machines, Languages, and DFA (Chapter 8).
- Week 5: NFA and Regular Expressions (Chapter 9).
- Week 6: Operations on Regular Machinery (Chapter 10), including illustrations using Grail.
- Week 7: The Automaton/Logic Connection, Symbolic Techniques (Chapter 11), including illustration using BED.
- Week 8: The Pumping Lemma (Chapter 12).
- Week 9: Context-free Languages (Chapter 13), including Consistency and Completeness proofs.
- Week 10: Push-down Automata and Context-free Grammars (Chapter 14).
- Week 11: Turing Machines (Chapter 15). Basic Undecidability Proofs (Chapter 16).
- Week 12: Basic Notions in Logic (Chapter 18).
- Week 13: Complexity Theory and NP-Completeness (Chapter 19).
- Week 14: Model Checking: Basics (Chapter 21).

• A graduate course on automata, relationships with Boolean methods, undecidability, NP-completeness, and modern formal verification methods may be taught along the following lines, taking advantage of graduate students' relative independence, as well as keeping their impatience with rote work in mind:

- Week 1-3: Quickly go through Chapters 2 through 6.
- Week 4-6: Review concepts from NFA, DFA, and RE (Chapters 7 through 10 and Chapter 12), suggesting that they use Grail to quickly note various results.
- Week 7: Chapters 11, Automaton/Logic Connection. Reachability using BDDs. It is possible to cover Chapter 20 (Presburger arithmetic) at this juncture.
- Week 8: Chapter 13, 14.
- Week 9: Chapter 15.
- Week 10: Chapter 16.
- Week 11: Chapter 17.
- Week 12: Chapter 18.

– Week 13: Chapter 19.
– Week 14: Chapter 21 through 23.

B

BED Solution to the tic-tac-toe problem

The reader may enter the following script into the BED tool and watch the display of all draws.

```
var turn turnp
    a00 a00p    b00 b00p
    a01 a01p    b01 b01p
    a02 a02p    b02 b02p
    a10 a10p    b10 b10p
    a11 a11p    b11 b11p
    a12 a12p    b12 b12p
    a20 a20p    b20 b20p
    a21 a21p    b21 b21p
    a22 a22p    b22 b22p ;

let init=!a00 and !b00 and   !a01 and !b01 and
         !a02 and !b02 and   !a10 and !b10 and
         !a11 and !b11 and   !a12 and !b12 and

         !a20 and !b20 and   !a21 and !b21 and
         !a22 and !b22 and   !turn ;

let initp=
  !a00p and !b00p and !a01p and !b01p and
          !a02p and !b02p and !a10p and !b10p and
          !a11p and !b11p and !a12p and !b12p and

          !a20p and !b20p and !a21p and !b21p and
          !a22p and !b22p and !turnp ;

let samerow0 = (a00=a00p) and (b00=b00p) and (a01=a01p) and (b01=b01p)
and (a02=a02p) and (b02=b02p);
let samerow1 = (a10=a10p) and (b10=b10p) and (a11=a11p) and (b11=b11p)
and (a12=a12p) and (b12=b12p);
let samerow2 = (a20=a20p) and (b20=b20p) and (a21=a21p) and (b21=b21p)
```

```
and (a22=a22p) and (b22=b22p);
let samecol0 = (a00=a00p) and (b00=b00p) and (a10=a10p) and (b10=b10p)
and (a20=a20p) and (b20=b20p);
let samecol1 = (a01=a01p) and (b01=b01p) and (a11=a11p) and (b11=b11p)
and (a21=a21p) and (b21=b21p);
let samecol2 = (a02=a02p) and (b02=b02p) and (a12=a12p) and (b12=b12p)
and (a22=a22p) and (b22=b22p);

let M00 = !a00 and !b00 and a00p and !b00p and !turn and turnp
and samerow1 and samerow2 and samecol1 and samecol2;

let M01 = !a01 and !b01 and a01p and !b01p and !turn and turnp
and samerow1 and samerow2 and samecol0 and samecol2;

let M02 = !a02 and !b02 and a02p and !b02p and !turn and turnp
and samerow1 and samerow2 and samecol0 and samecol1;

let M10 = !a10 and !b10 and a10p and !b10p and !turn and turnp
and samerow0 and samerow2 and samecol1 and samecol2;

let M11 = !a11 and !b11 and a11p and !b11p and !turn and turnp
and samerow0 and samerow2 and samecol0 and samecol2;

let M12 = !a12 and !b12 and a12p and !b12p and !turn and turnp
and samerow0 and samerow2 and samecol0 and samecol1;

let M20 = !a20 and !b20 and a20p and !b20p and !turn and turnp
and samerow0 and samerow1 and samecol1 and samecol2;

let M21 = !a21 and !b21 and a21p and !b21p and !turn and turnp
and samerow0 and samerow1 and samecol0 and samecol2;

let M22 = !a22 and !b22 and a22p and !b22p and !turn and turnp
and samerow0 and samerow1 and samecol0 and samecol1;

let N00 = !a00 and !b00 and !a00p and b00p and turn and !turnp
and samerow1 and samerow2 and samecol1 and samecol2;

let N01 = !a01 and !b01 and !a01p and b01p and turn and !turnp
and samerow1 and samerow2 and samecol0 and samecol2;

let N02 = !a02 and !b02 and !a02p and b02p and turn and !turnp
and samerow1 and samerow2 and samecol0 and samecol1;

let N10 = !a10 and !b10 and !a10p and b10p and turn and !turnp
and samerow0 and samerow2 and samecol1 and samecol2;

let N11 = !a11 and !b11 and !a11p and b11p and turn and !turnp
and samerow0 and samerow2 and samecol0 and samecol2;
```

```
let N12 = !a12 and !b12 and !a12p and b12p and turn and !turnp
and samerow0 and samerow2 and samecol0 and samecol1;

let N20 = !a20 and !b20 and !a20p and b20p and turn and !turnp
and samerow0 and samerow1 and samecol1 and samecol2;

let N21 = !a21 and !b21 and !a21p and b21p and turn and !turnp
and samerow0 and samerow1 and samecol0 and samecol2;

let N22 = !a22 and !b22 and !a22p and b22p and turn and !turnp
and samerow0 and samerow1 and samecol0 and samecol1;

let T = M00 or M01 or M02 or
M10 or M11 or M12 or
M20 or M21 or M22 or

        N00 or N01 or N02 or
N10 or N11 or N12 or
N20 or N21 or N22 ;

let atmostone =
 !(a00 and b00) and !(a01 and b01) and !(a02 and b02)
and
 !(a10 and b10) and !(a11 and b11) and !(a12 and b12)
and
 !(a20 and b20) and !(a21 and b21) and !(a22 and b22) ;

let wina1 = atmostone and
   (a00 and !b00) and (a01 and !b01) and (a02 and !b02) ;

let wina2 = atmostone and
   (a10 and !b10) and (a11 and !b11) and (a12 and !b12) ;

let wina3 = atmostone and
   (a20 and !b20) and (a21 and !b21) and (a22 and !b22) ;

let wina4 = atmostone and
   (a00 and !b00) and (a10 and !b10) and (a20 and !b20) ;

let wina5 = atmostone and
   (a01 and !b01) and (a11 and !b11) and (a21 and !b21) ;

let wina6 = atmostone and
   (a02 and !b02) and (a12 and !b12) and (a22 and !b22) ;

let wina7 = atmostone and
   (a00 and !b00) and (a11 and !b11) and (a22 and !b22) ;
```

```
let wina8 = atmostone and
   (a02 and !b02) and (a11 and !b11) and (a20 and !b20) ;

let winb1 = atmostone and
   (!a00 and b00) and (!a01 and b01) and (!a02 and b02) ;

let winb2 = atmostone and
   (!a10 and b10) and (!a11 and b11) and (!a12 and b12) ;

let winb3 = atmostone and
   (!a20 and b20) and (!a21 and b21) and (!a22 and b22) ;

let winb4 = atmostone and
   (!a00 and b00) and (!a10 and b10) and (!a20 and b20) ;

let winb5 = atmostone and
   (!a01 and b01) and (!a11 and b11) and (!a21 and b21) ;

let winb6 = atmostone and
   (!a02 and b02) and (!a12 and b12) and (!a22 and b22) ;

let winb7 = atmostone and
   (!a00 and b00) and (!a11 and b11) and (!a22 and b22) ;

let winb8 = atmostone and
   (!a02 and b02) and (!a11 and b11) and (!a20 and b20) ;

let allmoved = (a00 = !b00) and (a01 = !b01) and (a02 = !b02)
and
   (a10 = !b10) and (a11 = !b11) and (a12 = !b12)
and
   (a20 = !b20) and (a21 = !b21) and (a22 = !b22) ;

let draw = allmoved and
!wina1 and !wina2 and !wina3 and !wina4 and
!wina5 and !wina6 and !wina7 and !wina8 and
!winb1 and !winb2 and !winb3 and !winb4 and
!winb5 and !winb6 and !winb7 and !winb8 ;

upall init; upall samerow0; upall samerow1; upall samerow2;
upall samecol0; upall samecol1; upall samecol2; upall M00;

upall M01; upall M02; upall M10; upall M11; upall M12;
upall M20; upall M21; upall M22;

upall N00; upall N01; upall N02; upall N10; upall N11;
upall N12; upall N20; upall N21; upall N22; upall T;

upall atmostone;
```

```
upall wina1; upall wina2; upall wina3; upall wina4; upall wina5;
upall wina6; upall wina7; upall wina8;

upall winb1; upall winb2; upall winb3; upall winb4; upall winb5;
upall winb6; upall winb7; upall winb8;

upall allmoved; upall draw; view draw;
```

Here, be prepared to obtain a six-page BDD describing all possible draws in one fell swoop.

References

1. Agrawal, M., Kayal, N., and Saxena, N. "PRIMES is in P", August 2002. http://www.cse.iitk.ac.in/primality.pdf.
2. Aagaard, M. D., Jones, R. B., and Seger, C. -J. "Combining Theorem Proving and Trajectory Evaluation in an Industrial Environment". In Alan Hu, editor, *Design Automation Conference (DAC)*, 1998.
3. ACL2 Version 2.9. http://www.cs.utexas.edu/users/moore/acl2/.
4. Adiga, N. R., et al. "An Overview of the BlueGene/L Supercomputer". In *Conference on High Performance Networking and Computing: SC2002*, page 60, 2002.
5. Aho, A., Hopcroft, J.E., and Ullman, J.D. *The Design and Analysis of Computer Algorithms*. Addison-Wesley, 1974.
6. Aho, Alfred V., Hopcroft, John E., and Ullman, Jeffrey D. *Data Structures and Algorithms*. Addison-Wesley, 1983.
7. Andersen, Henrik Reif. "An Introduction to Binary Decision Diagrams" (tutorial associated with the bed tool), April 1998.
8. Ball, T., Cook, B., Levin, V., and Rajamani, S. K. "SLAM and Static Driver Verifier: Technology Transfer of Formal Methods inside Microsoft". In *IFM 04: Integrated Formal Methods*, pages 1–20. Springer-Verlag, April 2004.
9. Ball, T. and Rajamani, S.K. "The SLAM Toolkit". In *Proc. Computer-Aided Verif.*, pages 260–264, 2001. LNCS 2102.
10. Barringer, H., Goldberg, A., Havelund, K., and Sen, K. "Program Monitoring with LTL in EAGLE". In *International Conference on Parallel and Distributed Processing Systems (IPDPS)*, 2004.
11. Biere, A., Cimatti, A., Clarke, E.M., and Zhu, Y. "Symbolic Model Checking without BDDs". In *Tools and Algorithms for the Analysis and Construction of Systems (TACAS'99)*, 1999. LNCS 1579.
12. Brown, S. and Vranesic, Z. *Fundamentals of Digital Logic*. McGraw-Hill, 2003.

13. Bryant, Randal E. "Symbolic Boolean Manipulation with Ordered Binary Decision Diagrams". *ACM Computing Surveys*, 24(3):293–318, 1992.

14. Bryant, R.E. "Graph-Based Algorithms for Boolean Function Manipulation". *IEEE Transactions on Computers*, C-35(8):677–691, August 1986.

15. Cantin, J. F., Lipasti, M. H., and Smith, J. E. "The Complexity of Verifying Memory Coherence". In *Proceedings of the fifteenth annual ACM symposium on Parallel algorithms and architectures (SPAA)*, pages 254 – 255, San Diego, 2003.

16. Cantin, Jason. The Complexity of Verifying Memory Coherence. www.jfred.org/patpub.html.

17. Church, Alonzo. *Mathematical Logic*. Princeton University Press, 1956.

18. Clarke, E. M. and Emerson, E. A. "Synthesis of Synchronization Skeletons for Branching Time Temporal Logic". In *Logic of Programs*, LNCS 131, pages 52–71. Springer-Verlag, 1981.

19. Clarke, E. M., Emerson, E. A., and Sistla, A.P. "Automatic Verification of Finite-state Concurrent Systems using Temporal Logic Specifications. *ACM Transactions on Programming Languages and Systems (TOPLAS)*, 8(2):244–263, 1986.

20. Clarke, E. M., Grumberg, O., and Peled, D. *Model Checking*. MIT Press, December 1999.

21. Clarke, E.M., Grumberg, O., Jha, S., Lu, Y., and Veith, H. "Counterexample-guided Abstraction Refinement for Symbolic Model Checking". *Journal of the ACM (JACM)*, 50:752 – 794, September 2003.

22. Clarke, E.M., Kimura, S., Long, D.E., Michaylov, S., Schwab, S. A., and Vidal, J.-P. "Parallel Symbolic Computation on Shared Memory Multiprocessor". In *International Symposium on Shared Memory Multiprocessors*, Tokyo, Japan, April 1991. (Also CMU TR CMU-CS-90-182).

23. Clay Mathematical Institute. http://www.claymath.org/millennium/ P_vs_NP/Official_Problem_Description.pdf.

24. Collatz Problem. http://mathworld.wolfram.com/CollatzProblem.html.

25. Comon, H., Marché, C., and Treinen, R. *Constraints in Computational Logics, Theory and Applications*. Springer, 2001.

26. Continuum Hypothesis. http://mathworld.wolfram.com/ContinuumHypothesis.html.

27. Cook, Stephen A. "Computer Science Lectures, NP-Completeness". www.cs.toronto.edu/~sacook/.

28. Corella, F., Shaw, R., and Zhang, C. "A formal proof of absence of deadlock for any acyclic network of PCI buses". In *Hardware Description Languages and their Applications*, pages 134–156. Chapman Hall, 1997.

29. Cormen, Thomas H., Leiserson, Charles E., and Rivest, Ronald L. *Introduction to Algorithms*. McGraw-Hill Company, 1990.

30. Courcoubetis, C., Vardi, M.Y., Wolper P., and Yannakakis, M. "Memory Efficient Algorithms for the Verification of Temporal Properties". In *Formal Methods in System Design*, volume 1, pages 275–288, 1992.

31. Curry, H.B. and Feys, R. *Combinatory Logic (Vol. 1)*. North Holland, Amsterdam, 1958.

32. Davis, M., Logemann, G., and Loveland, D. "A Machine Program for Theorem Proving". *Communications of the ACM*, 5(7):394–397, 1962.

33. Davis, M. and Putnam, H. "A Computing Procedure for Quantification Theory". *Journal of the ACM*, 7(1):201–215, 1960.

34. Demaine, E.D., Hohenberger, S., and Liben-Nowell, D. "Tetris is Hard, Even to Approximate". In *Computing and Combinatorics (COCOON)*, pages 351–363, 2003.

35. DeMillo, R. A., Lipton, R.J., and Perlis, A.J. "Social Processes and Proofs of Theorems and Programs". *Communications of the ACM*, 22(5):271–280, 1979.

36. Dijkstra, Edsger W. "The Humble Programmer". *Communications of the ACM*, 15(10):859–866, 1972.

37. Dijkstra, Edsger W. *A Discipline of Programming*. Prentice-Hall, 1976.

38. Dill D. Comment made in keynote address offered at the Principles of Programming Languages (POPL) Conference, 1999.

39. Du, D.-Z. and Ko, K.I. *Problem Solving in Automata, Languages, and Complexity*. John Wiley & Sons, 2001.

40. Floyd, R. W. "Assigning Meanings to Programs". In J. T. Schwartz, editor, *Mathematical Aspects of Computer Science*, pages 19–32, Providence, RI, 1967. American Mathematical Society.

41. Formal Methods. Jonathan Bowen's Formal Methods Resource Page at http://www.afm.sbu.ac.uk.

42. Friedman, S. J. and Supowit, K. J. "Finding the Optimal Variable Ordering for Binary Decision Diagrams". In *Proceedings of the 24th ACM/IEEE conference on Design automation*, pages 348–356, 1987.

43. Fritz, Carsten. "Constructing Büchi Automata from Linear Temporal Logic Using Simulation Relations for Alternating Büchi Automata". In *CIAA*. Springer-Verlag, 2003. An excellent LTL to Buchi Automaton tool is available from http://www.ti.informatik.uni-kiel.de/~fritz/ABA-Simulation/ltl.cgi.

44. Garey, M. R. and Johnson, D.S. *Computers and Intractability, A Guide to the Theory of NP-Completeness*, chapter Appendix A5, pages 236–237. W.H. Freeman and Company, 1978.

45. Ginsburg, Seymour. *The Mathematical Theory of Context Free Languages*. McGraw-Hill, 1966. QA 267.5 S4 G5.

46. Gopalakrishnan, G.L., Yang, Y., and Sivaraj, H. "QB or Not QB: An Efficient Execution Verification Tool for Memory Orderings". In *Computer Aided Verification*, pages 401–413, 2004.

47. Gordon, Michael. "Why Higher-order Logic is a Good Formalism for Specifying and Verifying Hardware". In *Formal aspects of VLSI design*, 1986.

48. Gordon, Michael. *Programming Language Theory and Implementation*. Prentice-Hall, 1993.

49. Gurumurthy, S., Kupferman, O., Somenzi, F., and Vardi, M.Y. "On Complementing Nondeterministic Büchi Automata". In *12th Advanced Research Working Conference on Correct Hardware Design and Verification Methods*, Lecture Notes in Computer Science. Springer-Verlag, 2003.

50. Halmos, Paul R. *Naïve Set Theory*. Van Nostrand, 1968.

51. Har'El, Z. and Kurshan, R.P. "Software for Analysis of Coordination". In *Proc. Int'l Conference on System Science*, 1988.

52. Harrison, John. "Formal Verification of Square Root Algorithms". *Formal Methods in System Design*, 22(2):143–153, 2003.

53. Henzinger, T. A., Jhala, R., Majumdar, R., and Sutre, G. "Lazy Abstraction". In *Principles of Programming Languages*, pages 58–70. ACM, January 2002.

54. Hilbert's Problems. http://mathworld.wolfram.com/HilbertsProblems.html.

55. Hoare, C. A. R. "An Axiomatic Basis of Computer Programming". *Communications of the ACM*, 12(10):576–580, October 1969.

56. Hoare, C. A. R. Communicating Sequential Processes. *CACM*, 21(8):666–677, 1978.

57. Hoare, C.A.R. "The Emperor's Old Clothes". *Communications of the ACM*, 24(2):75–83, 1981.

58. Holzmann, Gerard, J. *Design and Validation of Computer Protocols*. Prentice Hall, 1991.

59. Holzmann, Gerard J. *The SPIN Model Checker: Primer and Reference Manual*. Addison Wesley, 2004.

60. Hopcroft, J.E., Motwani, R., and Ullman, J.D. *Introduction to Automata Theory, Languages, and Computation*. Addison-Wesley, 2001.

61. Hopcroft, J.E. and Ullman, J.D. *Introduction to Automata Theory, Languages, and Computation*. Addison-Wesley, 1979.

62. Joe Hurd. "Computer Science Lectures, Schröder Bernstein Theorem". http://www.cl.cam.ac.uk/~jeh1004/compsci/lectures/.

63. Introduction to Logic. http://logic.philosophy.ox.ac.uk/.

64. Jackson, D. In tutorial offered at the Third IEEE International Symposium on Requirements Engineering, Annapolis, Maryland, USA, January 6-10, 1997.

65. Jaffe, J. "A Necessary and Sufficient Pumping Lemma for Regular Languages". In *ACM SIGACT News archive*, volume 10, pages 48–49. ACM Press, 1978. Issue 2, Summer 1978, ISSN:0163-5700.

66. JFLAP. http://www.jflap.org/.

67. Jones, Neil D. *Computability and Complexity from a Programming Perspective*. MIT Press, 1997.

68. Kelley, D. *Automata and Formal Languages: an introduction*. Prentice-Hall, 1995.

69. Kepler Conjecture. http://mathworld.wolfram.com/KeplerConjecture.html.

70. Klarlund, N., Møller, A., and Schwartzbach, M. I. "MONA Implementation Secrets". *International Journal of Foundations of Computer Science*, 13(4):571–586, 2002.

71. Kozen, Dexter C. *Automata and Computability*. Springer, 1997.

72. Kurshan, R.P. *Computer-aided Verification of Coordinating Processes*. Princeton University Press, 1994.

73. Lamport, Leslie. "Proving the Correctness of Multiprocess Programs". *IEEE Transactions on Software Engineering*, SE-3(2):125–143, 1977.

74. Lamport, Leslie. "How to Write a Proof". *American Mathematical Monthly*, 102(7):600–608, August-September 1993.

75. Lamport, Leslie. "How to Write a Long Formula". *Formal Aspects of Computing*, 3:1–000, 1994.

76. Nancy G. Leveson. *SAFEWARE: System Safety and Computers*. Addison-Wesley, 1995.

77. Loeckx, J.and Sieber, K. *The Foundations of Program Verification: Second Edition*. John Wiley & Sons, 1987.

78. Manna, Z. *Mathematical Theory of Computation*. McGraw-Hill, 1974.

79. Manna, Z. and Pnueli, A. "The Modal Logic of Programs". In *ICALP*, pages 385–409, 1979.

80. Manna, Z. and Pnueli, A. *The Temporal Logic of Reactive and Concurrent Systems*. Springer Verlag, 1992.

81. Marques-Silva, João P. and Sakallah, Karem A. "GRASP - a New Search Algorithm for Satisfiability". In *International Conference on Computer-Aided Design*, pages 220–227, 1996.

82. Maurer, H. "A Direct Proof of the Inherent Ambiguity of a Simple Context-Free Language". *JACM*, 16(2):256 – 260, April 1969.

83. McMillan, Kenneth L. *Symbolic Model Checking*. Kluwer Academic Press, 1993.

84. Meyer, Bertrand. "Design by Contract: The Lessons of Ariane". *IEEE Computer*, 30(2):129–130, January 1997.

85. Milner, R. *A Calculus of Communicating Systems*, volume 92 of *Lecture Notes in Computer Science*. Springer-Verlag, 1980.

86. Milner, R. and Tofte, M. "Co-induction in Relational Semantics". *Theoretical Computer Science*, 87(1):209–220, 1990.

87. MiniSAT. http://www.cs.chalmers.se/Cs/Research/FormalMethods/MiniSat/.

88. Moret, B.M.E. and Shapiro, H.D. *Algorithms from P to NP*. Benjamin Cummings, 1991.

89. Morris, F.L. and Jones, C.B. "An Early Program Proof by Alan Turing". *IEEE Annals of the History of Computing*, 6(2):139–143, April-June 1984.

90. Moskewicz, M.W., Madigan, C.F., Zhao, Y., Zhang, L., and Malik, S. "Chaff: Engineering an Efficient SAT Solver". In *Proceedings of the 38th Design Automation Conference (DAC'01)*, 2001.

91. The MPEC Tool Release for Verifying Multiprocessor Executions. http://www.cs.utah.edu/formal_verification/.

92. Nelson, G. and Oppen, D.C. "Simplification by Cooperating Decision Procedures". *ACM Transactions on Programming Languages and Systems*, 1(2):245–257, 1979.

93. Owicki, S. S. and Gries, D. "Verifying Properties of Parallel Programs: An Axiomatic Approach". *Commun. ACM*, 19(5):279–285, 1976.

94. Owre, S., Shankar, N., and Rushby, J. "PVS: A Prototype Verification System". In *11th International Conference on Automated Deduction (CADE), Saratoga Springs, NY*, pages 748–752, June 1992.

95. PCI Special Interest Group–PCI Local Bus Specification, Revision 2.1, June 1995.

96. C. A. Petri. Fundamentals of a Theory of Asynchronous Information Flow. In *Proc. of IFIP Congress 62.*, pages 386–390. North Holland Publ. Comp., 1963.

97. Pnueli, Amir. "The Temporal Logic of Programs". In *Proceedings of the 18th IEEE Symposium Foundations of Computer Science (FOCS 1977)*, pages 46–57, 1977.

98. Pratt, Vaughan. "Every Prime Has a Succinct Certificate". *SIAM Journal on Computing*, (4):214–220, 1975.

99. Queille, J. and Sifakis, J. "Specification and Verification of Concurrent Systems in CESAR". In M. Dezani-Ciancaglini and U. Montanari, editors, *Fifth International Symposium on Programming*, Lecture Notes in Computer Science 137, pages 337–351. Springer-Verlag, 1981.

100. Rabin, Michael O. and Scott, Dana S. "Finite Automata and their Decision Problems". *IBM J. Res. Dev.*, (3):114–125, 1959.

101. Ramalingam, G. "The Undecidability of Aliasing". *ACM Transactions on Programming Languages and Systems*, 16(5):1476–1471, September 1994.

102. RSA Algorithm: 2002 A. M. Turing Award Winners Web Page. http://www.acm.org/awards/turing_citations/rivest-shamir-adleman.html.

103. RTI. "The Economic Impacts of Inadequate Infrastructure for Software Testing," Final Report, National Institute of Standards and Technology, May 2002. Research Triangle Institute (RTI) Project Number 7007.001. Accessed at http://spinroot.com/spin/Doc/course/NISTreport02-3.pdf (February 2006).

104. Rushby, John, M. "Disappearing Formal Methods". In *High-Assurance Systems Engineering Symposium*, pages 95–96, Albuquerque, NM, nov 2000. Association for Computing Machinery.

105. "SAT Live". www.satlive.org.

106. Savage, John E. *Models of Computation: Exploring the Power of Computing*. Addison-Wesley, 1998.
107. Schmidt, David. *Denotational Semantics: A Methodology for Language Development*. 1979. Out of print; copies of book available at http://www.cis.ksu.edu/~schmidt/text/densem.html.
108. Shepherdson, J. C. "The Reduction of Two-Way Automata To One-Way Automata". *IBM Journal of Research and Development*, 3(2):198–200. http://www.research.ibm.com/journal/rd/032/ibmrd0302P.pdf.
109. Shlyakhter, I., Seater, R., Jackson, D., Sridharan, M., and Taghdiri, M. "Debugging Overconstrained Declarative Models Using Unsatisfiable Cores". In *Proceedings of the 18th International Conference on Automated Software Engineering (ASE'03)*, 2003.
110. Shostak, Robert. "A Practical Decision Procedure for Arithmetic with Function Symbols". *Journal of the ACM*, 26(2):351–360, April 1979.
111. Sipser, Michael. *Introduction to the Theory of Computation*. PWS Publishing Company, 1997.
112. Spec Patterns. Discussion of LTL Patterns at http://patterns.projects.cis.ksu.edu.
113. Stanat, D.F. and Weiss, S.F. "A Pumping Theorem for Regular Languages". In *ACM SIGACT News archive*, volume 14, pages 36–37. ACM Press, 1982. Issue 1, Winter 1982, ISSN:0163-5700.
114. Stoy, Joseph E. *Denotational Semantics: The Scott-Strachey Approach to Programming Language Theory*. MIT Press, 1981.
115. Tauriainen, Heikki. "Automated Testing of Büchi Automata Translators for Linear Temporal Logic". Master's thesis, Helsinki University of Technology, 2000.
116. The Changing World of Software. http://www.sei.cmu.edu/publications/articles/watts-humphrey/ changing-world-sw.html.
117. Turing, Alan M. "On Computable Numbers: With an Application to the Entscheidungsproblem". In *Proceedings of the London Mathematical Society*, volume Series 2, November 1936. Number 42.
118. The Alan Turing Home Page. www.turing.org.uk/turing.
119. The Turing Machine and Computability. http://plato.stanford.edu/entries/turing/#2.
120. Vardi, M. Y. and Wolper, P. "An Automata-theoretic Approach to Automatic Program Verification". In *Proc. 1st Symp. on Logic in Computer Science*, pages 332–344, Cambridge, June 1986.
121. Whaley, J. and Lam, M.S. "Cloning-Based Context-Sensitive Pointer Alias Analysis Using Binary Decision Diagrams". In *Proceedings of the Conference on Programming Language Design and Implementation*, pages 131–144. ACM Press, June 2004.
122. Windows Driver Foundation. http://www.microsoft.com/whdc/driver/wdf/ default.mspx.

123. Wing, Jeanette M. "Computational Thinking". *Communications of the ACM*, 49(3):33–35, 2006.

124. Wise, David S. "A Strong Pumping Lemma for Context-Free Languages". *Theoretical Comp. Sci.*, (3):359–369, 1976.

125. Zhao, L. "Solving and Creating Difficult Instances of Post's Correspondence Problem". Master's thesis, University of Alberta, 2002. PCP Solver is available from http://web.cs.ualberta.ca/žhao/PCP/intro.htm.

(All web links were accessed on February 26, 2006 and found working.)

Index